Asia's Security

Asia's Security

Robert Ayson

BLOOMSBURY ACADEMIC
LONDON • NEW YORK • OXFORD • NEW DELHI • SYDNEY

BLOOMSBURY ACADEMIC
Bloomsbury Publishing Plc
50 Bedford Square, London, WC1B 3DP, UK
1385 Broadway, New York, NY 10018, USA
29 Earlsfort Terrace, Dublin 2, Ireland

BLOOMSBURY, BLOOMSBURY ACADEMIC and the Diana logo
are trademarks of Bloomsbury Publishing Plc

First published by Palgrave Macmillan, 2015
Reprinted by Bloomsbury Academic, 2024

Copyright © Robert Ayson, 2015

Robert Ayson has asserted his right under the Copyright,
Designs and Patents Act, 1988, to be identified as the author of this work.

All rights reserved. No part of this publication may be reproduced or
transmitted in any form or by any means, electronic or mechanical,
including photocopying, recording, or any information storage or retrieval
system, without prior permission in writing from the publishers.

Bloomsbury Publishing Plc does not have any control over, or responsibility for,
any third-party websites referred to or in this book. All internet addresses given
in this book were correct at the time of going to press. The author and publisher
regret any inconvenience caused if addresses have changed or sites have
ceased to exist, but can accept no responsibility for any such changes.

A catalogue record for this book is available from the British Library.

A catalog record for this book is available from the Library of Congress.

ISBN: HB: 978-1-1373-0182-6
PB: 978-1-1373-0181-9
ePDF: 978-1-1373-0183-3

To find out more about our authors and books visit www.bloomsbury.com
and sign up for our newsletters.

Contents

List of Maps	viii
Acknowledgements	ix

1 Introduction: How Should Asia's Security Be Approached?   1
    The importance of Asia's security   1
    Asia's evolution   6
    The extent of twenty-first-century Asia   9
    The meaning of security   14
    The meaning of Asia's security   19
    Asia's regional security   20
    The organisation of this book   22

2 Peace: Why Does Asia Seem More Secure?   26
    Interstate and external explanations of Asia's peace   27
    Peace at hand?   32
    Military power: balance or imbalance?   35
    Internal explanations   38
    Independence and Asia's postcolonial world   40
    Asia's interstate–intrastate nexus   45
    Conclusion   48

3 Power: Is It All About the USA and China?   51
    Great powers and Asia's security   52
    The USA as a great power   54
    Great power China?   58
    The perils and limits of great power status   64
    What about Russia, Japan and India?   67
    The importance of Asia's not so great powers   72
    Do the great powers matter that much?   75
    Conclusion   78

4 Money: Can Economic Interdependence Keep Asia Safe?   80
    Asia's economic interdependence   81
    The economic costs of Asian conflict   84
    Peace through interdependence?   87
    The political uses of economic relationships   90

    Economic coercion: sanctions in Asia   94
    Energy interdependence and insecurity   95
    Uneven effects on domestic security in Asia   98
    Conclusion   101

5 **Guns: Will Military Technology Lead to New Conflict in Asia?**   103
    The nuclear dimension of Asia's security   104
    Could nuclear war really happen in Asia?   108
    The competition for conventional military advantage in Asia   111
    Growing power projection capabilities   115
    High-tech: Asia's cyber challenges   118
    Low-tech: the problem of small arms   121
    A proliferation of drones?   123
    Conclusion   125

6 **Rivalry: Will Territorial Competition and Nationalism Ruin Asia's Peace?**   127
    The importance of territory   129
    Historical complexities   135
    Lighting the fire: nationalism   141
    No real solutions?   145
    Conclusion   148

7 **Fragmentation: Are Asia's Main Security Problems Domestic Ones?**   150
    Force and politics in Asia   152
    Asia's insurgencies   156
    Failed states and human security   160
    The spread of internal conflict?   165
    The decline of internal conflict?   168
    Conclusion   172

8 **Hazards: Will Non-State Actors and Transnational Challenges Overtake Asia?**   175
    Transnational terrorism   176
    Piracy in Asia   183
    The drugs and guns trades   186
    The flow of people   189
    Are transnational plagues regional security issues?   192
    Climate change and Asia's security   194
    Conclusion   197

9	**Interference: Can Intervention Work in Today's Asia?**	200
	Understanding intervention	201
	From peacekeeping to intervention?	203
	Regional interventions: East Timor and Solomon Islands	206
	Counter-insurgency: the Afghanistan experience	209
	Justifying interventions: a changing canvas	214
	Intervention without boots on the ground?	216
	Mediation	219
	Conclusion	222
10	**Solidarity: Can Asian States Work Together On Security?**	224
	Understanding cooperation	225
	The ASEAN family of groupings	230
	Coherence: a fruitless quest?	236
	Beyond Southeast Asian regionalism	238
	An Asian security community?	241
	Conclusion	244
11	**Division: Will Alliances and Partnerships Split the Region?**	246
	US alliances in Asia	247
	Alliance benefits and doubts	250
	The surprising revitalisation of Asia's alliances	254
	Alliances and security dilemmas	258
	A concert of powers in Asia?	262
	Conclusion	267
12	**Conclusion: Towards a New Asian Security?**	269
	A pecking order for Asia's security?	270
	Beyond state-centric confines	273
	Asia's major power security problem	275
	How do we get deeper cooperation?	278
	How important is security for Asia?	282
Bibliography		284
Index		302

List of Maps

1.1 Asia and its borderlands   13
2.1 Maritime territorial claims in the South China Sea   133
2.2 Maritime territorial disputes in North Asia   136

Maps 1.1, 1.2 and 1.3 are copyright of CartoGIS at the Australian National University, and the author and publishers are grateful for the institution's permission to reproduce them in this volume.

Acknowledgements

The first draft of this book was completed during a period of research and study leave granted by Victoria University of Wellington, New Zealand for which I am very grateful. I also owe a debt of thanks to Brendan Taylor and his colleagues at The Australian National University's Strategic and Defence Studies Centre, Canberra for providing an ideal venue to draft many of the early chapters, and to Michael Wesley for also making this possible. I wish to acknowledge the CartoGIS team at ANU's College of Asia and the Pacific for their careful translation of my ideas for three images into the maps that appear in this volume.

A number of my VUW colleagues and friends at the ANU and beyond have encouraged me along the way towards the completion of this project. In this regard I especially wish to thank David Capie, Manjeet Pardesi, Hugh White, Desmond Ball, Amy King, Marc Lanteigne and Christine Leah. I am especially grateful to Rosemary Foot, who, as the 2014 Sir Howard Kippenberger Chair at Victoria, kept reminding me about the importance of focusing on this book, and to Lawrence Freedman for welcoming the chance to read early chapters and for his constantly reassuring confidence in my work. James Hurndell provided some very helpful research assistance at a crucial stage of the book's evolution.

There is one person above all who made this book a possibility. Steven Kennedy first approached me in Canberra about writing this book about seven years ago. When he visited New Zealand a few years later, his enthusiasm for the project meant that I could not say no a second time. I owe him an apology for not completing the book prior to his retirement from Palgrave Macmillan, since which time I have been very grateful to Stephen Wenham for helping to guide the book project in its later stages, and to Madeleine Hamey-Thomas for providing an ongoing point of contact. The comments from two anonymous reviewers have also made an important difference to what readers will see in the following pages.

As with all of my writing, this book could not have been completed without the understanding and loving support of my very patient wife Catherine, and the members of both our families.

Wellington   Robert Ayson

Chapter 1

Introduction: How Should Asia's Security Be Approached?

The appearance of this book presupposes that its subject matter is sufficiently important to justify the writing of it. In my view, no region matters more than Asia to the world's security. But this position needs to be explained and substantiated, which is the first task of this introductory chapter.

Moreover, if Asia's security is indeed that important, how far and wide does *Asia* extend? This is a challenging question because of the region's political, social and geographical diversity. Should security analysts attempt to identify a core of Asia that matters more than peripheral areas, or should they take a more inclusive approach? This volume takes the latter approach, but this choice also requires an explanation.

Additional queries follow. If a comprehensive understanding of Asia is justified, it is still not clear what the *security* of Asia involves. This means engaging with the lively debate about the nature of security against the competing dynamics of the modern world. Today security means rather more than the simple absence of war, but this does not mean that the control of violence is unimportant, especially for a part of the world such as Asia, where the military capabilities of many important countries are growing.

Even this does not complete the initial picture. Some sense also needs to be made of what the security of Asia as a *region* involves. Here, the argument being posed is that Asia's security is distinct from the sum of the security conditions experienced in the various parts of the region. But this in turn requires the identification of some broad themes that affect Asia's regional security as a whole. In other words, asking the right questions is going to be crucial here.

The importance of Asia's security

Asia's security affairs are inherently important for any consideration of the state of the world. While it may be thought that preparations for armed conflict have become an increasingly rare part of international politics in some parts of the world, that proposition is being tested in Asia. Part of this test comes from the way the region is responding to the rise of China, a phenomenon of remarkable proportions. China's rise has immediate and important security implications as its leaders have access to increased regional influence: a growing economy, for example, supports more advanced military capabilities.

Something similar can be said for the relationship between China, the main rising power of Asia and the entire world, and the United States, which has until recently enjoyed regional predominance and remains the globe's most powerful country. Their relationship, the most important of any in Asia, has a very significant security dimension. One of the main reasons that many of China's neighbours value America's involvement in Asia rests with the large military presence that Washington alone is able to maintain in Asia. How and to what extent China's growing military capabilities are able to undermine America's predominant position in Asia rates among one of the most important trends that the rest of the region is watching very closely. Many countries feel their own security rests primarily on this correlation of American and Chinese forces. As Thomas Christensen (2001) argued over a decade ago, China does not have to catch up with the United States to pose challenges to its predominance.

Moreover, if the proposition that major powers no longer go to war with one another is to be tested anywhere in the world, this test is more than likely to occur in Asia. This is about much more than US–China relations directly. It is about whether they still might find themselves in a serious crisis over the Korean peninsular (as North Korea issues nuclear threats of one kind or another, or collapses) or over the future of Taiwan. It is also about the very tense standoff between China and Japan, and the mix of strategic competition and cooperation in the relationship between India and China. Protracted territorial disputes complicate the picture in both of these cases and leave the region with some of its most significant security challenges. Talk of a replay of the war through misunderstanding that began in Europe in 1914 often raises the possibility of an escalation of the competition between China and Japan over islands in the East China Sea. Many Southeast Asian countries are also paying careful attention to China's growing ability to enforce its extensive claims in the South China Seas, where a number of countries have competing territorial claims. Territorial disputes have a proven record of generating armed conflict in Asia, as the series of limited conflicts between India and Pakistan over the future of Kashmir testify.

This is no hypothetical discussion of security and insecurity. It is the real thing. These concerns about the future of Asia's interstate security relations are amplified by the concentration of military power in the region. Of the handful of countries in the world known to possess nuclear weapons, four are found in Asia (China, India, Pakistan and North Korea) and the two largest holders of such weapons (the United States and Russia) continue to have an influence on Asia's overall strategic balance. Much of the world's military modernisation is occurring in Asia, generating some of the most lucrative international markets for arms manufacturers, and in particular for advanced maritime weapons systems including submarines, advanced fighter aircraft, missiles, and advanced command and control systems. If arms races genuinely exist anywhere in the world, and if they threaten security by making war more likely, they are most probably to be found in and around Asia.

Security is vitally important, not just for relations between the states of Asia. It is also not difficult to find security challenges playing a prominent role in the domestic political affairs of a number of Asian countries. A number of unresolved internal conflicts are scattered throughout Asia, including in Thailand, the Philippines and Myanmar (Burma). The role of the armed forces in domestic affairs has declined markedly, but it is still a factor in quite a few cases. One of these is Pakistan, whose integrity and cohesiveness always seems to be in doubt, with continuing violence by militant groups challenging the authority of the central government, and it is not many years since Sri Lanka was convulsed by a bloody and tragic civil war. China's leaders continually worry that the legitimacy of the Community Party's rule is being challenged by separatist violence in Xinjiang. India continues to be faced with a troubling insurgency conducted by Naxalite guerillas, and much smaller countries on the other side of the region have experienced serious internal conflicts. These include civil wars that were occurring not so long ago in parts of Papua New Guinea and Solomon Islands.

To this assortment of interstate and domestic security challenges can be added a number of important transnational (or 'non-traditional') security concerns in Asia. Indeed, the region is an important testing ground for the intriguing proposition that the transnational security agenda has become the dominant one internationally. For example, concerns that various terrorist movements can develop transnational networks have tended to include Southeast Asia high on the list (including groups that have operated in Indonesia, Malaysia, the Philippines and Thailand). Maritime Southeast Asia has also been home to some of the most persistent examples of piracy. The international trade in illicit drugs includes Asia as a vital centre: the flows of opiates from Myanmar and Afghanistan into neighbouring countries have had serious social effects.

A number of Asia's most populous countries also feature as particularly susceptible to the effects of climate change, which include rising sea levels. The low-lying and highly populated delta portion of Bangladesh, for example, is likely to witness serious problems of this sort. Some small Pacific Island countries are likely to disappear entirely from the map. Extreme weather events, including prolonged droughts, seem almost certain to add to the social and political challenges of already water-stressed parts of India and China. If climate change is likely to exacerbate tensions anywhere in the world, Asia will provide some of the most important potential examples. The notion that natural disasters (including typhoons and other severe storms) constitute an emerging variety of a new security challenge also directs attention to Asia.

This initial depiction of Asia risks giving the impression that the region is a hotbed of bad news and distress. But the record over more recent decades suggests that this would be a very inaccurate portrayal. Asia has been booming. In many ways the outstanding good news story of the late-twentieth and early-twenty-first centuries is the way so many parts of Asia have been able to lift hundreds of millions of people into higher standards of living. This phenomenon has been most impressive and obvious in China, whose economic transformation is little short of a modern miracle. But it was Japan in earlier decades which forged a model of export-led economic growth that many Asian economies were able to replicate.

If the twenty-first century is an Asian one, it is largely a result of this impressive economic performance, which has rightly attracted global attention. Producers and suppliers of energy and other goods find that Asia offers the most vibrant markets for their products. China has become the number one trading partner, not just for a slew of countries in its own Asian neighbourhood, but for a range of countries in Europe, the Middle East, Africa and the Americas. China is also at the centre of a vast interlocking network of production chains as components from different national origins are brought together and assembled using supply chains that criss-cross around Asia and beyond.

As they grow richer, the Asian economies are not operating in a vacuum but are dependent on each other. Global prosperity also depends increasingly on Asia's economic dynamism and the strong links between many of Asia's economies. To some degree at least, Asia's prosperity is important for the whole world. But this also provides a further reason for everyone to take an interest in Asia's security.

In the first instance, increasing wealth in many parts of Asia depends in part on stability in the region, including the relative absence of large interstate wars and of constant violent threats, which can upset domestic politics and generate major uncertainty. Asia does not have to be completely peaceful to be rich. But it can be hard to achieve economic growth when constant threats of war and revolt abound. This means that unless the region's leaders do not think economic prosperity is an important goal (and perhaps only North Korea acts as if this might not be the case), they are bound to see at least some merit in maintaining a positive security environment. In the second instance, this also means that Asia's security (and insecurity) matters to the rest of the world. Any country whose prosperity depends in part on Asia's rising wealth will also have an interest in a secure and stable Asia, even if it has a very limited ability to affect those security conditions.

This also means that there is an increasing global dependence on political relationships within Asia which, along with economic growth, bring important security implications. While many of Asia's main countries are increasingly prosperous and politically confident, they are also nervous about what their future will hold. In part, their growing confidence can become a dual-edged sword, because nationalist sentiment can easily become a security problem in relations between states. Larger economies allow military capabilities to be developed that can in turn threaten the neighbourhood in which they operate. Old enmities which may have been hidden while Asian countries were weaker are now coming to the fore as new possibilities for influence become available.

The pace of change in Asia, economically and socially, has challenged traditional institutions and patterns of political order in a range of Asian countries. The region is known much more for its political diversity than for its political unity. Asia is home to the world's largest democracy (India) and the world's largest Muslim country (Indonesia), which itself has been democratic for over a decade. China and Vietnam are both fascinating examples of large countries that have retained one-party forms of government led by communist parties but which have also developed market economies. North Korea has retained the former and not engaged in the latter, whereas South Korea is a leading example of a successful transition from military authoritarianism to liberal democracy.

This is not to say that regional cooperation is infeasible. Asia is also home to one of the most impressive attempts at regional cooperation amid political diversity. The Association of Southeast Asian Nations (ASEAN) is the hub of a wider array of multilateral organisations in Asia. But it is not clear whether any of these groupings can really make a difference when security crises arise. This creates space for the United States to play a role as a potential security guarantor. But while a number of Asian countries are long-standing American allies and quite a few are emerging security partners, others remain suspicious of Washington's intentions. With or without that American role, few Asian countries are confident that the natural tendency of their neighbours will be to cooperate when rumours and threats of serious conflict arise. That lack of confidence should also mean that a wider world more dependent upon Asia is more vulnerable to the shortages in political trust within the region.

Asia's evolution

People from the Western world have been talking about Asia for thousands of years. The ancient Greek historian Herodotus used the term Asia to refer to the lands to the east of the Aegean Sea. When the Christian apostle Paul and his companions were spreading their message 2,000 years ago, their destinations included parts of what was then called Asia Minor. This area is now regarded as part of today's Turkey, and the area then known as Asia Major is now called Iraq. Indeed, for many observers today, Turkey remains on the cusp between Europe and the Middle East. The idea of a connection between Europe in the west and Asia in the east has also been reflected in the notion of a vast 'Eurasian' area, the control of which was considered by geopolitical thinkers such as Halford Mackinder (1904) and Nicholas Spykman (1944) to be the key to determining world history.

It was the growth of European empires that brought more modern (and more eastward) notions of Asia into being. Indeed, for much of its history as an idea 'Asia' remained a largely Western conception. Some earlier connections between Europe and Asia, including the Silk Road used by merchants in Marco Polo's time, were developed by land. But a new age opened up at the end of the fifteenth century when Portuguese vessels began rounding the southernmost tip of Africa, first connecting Europe to India by sea and then moving further eastwards to what would be named the Malacca Strait area (near Indonesia, Singapore and Malaysia). Partly because of commercial interests, the core of Asia moved eastwards too. But Asia was not a unified political area. Instead, it was divided into the territorial claims of competing European empires. The objects of this competition included China (extending to Macau, Hong Kong and Taiwan) and the Korean peninsula, both parts of what is now called East Asia. Also included in this competition were the maritime archipelagic territories in the Malay world and the Philippines, and the mainland parts of what is now called Southeast Asia, including Vietnam and Burma. European imperial territories included the lands in and around the Indian subcontinent, including what are in the present-day India, Pakistan, Bangladesh and Sri Lanka. Most of these places were once part of British India but are now sovereign states collectively referred to as South Asia. Afghanistan, where England and Russia played a great game of imperial competition, is in turn part of a stretch of lands now referred to as Central Asia.

With so many of these countries divided up among the competing European empires (and in China's case, some of its coastal areas divided between them), the idea of Asia as a single region was as much a conception as a reality. As a concept Asia probably made more sense in cultural and religious rather than political terms. It was a place of historical civilisations (Chinese, Japanese, Vietnamese, Javanese and others) and major faiths (including Hinduism and Buddhism). The spread of Islam as far eastwards as the Malay archipelago (where it is still evident today as the majority religion in Indonesia, Malaysia and Brunei) was also an important piece of earlier 'glue' that held a fair bit of today's Asia together. It must be said, however, that trading links with the Arab world provided many of the sources of this connection. As has often been the case in world politics, economic power and religious influence went hand in hand.

Asia made little sense as a political concept, partly because there were few local rivals to the powerful European empires. The strongest Asian country of all time, imperial China, had previously presided over a hierarchical system of relations in its surrounds in which it was firmly on top. This involved a China-centric form of interstate relations with neighbours who, as Fairbank and Ch'en (1968) have shown, paid tribute to Peking (now Beijing). But China was in fact getting weaker just as the European empires were getting stronger, and a number of these were to carve up parts of coastal China for their own trading interests. India, the other largest Asian country by population, was neither unified nor united for much of this time, but was instead a patchwork of smaller principalities. Like most of the rest of the world, the Asian countries were not included as independent political actors in the European-dominated system of sovereign states that had been evolving since the Peace of Westphalia had ended the era of European religious wars in 1648.

This exclusion began to change in the late nineteenth century, when a rapidly modernising Japan (which had evaded Western colonial conquest) forced its way into that expanding states system. Japan's defeat of Russia in 1905 marked a significant change in Western perceptions of Asia, for all-too-obvious reasons. This was a crucial moment for what Hedley Bull and Adam Watson (1985) call the expansion of international society, as non-European states began to participate in the international system as independent actors. The further growth of Japan's influence by military means was one of the formative moments for Asia as we know it today. Japan's conquests were many decades in the making. At the start of the twentieth century these included Korea and Taiwan. Then came Manchuria as part of the bitter imprint of Japanese militarism in China, followed by the development of the deceptively entitled 'Greater East Asia Co-Prosperity Sphere'. As a consequence, Japan effectively held sway over a great deal of what is now called East Asia. Further south, Japan's forces controlled parts of the Philippines, the Malay world and New Guinea, from which they appeared also to threaten Australia. Further west, Japan's position in Burma connected Southeast and South Asia, threatening India, the crown in Britain's empire. To the east, Japan had also brought the United States into the war through the infamous 1941 Pearl Harbor attack on Hawaii, a group of islands lying thousands of kilometres into the Pacific Ocean. As much by accident as by design, Japan's military expansion made sense of Asia as a connected entity stretching both from east to west and from north to south.

Opposition to Japan's control also generated unity of a political sort among many of the local Asian elites, who were also determined that their countries should not return to European tutelage as colonial possessions. The political map of today's Asia is largely a product of this post-war process of sovereign independence, exploding the international ranks of nation-states and United Nations members. The most important of these transitions in Asia came in India, whose independence from Britain in 1947 foretold the end of what had been the strongest and most widespread European empire in Asia. India was initially keen to build bridges with other newly independent Asian countries, but became rather inward-looking in its foreign policy, preaching self-sufficiency rather than mutual economic reliance.

While this did little to knit Asia together as a regional system of states, India and Indonesia worked together to give a new grouping of countries a strong Asian component. The non-aligned movement, which began formally with the Bandung conference held in Indonesia in 1955, was an attempt to escape the shackles of the Cold War competition between the Eastern and Western blocs led by the Soviet Union and the United States, respectively. Originating in the division of Europe after the Second World War, the Cold War had very clearly spread to Asia by 1950, when North Korea attacked South Korea. Stalin's Soviet Russia was not the only backer of Kim-il Sung's North Korea in this unprovoked aggression: neighbouring China also offered political and then military support, having been the location for Mao Zedong's successful communist revolution in 1949.

The subsequent view that Asia comprised a series of dominoes ready to fall under communist influence, which helped to justify American intervention in Vietnam in the following decade suggested an increasingly interconnected Asia (albeit of a negative variety). So too did evidence that Mao's China, which clashed with India in the early 1960s, was seeking to export revolution by assisting communist parties in a range of Southeast Asian countries. But in another ironic twist, the fear of mutual interference in each other's fragile domestic politics drew a number of these same countries together in the formation of ASEAN in 1967. Only a small group at first, this development was a clear sign of the potential for common endeavour between some of Asia's young, independent and sovereign states (Indonesia, Malaysia, the Philippines and Singapore) as well as Thailand, which had managed to elude colonial control.

If the original ambition shared by Asian governments of all stripes was political autonomy from external domination, the second was the pursuit of economic prosperity as a pathway to national development. In their quest there was an undoubted regional example in Japan, which for many post-war years served as both a model for export-led growth and as an engine (as a market and source of investment) for much of Asia's growing dynamism. Japan also changed wider international views of Asia by becoming a leading trading partner of so many countries in other parts of the world. The similar export-led trajectories of a series of other market economies in East Asia, including South Korea, Taiwan, Hong Kong and Singapore meant that after the mid-1970s Asia was becoming better known for economic growth than for political revolution and violence. The trump card in this sequence was the decision of the Communist Party of China, under the new leadership of Deng Xiaoping, to take advantage of capitalism and open up to the international economy. Alongside the 1970s rapprochement between the United States and China, this brought China fully into the modern states system.

While it would take many years to occur, China eventually surpassed Japan as Asia's leading economy and the main engine for growth in the region. China's massive appetite for resources, including for energy, when coupled with the continuing demand from Japan, a rapidly growing India, South Korea, and others, has helped to create new supply lines that connect Asia to the Middle East, raising the wider profile of the region. Maritime Southeast Asia, including the historically significant Malacca Strait, sits in these projections as a gateway between the Indian and Pacific Oceans. Whether or not the sense of regional transformation that has been presented by Robert Kaplan (2010) really holds, there seems to be little doubt that an increasingly interconnected and wider Asia is in the process of being formed.

The extent of twenty-first-century Asia

This all leads to an important contemporary question about the extent of Asia: what comprises today's Asia? Before responding to this immediate question, it is necessary to make one point that should seem clear from the preceding analysis. Understandings of Asia need to be dynamic and adaptable, because this is exactly how Asia has been as an area in recent decades, even centuries. But the fluidity and subjectivity of these approaches does not suddenly make geographical considerations irrelevant.

Perhaps it is best to start with those countries that are regarded as part and parcel of Asia, no matter who is setting the boundaries. Here the obvious starting points are the countries of East Asia, which can be divided into two increasingly connected portions. There are the countries of North Asia: China, Japan, North Korea, South Korea and Taiwan (all of which, apart from the last one listed, are recognised as sovereign states). Landlocked Mongolia, often forgotten, should also be included in this group. Then there are the countries of Southeast Asia. There is a maritime group comprising Indonesia, Singapore, Malaysia, the Philippines, Brunei, and the more recently independent sovereign state of Timor Leste (East Timor), and the continental Southeast Asian countries: Vietnam, Thailand, Laos, Cambodia and Myanmar. All of these, with the exception of Timor Leste, are members of ASEAN, which is the hub of regional institutional cooperation for Asia as a whole.

It is tempting to stop there. To consider Asia simply as East Asia (and Asian Security as East Asian Security) is not as restrictive or silly as it may sound at first. Included here are almost all of Asia's booming economies. First, the two historical great powers (China and Japan) around which so much of Asia's destiny has been shaped. Then there are the gateways between the Indian and Pacific Oceans in the Southeast Asian maritime archipelago. As Asia matters increasingly to the distribution of world power, it is principally in East Asia that this is occurring. East Asia is where the main thrust of American influence in Asia is directed, and it is where the future of the US–China relationship will more often than not be played out. And because China's rise is such a dominant feature for this part of the world, China's own physical location tends to encourage an East Asian focus that squeezes out other, more expansive, notions of the Asian region.

This view also features in some of the leading debates on Asia's security in the academic literature. In the leading example of the pessimistic argument that Europe's violent past could be Asia's future, for example, Aaron Friedberg (1993/4) does mention India, but the focus of so much of his analysis falls squarely on East Asian dynamics. In trying to debunk Friedberg's thesis a decade on, and in arguing that an understanding of Asian security requires different (non-European) models suggesting that a less competitive region is possible, David Kang (2003) deliberately and explicitly restrains his area of emphasis to North and Southeast Asia. It is the rise of China, and what this rise means for the countries that surround it, and for American policy, that is the main reason for this debate in the first place.

East Asia will remain the bedrock of the conception of Asia used in this book. But it is still necessary to develop a more inclusive conception of the region to guide the security analysis is present in the coming chapters. First, while sometimes seeming to exist as an almost separate enclave from East Asia, parts of South Asia deserve to be included in the mix. The principle reason for this is India, which has the potential to be the second Asian world power (after China) in the next fifty years. There are few more important relationships between two Asian states than that between China and India (who, as later chapters will show, are strategic competitors). As India's local rival, and as an ally of China, Pakistan is part of this equation, as are to a lesser degree Nepal and Bhutan, which both sit between Asia's two giants. If South Asia is to be considered part of Asia, this also means including Sri Lanka and Bangladesh, which have both faced significant internal security challenges of their own and both have close historical connections to India.

How far further westwards and northwards one's conception of Asia should go is a matter of conjecture. There is a range of countries that are commonly regarded as constituting 'Central Asia', a misleading term because they are seldom very central in the way that many analysts approach Asia as a region. As the most prominent of these countries, an often troubled Afghanistan cannot be completely ignored in any comprehensive assessment of Asian security. But the main reason to include it is because of what developments inside and across Afghanistan's borders mean for India, Pakistan and China. The remaining Central Asian republics are all former parts of the Soviet Union. Kazakhstan, Kyrgyzstan, Tajikistan, Turkmenistan and Uzbekistan (whose combined populations are greater than 60 million) do not often feature in mainstream analyses of Asian security. But it is necessary to keep them in mind as parts of what this book will refer to as 'Asia's borderlands'. Many of them belong, with China and Russia, to the Shanghai Cooperation Organisation. All are finding that their economic destinies rely increasingly on their growing links with China. They are, at the very least, part of Asia's borderlands.

This borderlands idea also applies in other directions. The Russian state symbol from the imperial era has two eagles. One looks to the west to Europe and the other to the east, and what was once called the 'Asiatic' part of Russian national sentiment has sometimes been regarded as the dominant influence. Russia has a long border with China, contests a number of northern islands with Japan, and is also a neighbour of North Korea. By virtue of its territorial extent, Russia has as at least as much claim to be a Pacific power as the United States. Moreover, as was very evident in the Cold War (as well as beforehand), Russia has been a significant strategic influence in East Asia, even though so much of recent attention has been rightly devoted to Moscow's intervention in Ukraine that began in early 2014. This is not to say that Russia is part of Asia. But it is to say that it is unwise to study Asian security without some reference to the part that Russia has to play.

The latter formula applies even more, it would seem, to the United States, though less so for geographical and territorial reasons, and more for military, economic and diplomatic ones. In both Guam and the state of Hawaii, the United States clearly has territory close to Asia. But perhaps even more than these factors, the 7th fleet of the United States Navy, along with US forces based in Japan and South Korea, speak of a power-projection capability into Asia that no other country in the world can match. Moreover, in a collection of security alliances that have operated since the 1950s (including those with Japan, South Korea, Thailand and the Philippines) the United States has been at the centre of a unique Asian security system. US–Asian economic connections are very strong (including the mutual dependence that links China and the United States) and almost every other country in the region has been keen to see the United States included in Asia's regional diplomatic forums. If the USA is not part of Asia in a strictly cultural or geographical sense, it cannot be excluded from calculations of where this part of the world is heading. The same cannot really be said for Canada, Mexico and Chile, which also have Pacific Ocean coastlines, but which do not have much of a profile in Asia.

But some other Asia-Pacific countries do need to be included in the depiction of the border zones of Asia that feature in this volume. This includes Australia, an island continent whose northern reaches are only a few hundred kilometres from Indonesia. Like its neighbour, New Zealand, Australia has become increasingly integrated with many of East Asia's economic and diplomatic processes. Papua New Guinea, which shares a land-border with Indonesia, should definitely be part of this extended list, as should at least some of the Pacific Island countries (including Solomon Islands and Fiji) whose destinies will be shaped more by Asia in future years and less by their traditional Western points of contact.

In sum, this offers a considerably expanded understanding of Asia, in contrast to some of the standard approaches that limit themselves to an East Asian focus. As Map 1.1 illustrates, the understanding of the region being used in this book involves a largely East Asian and South Asian core, surrounded by Asia's borderlands, which include parts of Central Asia, the Pacific territory of the United States and Russia, as well as Australasia and the South Pacific. Readers will find that this broader notion of Asia will introduce some approaches to security that could more easily be omitted in an exclusively East Asia vision. The diversity of the regional security experience will increase as a result. But to leave these factors out would make for an incomplete, if simpler, volume.

The extent of Asia is a work in progress, and will continue to be so as economic, political and military relationships evolve. The notion of a broader Asia is not a way of avoiding the importance of the central role played by the East Asian core. The wider connections into the borderlands are often being shaped by the dynamism of the East Asian countries and, in particular, the ties that are being developed as China has been rising economically, diplomatically and militarily. These dependencies may increasingly bring about the wider region that Barry Buzan and Ole Waever (2004) have referred to rather mechanistically as an Asian 'security super-complex'. But an increasingly integrated (and larger) Asia will not displace the political, cultural and geographical diversity of the region, which makes it both challenging and interesting to work out what Asia's security really means.

13

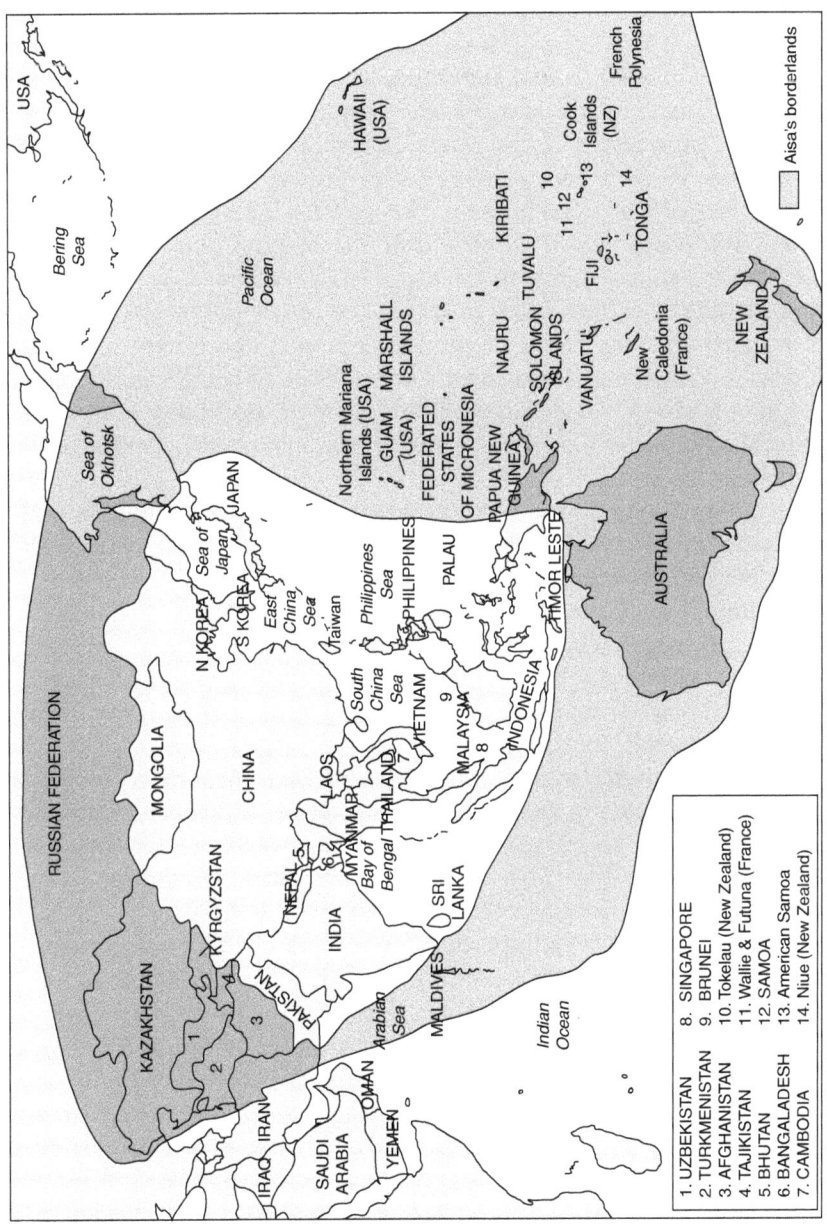

Map 1.1 *Asia and its borderlands*

The meaning of security

It should not be any surprise to learn that scholars frequently disagree about the issues that shape Asia's security. Some believe that Asia's security is driven by the relationships between the major powers: first, between China and the United States, but also between Japan, China and the USA, and increasingly between China, the USA and a rising India. Others focus on a series of hotspots or flashpoints in relations between states in Asia. These include the Korean peninsula, which is still technically in a state of war; Kashmir, a territory contested by India and Pakistan; and the potential for conflict over the South China Sea or East China Sea. For some country specialists, Asia's security depends less on the quality of the relations between states, and rests more on the possibility of internal disorder. Thailand and the Philippines both have southern insurgencies, Pakistan is regarded by some as close to being a failed state, and China's leadership is sometimes petrified by the possibility of internal disunity. Should one of Asia's giants collapse or splinter, the consequences could be very troubling.

Other analysts point to the emergence of transnational problems and issues that appear to pose broader security challenges to all, or at least most of, the countries of Asia. These include transnational terrorist networks, criminal syndicates and the problem of piracy. They also include the extent to which climate change, by increasing extreme weather events, and environmental degradation, could generate tensions between Asian countries over shared and scarce resources, and generate large movements of people across borders.

All of these features of the Asian security environment will be discussed in the chapters to come. But their impressive diversity does not render it impossible to have a meaningful understanding of security as a concept for the analysis that occurs in this volume. It is certainly permissible to have severe disagreements about what is Asia's most important security problem. It is important for readers of this book to query whether some security issues are really as serious and important as they sound. But it is still possible to find a basic agreement that security is about safety and protection in the face of potentially harmful factors. Security is almost always about the relationship an actor has with its surrounding environment. Security is defined and clarified as much by its absence as its presence. Exposure to threats of harm, which insecurity is all about, quickly serve as reminders of the importance of a more secure environment. Security, then, is a condition or quality of the environment in which an actor exists.

One immediate question is the extent to which security and insecurity are qualities of existence that are real or imagined. This dual notion of security, as a condition that is experienced psychologically as well as physically, is summed up in perhaps the most influential definition of the concept that has been offered to date. This comes from an article by the political scientist Arnold Wolfers (1952, pp. 484–5), who observed that 'security, in an objective sense, measures the absence of threats to acquired values, in a subjective sense, the absence of fear that such values will be attacked'. In other words, judgements about security can in part be statements about what is really out there: objective assessments about the existence of things that might do harm to states, communities and peoples. But these judgements are also likely to involve a significant subjective element: how actors feel about the environment around them, and what their view of the world encourages them to notice and overlook.

The relationship between these objective and subjective elements of security analysis plays a crucial role in this book. Sometimes, for example, actually existing problems are magnified to a level much greater than they might deserve. A build-up of armaments (which actually exists) can be read as a threat (which it might or might not be). Or sometimes a dangerous problem is completely missed or ignored. This raises the crucial role of the possibility of misperception, which as Robert Jervis (1976) has illustrated, has a very significant role in many international conflicts. The relationship between the objective and subjective elements of security can also be exploited. Terrorists, for example, are seldom strong enough to force states to do things (they are in that sense objectively weak). But in intimidating civilian populations with acts and threats of violence, they seek to build up pressure from below (they are in that sense subjectively stronger). Perceptions count.

This means that *Asia's Security* is about much more than material factors. It is about the psychology of perceptions. How China and the United States relate to one another (and whether they regard one another as partners or adversaries or somewhere in between) relies in part on the military and economic resources they understand one another to possess. But it also relies on how they are inclined to view each other and interpret those resources.

Views of what security really involves as a concept tend to carry marks of their origins. The definition being discussed here comes from the height of the Cold War, and from a Western country (the United States) whose main concern was the external threats posed to the West by the Soviet Union. As Wolfers (1952, p. 484) mentions earlier in the same article, in its 'common usage' at the time, 'security rises and falls with the ability of a nation to deter an attack, or to defeat it'. Clearly, that common usage presumes that the threat comes from outside and from another state or group of states, and that it consists of their capacity to use force against you.

Some, but by no means all, of the Asian security issues and strategies this book investigates fit into this pattern. Some of the problems come from within the state, and others do not feature threats of organised violence. How much one is willing to accept the broader understandings of security that follow is a matter for individual readers to determine for themselves as they work their way through this volume. But in treating security in terms of the management of threats to 'acquired values', Wolfers does all students of Asia's security a considerable service. One way of interpreting this statement is to argue that security involves the protection of those things that a particular society has learned to hold especially dear. These are its acquired values. Security in its rawest form is about the very survival of these things that are so important, and indeed vital, to any particular social group. What, then, are these values, who gets to define them, and how might they come to be threatened? These are questions around which so much of the security debate occurs.

Survival might be regarded as one of the core values of any group or individual, which needs to be promoted if anything else is to be possible; the quality of life makes little sense without life itself. The famous depiction of the state of nature by the English philosopher, Thomas Hobbes (1651, p. 186), as a life that is 'nasty, brutish and short' points to a pervasive insecurity of a very fundamental variety: the insecurity of life itself. Hobbes was reflecting the chaos and uncertainty that England's civil war had wrought in the mid-seventeenth century, but this principle can be taken as a general one. The ability to persist is itself a core value that needs protecting if security is to make much sense at all. And it is not surprising, then, that an act of war, which by its nature can put at risk the very existence of individuals and groups, has often been seen as the primary problem around which our understanding of security has evolved. Indeed, immediately before his famous statement about the 'naturall [sic] condition of mankind', Hobbes refers to 'continuall feare [sic], and danger of violent death'.

Whether war is increasingly obsolete or an enduring part of the interactions between human groups, including in Asia, it is difficult to think of an event that could do more harm to so many of the things that are valued. And the quest to avoid war, or at least to manage it, and sometimes to use it against others, has been a very important part of security analysis down the years. It is still at the centre of much of the analysis of Asia's security: this is a region where war may now be less frequent than in the not too distant past, but it has by no means been made impossible or unthinkable. Warlike acts (including threats of organised violence), and war itself, are still features of regional politics.

Alternatively, if climate change is regarded as a potential threat to the survival of human communities, then this emerging problem may be regarded as a direct security challenge (including in Asia) for rather similar reasons. But even a modicum of climate change, the whispers of a potential revolution, or the fears that may come from the risk of a distant war can also affect the *conditions* of existence. Here too there are acquired values that communities wish to protect and promote. These may include the freedom to go about one's daily life without the threat of state coercion. They may include the ability to exchange goods and services with other countries without fear of a violent disruption to trade and commerce. They may include the integrity of territorial claims under international law against the intrusion of predators. Security is about more than the possibility or impossibility of survival itself.

In each case, a social group is seeking to secure things that matter dearly in the face of threats that already exist or are potentially to come, objectively real and subjectively felt. Security may seem a simple business because this logic so often repeats itself, but it is also complex because this logic repeats itself in so many different circumstances. This is certainly the case in Asia. Some countries will regard their number one security challenge as the danger of attack from a hostile neighbour. It seems eminently fair, for example, that South Korea should regard its security environment in this way, given the armed forces that North Korea has amassed and the angry rhetoric that often comes from its regime. But in other countries there will be a greater concern about security challenges from within. The Philippines government, for example, may be increasingly worried about China's claims over disputed reefs and islands, but it remains deeply concerned about a Muslim insurgency in the south of the country. Australian politicians are often exercised over boatloads of unauthorised asylum seekers who have set sail from Indonesia. Their insecurity may be more imagined than real, but a great deal of effort in Canberra has gone into responding to this non-state challenge.

This last example is important because it is a reminder of the main political actors involved in almost all of the security issues that occur in Asia. These actors are sovereign states which sometimes are also referred to as nation-states (and sometimes more loosely and simply as states, as in the *states parties* who sign and are bound by international treaties). It is useful to break down the nation-state combination into its component parts. The 'state' part of the term refers to the permanent institutions that define a given sovereign political unit. These institutions include the armed forces, but they also include the rules by which the armed forces and other features of the state are used and organised. The 'nation' part of this formula refers to the people and territory to which the sovereignty (or legitimate authority) of those state institutions extend.

In reality, of course, these features intersect and can be difficult to distinguish, but they add up to two important considerations for security.

The first is that the daily working of international relations (including in Asia) would not be possible if these separate nation-states did not recognise the sovereign independence of one another. Not doing so is a potential challenge to mutual co-existence. If sovereignty is seen as a value bound up in the survival of a nation-state, challenges to sovereign independence can rank among the most worrying of security problems. The second is that the daily workings of domestic political affairs are made especially challenging when the sovereign authority of the state is challenged from within. Breakaway movements, coups and civil wars are all reminders of the linkages between security and the domestic aspects of sovereign statehood. One challenge here is that in a number of cases, including in Asia, there is more than one group that believes itself to be a separate nation within the boundaries of an existing single state.

These issues will arise in a number of the coming chapters. The important point to note at this early stage is that the different levels in which security almost always presumes sovereign states to be the most significant political actors. This is certainly the case when scholars are referring to international security: the security of the system of sovereign states which can be threatened by one or more of its members. An important part of international security (which is, more properly speaking, interstate security) occurs in Asia, and is the focus of a number of the chapters of this book.

At other times scholars will be evaluating national security – the security of an individual sovereign state. This is sometimes in the context of threats from other nation-states, but it also often focuses on real and perceived threats from within. Concerns about domestic stability that have been prevalent in much of Asia and its borderlands testify to the significance of this type of security analysis in the region on which this book is focused. Again, the state is the main actor: often as the unit whose authority is challenged, but also often as the main responder to those challenges.

There is a significant and emerging body of scholarship on human security which deals with the security interests and concerns of individuals and groups. This may sound as if the nation-state is being left out of the picture, and human security theorists often argue that their approach is a response to an overly state-centric set of assumptions in the existing security literature. But the sovereign state still plays a very big role. Sometimes this is in a negative sense: often human security measures the vulnerability of groups and individuals to the practices of the military and police forces of the state (which have become threats to the citizens they are supposed to protect). Sometimes this is in a more positive sense: states are expected to protect and promote human security (and if they do not, it is not clear who would). As later parts of this book will show, human security arguments are a significant part of a comprehensive understanding of Asia's security.

The meaning of Asia's security

The different but often overlapping concerns that these various approaches to security reflect are not distributed evenly across Asia and its borderlands. In different parts of the region, security makes sense in quite different circumstances. This leads to an important question: is Asia's security simply the sum of these different and various parts? Similarly, is Asia's *insecurity* simply a combination of the security problems that are being experienced across the region?

The answer to these questions contained in this book is a resounding no. There are two reasons for this negative response. The first is impracticality. It is simply impossible to appreciate all the security conditions and challenges across the very extensive part of the world that Asia and its borderlands represents. This would be a truly herculean task, and would be so even if this book had restricted itself to East Asian security alone. In fact, even to consider *all* the security concerns in the many parts of China or India alone would be beyond this or any volume. This book takes a comprehensive look at Asian security and understands Asia in a broad sense, as has already been noted. But it cannot claim to cover all of the issues everywhere in Asia and its borderlands. The second reason for the negative answer is analytical. If a complete understanding of all of this region's security issues was possible, bringing them together would not offer a meaningful understanding of Asia's security. The security of a region is not the sum of all the various parts. It is something distinctly different.

A sound appreciation of Asia's security involves differentiation and prioritisation. Any security problem anywhere in Asia can certainly harm things of value to some individual or group. For example, tribal violence is fairly common in some of the southern highlands of Papua New Guinea. This violence is a threat to the security of local citizens and communities, and it sometimes attracts the attention of the national government in the capital city, Port Moresby. But these occasions of violence do not generally reverberate around Asia. Similarly, while attacks on New Zealand and Afghanistan forces in the northern parts of the isolated Bamiyan province certainly affected security in that part of the country, they did little to alter the overall security picture for Afghanistan, and they made very little contribution to security in the rest of the Asian region and its borderlands.

This is not to suggest that small events cannot have big, and even catastrophic, consequences: the spark that ignited the First World War in 1914 was set off by a Serbian nationalist in Sarajevo. But the assassination of the Austrian Archduke Franz Ferdinand began a chain of military mobilisation that was to result in a catastrophic war affecting the whole of Europe and beyond. While seemingly small events can and do have wider systemic effects, many local crises do not have a wider impact. Every security problem in Asia and its borderlands is of concern to some group, and in a number of cases, to a state and some of its immediate neighbours. But often the rest of the region remains largely unaffected.

By contrast, there are some security problems that can almost be guaranteed to have wider regional implications. Should Japan and China be unable to contain their differences and find themselves in a major crisis over disputed islands in the East China Sea, the rest of Asia should probably be watching carefully. An escalation to war could bring in Japan's major ally, the United States, and it is not clear how and when such a conflict would end. Everyone in Asia, including those simply having important trading links with Japan and China, could be affected negatively. The security of Asia (and not just security in a particular part of the region) would be at stake. It does not have to be an interstate security problem to have these wider effects. If one of Asia's largest countries found itself challenged by substantial political instabilities and civil war, the spillover effects for others could be very difficult. Indeed, for analysts such as Gordon Chang (2010), the spectre of a weak China is much more worrying than the implications of a China that is becoming too strong. A serious collapse in America's economy or its willingness to remain a major military player in Asia would also be very likely to have ripple effects that almost every country in the region would notice.

Asia's regional security

This book is primarily interested in the security issues that can cause these wider effects or ripples across Asia (or at least ripples that affect a good many places in Asia). Another way of saying this is that the focus is on Asia's *regional* security. But what does it mean to treat Asia as a region? (And does Asia in fact qualify?) A region is partly a geographical concept. The component countries of a region are normally located next to one another: *proximity* counts. But a region makes sense also because of the relationships between its proximate components. The countries of a region normally share something in their relations with one another that they do not share to the same intensity with those outside it. It might seem that these relationships are about *similarity*. In this case, Asia could be understood as a region if it consisted of a group of countries with similar cultures and ways of organising themselves, both socially and politically. To some extent this seems to apply to parts of Asia. For example, Southeast Asian elites sometimes refer to an ASEAN way of doing business that emphasises consensus and informality. Moreover, traditions that insist on the obligation of the individual to the collective seem to shape life in a range of East Asian countries.

But if Asia is a region, it is simply too diverse for similarity to be its most meaningful characteristic. Instead, there is a need for some sort of transmission belt that links phenomena in the various parts of Asia and its borderlands. This approach draws attention to relationships of *interdependence* between groups of countries which cannot properly seek their security objectives without paying heed to what their regional neighbours and partners are expected to do. They and their security are dependent on each other. At times this security interdependence is particularly intense in parts of Asia. The security of Pakistan, for example, cannot be understood without some understanding of how it relates to the security of India, and vice versa. There is also close security interdependence between Malaysia and Singapore, North Korea and South Korea, Australia and Indonesia, and Japan and China. These pairings can also be interdependent with one another: what happens in the security relationship between Japan and China can easily effect the security outlooks of North and South Korea. The ripple effect referred to earlier is a metaphor for security interdependence on a grand scale, where an event in one part of Asia has security effects throughout the region. As will become clear in various parts of this book, some issues and actors have a greater tendency than others to generate region-wide security effects.

Viewing Asia as a region of interdependent parts allows for an analytical framework to recognise those issues that affect the region as a whole. Indeed, it might be possible to go as far to speak of Asia and its borderlands as a regional system. According to the American scholar, Robert Jervis (1997, p. 6):

> We are dealing with a system when (a) a set of units or elements is interconnected so that changes in some elements or their relations produce changes in other parts of the system, and (b) the entire system exhibits properties and behaviors that are different from those of the parts.

The first of these criteria is talking about the ripple effect, a metaphor that Jervis (1997, p. 9) himself introduces and uses. The second confirms the earlier observation that it is not plausible to measure Asia's security simply by adding up as many of the security issues, problems and threats as can be identified. It means that Asian regional security has a quality that is distinct from the specific security concerns and issues that might be found in any individual part of it. Moreover, it is quite possible that insecurity for one part of Asia can exist at the same time as the overall security assessment for the region is much more positive. In the late

1970s, for example, Vietnam was regarded as a security threat by a number of other Southeast Asian countries, and this common concern generated a sense of sub-regional unity. The reverse might also apply: Pakistan tends to feel insecure if India is feeling much more positive about its own situation. A strong and secure India can appear as a threat to Pakistan's future. Measuring regional security in this situation cannot just be a matter of adding up the respective security perceptions of the component countries of Asia.

Even so, the capacity for shared security concerns to unite regional neighbours suggests that, in addition to proximity and interdependence, a regional security system can also feature a degree of security *integration*. Here, the component actors of a region come to act increasingly as if they constitute a single unit – which again would amplify the differences between the security of a region and the security of its component parts. This is no small matter, because the nation-states that are normally seen as the main political actors in any region in the world attach high importance to their sovereign independence. It may be possible one day, as Barry Buzan (2000, p. 3) has written, to see the transformation of 'an anarchic system of states to a single large actor within the system'. Yet for the time being at least this seems rather optimistic, especially for Asia, where states seem unwilling to sacrifice their sovereign interests into a wider pool. Yet a more integrated Asia may offer the prospects for a genuinely regional response to the problems that ripple effects have a habit of sharing around.

As will become clear in subsequent chapters, the patterns of security integration in Asia are irregular and inconsistent. It is too much to expect a genuinely united or unified approach on many issues. Some interdependencies can make agreement harder to come by, even if the countries involved would both be better off by sharing a common approach. Two or more regional neighbours who are connected by their claims over the same piece of territory will often find that the thing that connects them is more likely to divide than unite them. That is certainly the experience of Japan and China over territorial claims in the East China Sea. They may have a common problem, but each will probably see it as the other's fault.

The organisation of this book

The focus in this book is squarely on Asia's regional security. This volume seeks to provide an *overall* assessment of Asia's security which appreciates the region's diversity but also anticipates regional security interdependence. For this reason, a sum-of-the-parts approach is unlikely to work. That means resisting the temptation to take the reader through a tour of security in different parts of Asia, with a section on North Asia, one on Southeast Asia, another on South Asia and so on. A country-by-country security analysis would fall down for the same reason. As a result, a thematic approach that draws on examples of security issues, problems, perceptions and interactions from across the region is going to work better. Consequently, this book poses a series of big questions about Asia's security that allow for an ongoing debate about the subject. The first such question has been the focus of this introductory chapter, which has been asking 'How should Asia's security be approached?' Subsequent chapters will deal with eleven more questions.

Chapter 2 asks a question that is sometimes overlooked but is necessary to investigate in any assessment of Asia's security. It has already been argued that a study of security involves an assessment of threats to acquired values. Traditionally, threats of armed violence have often been regarded as central to this assessment. But what is striking about Asia in recent years is the infrequency of war, at least of the major interstate variety, and in terms of the frequency of major civil conflicts. The same could not be said about Asia 50 years ago. Hence the next question is a deliberately provocative one: 'Why does Asia seem more secure?'

This is followed by three chapters asking questions around the changing distribution of power in Asia, one of the most significant factors in regional security. Chapter 3 investigates the relationship between the region's two leading or great powers that continues to attract very high levels of attention (and at times concern). Hence the question: 'Is it all about the USA and China?' As readers might expect, the answer is a qualified one, because while these two large countries are fundamentally important to Asia's security, it would be risky to neglect India and Japan, and there may be other players that also need to be considered.

Chapter 4 is about the security implications of the main driver of the changing power equation in Asia that is affecting all of the region's countries, large and small. This is the economic prosperity of many parts of the region which through trade, investment and production has produced much of the interdependence which now exists in Asia. This shared prosperity is often regarded as a major cause for security optimism. This proposition needs to be tested, hence the question here is 'Can economic interdependence keep Asia safe?'. This fourth chapter will also look at the way that economic growth contributes to growing military and political influence, something that is itself not always a guarantee of peaceful relations.

The last of the power questions follows from this. Thanks in part to their expanding economies, a range of Asian countries are able to acquire increasingly advanced military capabilities. A few in the region also have nuclear weapons, and many are investing in cyber capabilities. Almost all will say that they are doing this for peaceful purposes, but there are real doubts about whether this growth in armaments is as safe as individual countries are suggesting about their own forces. Chapter 5 therefore asks 'Will military technology lead to new conflict in Asia'.? One of the features of this chapter is an analysis of the argument that there is a dangerous arms race occurring in Asia which imperils regional security.

The proposition behind the last question is challengeable on important theoretical grounds: armed forces are not generally sent into war without a political aim in mind. When analysts try to predict what might cause the end of Asia's period of relative peace they are likely to point to the growing rivalry between neighbours over old territorial disputes. These difficulties are in turn inflamed by nationalist sentiments that are evident in a number of East Asian and South Asian countries. Some seem convinced that Asia's next conflict will occur over contested territories in the South China Sea or East China Sea. But, like all alternative propositions, this argument also needs testing. Hence the question for Chapter 6: 'Will territorial competition and nationalism ruin Asia's peace?'

By this stage of the book, a good deal of the analysis will have been about the interactions between increasingly strong states where that strength can be evident politically, economically and militarily. But as is already clear from this introductory chapter, Asia's security, and especially its insecurity, is often connected to the problems and challenges of state weakness. Threats to human security, for example, often occur when governments become fearful that they are losing control of a disintegrating country, or when they are simply unable to provide protection to their citizens. This makes for a very important question for Chapter 7: 'Are Asia's main security problems domestic ones?'

It is also necessary to consider those security issues and challenges in Asia that appear to have the potential to bypass states as security actors and that might even threaten to make the power of states irrelevant. These come in two main forms: actors other than states which bring with them their own security agendas, and transnational security issues which cross borders and boundaries with seeming impunity. Hence Chapter 8 asks: 'Will non-state actors and transnational challenges overtake Asia?'.

At this point a significant array of Asia's security problems and challenges will have been considered. But what is to be done about them? What security responses are available, and possible? Chapter 9 looks at a range of ways in which the governments of Asia might step in to seek answers to existing security problems. 'Can intervention work in today's Asia?' is the question here. This will mean an evaluation of intervention in its different forms, from crisis management to peacekeeping in the region. But it also means understanding the generally high level of sensitivity among Asian countries to ideas of external interference in domestic matters.

The next chapter on potential responses to Asia's security problems takes an even broader view. The question for Chapter 10 is 'Can Asian states work together on security?'. It has sometimes been asserted (often from outside the region) that Asia lacks traditions and institutions of cooperation to tackle security problems. Is this an accurate depiction, or are there reasons to be hopeful? Does ASEAN, for example, offer a way forward? Might Asia be on the cusp of regional responses to regional security problems? Or are the critics right?

Chapter 11 examines a different response: a pattern of security cooperation in Asia that has been much more selective in its membership. This involves the security alliances and partnerships in Asia that are often under the leadership of the United States. Some say these relationships have been guarantors of security in Asia. Others find them divisive and wonder if they create adversarial relationships with those they do not include. This is a live issue, because a number of these relationships have been strengthened in response to the rise of China. Hence the question: 'Will alliances and partnerships split the region?'. In the quest for greater regional security, will these arrangements in fact lead to even even less?

That leaves Chapter 12 to round off the book, and it seems only fitting to use it to consider where Asia's security might be heading. This chapter ponders whether the region is moving 'Towards a new Asian security?'. One part of this closing chapter takes stock of the answers to the various questions that have already been addressed. In particular, this will be a matter of identifying those factors that seem most likely to cause the most significant ripple effects for Asia's security as a whole. The second part of the conclusion is to examine whether a new kind of regional response might be on the table to deal with these bigger regional challenges.

Identifying that response will involve a return to one of the main features of this introduction: the idea that security is about the absence of threats to things that are valued. The issue here is whether the major contributors to Asia's security can develop some shared values around which a new Asian security could be fashioned. This may seem an ambitious idea, but for such an important region as Asia, ambition might be exactly what is required. One of the obvious signs of Asia's importance is the increasing shift of global power in its direction. It is therefore only appropriate to move now to consider what the changing distribution of power to and in Asia means for its security.

Chapter 2

Peace: Why Does Asia Seem More Secure?

Scholars who look at security issues in Asia are generally not optimistic people. As the next seven chapters of this book will confirm, there are still plenty of security challenges in the region that give reason for concern. But today's Asia is not a region riven by major armed conflicts, at least not major conflicts between states. In particular, it has been some time since Asia and its borderlands featured a major armed clash between the larger powers. In 1979, China attacked Vietnam in what was to become a rather short and limited war. More recently, in 1999, India and Pakistan, who had both tested nuclear weapons just a year earlier, conducted a brief and even more limited war in Kargil, a district of the disputed Kashmir area. But this sort of behaviour, while worrying, falls considerably short of the major wars that took place in Asia in earlier decades, including the Korean War of the early 1950s. In one study, Timo Kivimäki (2011, p. 58), suggests that battle deaths from interstate wars involving East Asian countries between 1980 and 2005 had dropped by as much as 99.5 per cent in comparison to the period between 1946 and 1979.

Major wars in Asia are now very rare. The nation-states that make up the region have enjoyed mainly peaceful relations with one another for many years. This does not mean they lack points of significant disagreement. In fact, quite serious periods of tension between a number of the region's leading sovereign states occur from time to time. But, at the time of writing at least, it has to be observed that these significant differences of opinion have not escalated into outright military hostilities. It might be argued that, instead of interstate war, Asia has seen a proliferation of internal (or intrastate) armed conflict. There are certainly recent and current examples of this type of behaviour, and a number of these will be considered in Chapter 7 of this book. Internal violence has been a fact of life in parts of Myanmar, in the southern portions of the Philippines, and in Pakistan and Afghanistan, for example. But even these disturbances to the peace are not as frequent or as significant as they once were. There was a time, for example, when many Southeast Asian countries faced violent insurgencies of one sort or another. A number of them, including Indonesia, the largest and most diverse and complex Southeast Asian country, appear to have become more peaceful internally in recent decades.

It is rather too restrictive to define security simply as an absence of armed conflict. Insecurity is quite possible when war is absent. As was discussed in the introductory chapter, there may be other threats, including to a community or country's very survival, which can arise from non-military sources. But a general scarcity of armed conflict would still seem to be a good starting point if a security environment is to be regarded as desirable. A relatively peaceful regional environment may not always be a sufficient condition for security to exist, but it would seem to be a necessary one.

This makes it even more important to explain the relative peace, and with it a relatively good security environment, that large parts of Asia appear to have enjoyed in recent decades. It is necessary here to wonder what has changed in Asian security to make war both within and between states less likely. These are the tasks of this chapter. As these issues are considered it should not be surprising if it is found that a mix of factors are responsible. Later parts of this book will examine whether these factors are likely to continue to have a similar effect. For now, however, the focus is on Asia's security in recent decades. This will be undertaken first by considering external factors (mainly in terms of the relations between states and the distribution of power between them) that may be responsible for Asia's relative peace. This will be followed by an analysis of the security effects of the evolution of political systems that has been taking place in a number of Asian countries.

Interstate and external explanations of Asia's peace

It is hard to imagine any event that could cause greater region-wide insecurity in Asia (and thus a nasty variety of the ripple effects that reach out to the entire system) than a major interstate war in the region. This was certainly Asia's experience in the Second World War, and it certainly would have been Asia's experience had the nuclear-armed superpowers gone to war with each other in Asia during the Cold War. The search for explanations for the absence of catastrophic war between Asia's major powers is an important and interesting one. A natural place to look first is to consider the nature of the relationships between these countries.

Some of the interstate factors that have reduced the appearance of armed conflict in Asia may be generic and global rather than specific to the region being explored in this book. Indeed, over the past several decades, and certainly since the 1960s, major interstate wars have been fairly rare globally. That is not to say that war between states has been abolished. The invasion of Kuwait by Saddam Hussein's Iraq in 1990 and then the swift and limited war that followed, and the resumption of more serious hostilities as the United States led a much more controversial invasion of Iraq in 2003, are evidence from just one part of the world that war is still eminently possible. But it is still necessary to find a way to explain why war appears to have become much less likely, even if it has not become impossible or inconceivable.

There is a case to be made for the argument that, for most countries, war has ceased to be a valid policy option in their relationships with other members of the international system of states. To the extent that this is a global phenomenon it ought to apply relatively equally to Asia as to other parts of the world. To say that war is generally out of bounds as an instrument of policy is to make conceptual as well as practical observations. Conceptually, it questions the dictum of the early-nineteenth-century Prussian philosopher of war, Carl von Clausewitz (1976), that war amounts to an extension of diplomatic relations between states. This argument was widely held to be valid for many ensuing generations. The practical side of this issue is that among the governments which had held to this argument in its many variations were those who fought the horrific world wars of the first half of the twentieth century. Those wars may provide one major reason for the increasingly unacceptability of war itself. The Second World War (in which tens of millions of soldiers and civilians, including many in Asia, lost their lives) was widely seen as proof that major war had simply become too costly to tolerate.

It was probably in Europe rather than in Asia that the horrendous costs of the Second World War had the greatest effect in terms of the unwillingness of states to resort to major conflict as a future act of policy. An underlying commitment to avoid war lies behind the growth of Western European cooperation that is now reflected in the much larger and more concrete European Union. The first stage of this cooperation, the European Coal and Steel Commission, was designed to make it practically impossible for Germany or France to consider war as an option in their relationship by preventing the concentration of this vital natural resource in the hands of just one European government. However, no such post-war agreement, informal or formal, seems to have appeared in post-war Asia. Instead, a quite different story – the unravelling of European colonial power – was taking place in the region covered by this book.

Similarly, while there was no major interstate war in Europe after the surrender of Nazi Germany in 1945, the same did not apply in Asia after Japan's surrender later that year. In a significant Asian conflict that broke out in 1950, North Korea's assault on South Korea was then repelled by a US-dominated United Nations force that eventually precipitated China's military intervention. This was the most serious and worrying armed conflict of its type anywhere in the world in the early post-war period. At the same time, France was struggling to hold on to its possessions in mainland Southeast Asia, a process turning especially violent in Vietnam. The crucial year of 1954 saw not only France's defeat in Vietnam but also an attempted settlement in Geneva involving the great powers which meant, in the words of A. J. Stockwell (1992, p. 373), 'the formal entanglement of Indochina in the international relations of the Cold War'.

As time went on, the United States became increasingly concerned that the vacuum created by France's weakness and then absence would allow for communist political movements and revolutions to spread throughout that part of the region. By the mid-1960s, the United States was increasingly involved in a substantial war in mainland Asia. This conflict took the form of a civil war as the North Vietnamese government sought to undermine its anti-communist counterpart in the South. But it also took on the appearance of an interstate conflict between North Vietnam (backed variably by the Soviet Union and China) and the United States and some of its regional allies, including South Korea, Thailand, the Philippines, Australia and New Zealand, as well as the forces of the rickety, unpopular (and eventually overwhelmed) South Vietnamese regime.

This was not the only part of Asia where war had quite clearly remained a form of international intercourse. Relations between South Asia's main countries were hardly placid in the post-war period, and as with the case of Vietnam (and unlike the case of the Korean War), these wars lay in the unravelling of European power in the region. India may have achieved independence from Britain in 1947, but it did so as the result of a partitioning of British India that also gave rise to Muslim-majority Pakistan as a fellow independent state consisting of two non-contiguous eastern and western portions. That partition did not occur without violence on a massive scale, which brought extraordinary human suffering. Their leaderships adopted very different ideological positions, which, as Sumit Ganguly (2002, p. 10) has argued, 'embodied competing visions of nationalism and state-building', further violence was likely. As well as a war over the disputed Kashmir region in the immediate post-independence period, India and Pakistan fought again over the same question in 1965 and again in 1971, when East Pakistan sought to split itself off. In the latter case, the resulting newly independent state of Bangladesh came into being only after a short but severe war between Pakistan and India. Territorial competition and fears also characterised the short and limited but very significant war between India and China in 1962. As John Garver (2001, pp. 59-62) has argued, China's main reason for beginning this brief war was as a result of Mao's concerns that

India was seeking to reclaim a predominant position in Tibet. While often overlooked, it remains a permanent reminder of the potential for violent strategic competition between these two very large neighbours.

This does not mean that all of Asia was wracked by constant interstate warfare in the two decades after the defeat of Japan. But it is evidence enough to suggest that Asia's leaders did not assume that the costly experience of the Second World War had made regional wars unthinkable as instruments of policy. That fact may be doubly surprising to us when we consider that in August 1945 the biggest sign to the world community of the vastly increasing costs of industrial-age warfare had occurred in Asia. This was the atomic bombing of Nagasaki and Hiroshima by the United States, which constituted the last main violent act in the twentieth century's second global conflict.

It has to be said that since that time, nuclear weapons have had an ambivalent affect on Asia's security. This might seem an odd thing to say about the arrival of the most powerful weapon ever known. America's atomic attacks on Japan had an immediate effect on at least one of America's best strategic minds, Bernard Brodie (1946), leading him to conclude very quickly that armed forces could no longer be developed to wage war. From then on, he suggested, their only rational purpose was to prevent it. This was in effect a logic for nuclear deterrence, which became even more necessary by 1949, when the American nuclear monopoly was broken by Stalin's Soviet Union. Indeed, as the superpower arms competition really got into its stride, it became increasingly important to recognise that nuclear deterrence had become a two-way street. But Brodie's arguments took some time to gain favour. And nothing, including America's nuclear weapons, prevented the North Korean attack on South Korea in June 1950 that precipitated America's military intervention on the peninsula and signalled a new era of US strategic engagement in the region.

This meant that Asian wars had not suddenly been rendered obsolete by the coming of nuclear weapons. But their proven capacity for immense destruction raised the stakes so much that escalating conflicts that risked a costly exchange between nuclear-armed states became increasingly unwise. Stalin had backed North Korea in 1950, but he did not authorise Soviet forces to engage in the conflict against their American counterparts. Yet even without the use of nuclear weapons in Korea, the human costs of war were still ferocious. These were especially so for both sides of the Korean peninsula where, as William Stueck (1995, p. 361) recorded, 'the total number of people killed, wounded, and missing approaches three million, a tenth of the entire population. Another ten million Koreans saw their families divided; five million became refugees'. The human costs were also significant for some of the intervening powers, including among the poorly equipped forces of the People's

Liberation Army, which joined the war once US forces had pushed too far north for China's comfort. This did not prevent Mao and his colleagues from pushing on. The conclusion on this point from Zhang Shu Guang (1992, p. 282) is that China's 'leaders seem to have consistently believed in the necessity of sacrificing human lives to gain a final victory'.

Yet in a strict sense the resulting stalemate on the peninsula left no one in a winning position, reinforcing the place of the Korean War and its aftermath as the main Asian representation of the Cold War divide. For the superpowers at least, it was a limited conflict, and the United States did not use all of the military options it had available. Japan in 1945 would therefore remain the one and only place and time that nuclear weapons were used in anger in Asia or anywhere else. However, this did not mean that the presence of weapons of mass destruction could not be exploited for political ends. This was very clear in America's response to a series of crises in the Taiwan Strait in the 1950s as China seized and bombarded islands held by its nationalist adversary regime in Taipei. As the main protector of Taiwan (which was in the process of becoming a formal ally), the United States issued a series of explicit threats of possible nuclear strikes against China, as well as placing nuclear-capable systems in Taiwan.

Knowledge of these threats encouraged Mao to join the nuclear weapons club as quickly as possible, not least because China's communist leaders had deep concerns about their country's vulnerability to both conventional and nuclear attack. They were, in other words, determined to ensure that China should share in the deterrence possibilities that nuclear weapons were seen to provide. The big breakthrough in this regard came in 1964, when China tested its own nuclear weapon. This left the United States with another temptation: to use force to remove China's emerging arsenal before it was too late to do anything about it. Eschewing this option not only meant the avoidance of a major Sino-American war in the middle of the 1960s, but it also signalled a begrudging acceptance of China's place as an emerging world power, even though by this stage China and the United States did not enjoy formal diplomatic relations. But the nuclear implications for Asia did not stop at this point. The arrival of China's nuclear arsenal just two years after the brief Sino-Indian war encouraged India to move in a similar direction, culminating in what New Delhi would call a 'peaceful nuclear explosion' in 1974, but which was widely recognised as a rather more serious development. This in turn would provide Pakistan with a reason to pursue its own nuclear weapons. At least in its early years Pakistan's programme was supported by India's rival, China.

These latter developments in South Asia would take place gradually and away from the spotlight of world opinion. Much more obvious in the second half of the 1960s was the increasing involvement of the United States in the Vietnam War. By early 1969, over half a million of its soldiers were fighting there. And an initially secretive cross-border bombing campaign launched by the Nixon administration in 1969 helped to spread the conflict into Cambodia. According to critics such as William Shawcross (1979), this also helped to pave the way for the murderous rule of the Khmer Rouge, which traumatised Vietnam's south-eastern neighbour just a handful of years later. But the United States did manage to ensure that its two large communist rivals were not brought directly into the fighting. China and the Soviet Union instead provided military and economic assistance for Ho Chi Minh's Vietcong, who were pitched against the US-backed South Vietnamese regime. But their forces did not ultimately find themselves on the same battlefield as the Americans in any significant fashion, though as Jian Chen (1995, p. 360) has indicated, by the mid-1960s China was prepared to replay its Korean War intervention had the United States sent ground forces in North Vietnam. One of the reasons for the absence of this escalation was the fear that such a major power conflict could easily run completely out of hand, ruining much more than the lives of millions of Vietnamese (and Cambodian) people. In that sense at least, there was some degree of mutual interest between the major powers in avoiding a catastrophic conflict among themselves. This meant some measure of security in major power relations despite the dreadful experiences in mainland Southeast Asia for so many of the people there.

Peace at hand?

The end of the tumultuous and violent 1960s marked something of a watershed in Asia's security. From then on, with very few exceptions, Asia began to give the outward impression of being a much more peaceful place. It certainly became a more business-friendly place as the dynamism of Asia's growing economies stole the limelight from the receding military conflicts. In other words, Asia became a good news story: a cause for opportunity rather than a source of threats. Regardless of a few bumps and reversals, this period of relative security and stability still continues today. In a book such as this it is therefore necessary to identify those factors for peace which appear to have had less of a say in Asia's security during the first two decades of the post-war period, but which seem to have been more influential in the relatively peaceful decades since.

One possible answer is that the major powers had learned some Cold War rules in Asia, becoming less likely to provoke wars and finding themselves able to stabilise their relationship. In 1950, when North Korea went to war across the 38th parallel, the United States had not signalled very clearly that it would be prepared to defend South Korea in a Cold War contest. Unlike the situation in Europe, where the Cold War lines had been drawn starkly between East and West, Asia always presented a more fluid and confusing picture and it took some time for things to settle into recognisable patterns. Indeed, one of the reasons for that lack of clarity was the view that the United States did not regard the Korean peninsula as a vital security interest in the same way that it clearly regarded its Western European allies, who had just become part of the North Atlantic Treaty Organization (NATO).

But there was much more to Asia's complexity than this. Europe had NATO and the Soviet response took the form of the Warsaw Pact, reflecting the division of the continent into two blocs led by the two superpowers. This was the clearest representation anywhere in the world of the bipolar characteristic of the Cold War. But Asia increasingly became an arena for competition between three major powers, with China involved here alongside the United States and the Soviet Union. According to Michael Yahuda (2011), the leading scholar of the history of Asia's international politics, this tripolarity was particularly evident from the early 1970s until the collapse of the Soviet Union at the end of the 1980s.

This set up a more complex mix, with each of the three bilateral relationships (US–Soviet, Sino-Soviet, and Sino-US) moving sometimes in differing directions, each influenced by the two others. By the late 1960s, for example, the United States and the Soviet Union were relaxing their long-standing tensions in a process known as détente. But by this time the Soviet Union and China were barely on speaking terms, having experienced a major split in their relationship on ideological grounds. In a landmark study of this disagreement, Lorenz Lüthi (2008, p. 11) placed much of the blame for the situation on Sino-Russian economic differences in light of Mao's radical experiments with rapid industrialisation. The wider regional impact of this dispute cannot be exaggerated, because knowledge of the unshackling of China and the Soviet Union would eventually break the myth of a unified and monolithic communist bloc working in unison to encourage revolution in Asia. Clear empirical evidence for that divide became available in 1969, when a limited border conflict took place between Chinese and Russian forces. As one translated study by Kuisong Yang (2000) has suggested, Mao received more than he bargained for by encouraging this limited exchange as China's leaders feared full military escalation by the Soviet Union, including an attack on China's still-young nuclear weapons facilities.

This intra-communist discord paved the way for a marriage of convenience between China and the United States, which shared an interest in restraining Soviet influence in Asia. The formal breakthrough in US–China relations came when President Richard Nixon visited Chairman Mao in 1972. As Margaret MacMillan (2007) concluded in a very approachable book on this breakthrough, this meant that China was now seen by the United States as a legitimate member of the international system of states. Rather ironically, it was stage-managed partly by Nixon's Secretary of State, Henry Kissinger, who many still hold accountable for spreading the Indochinese war to Cambodia. Kissinger's geopolitical lenses always sacrificed the interests of smaller powers to the relations of the larger, namely the United States, China and the USSR. Because the last two of these could now be counted on to check each other because of their enmity, America's long-held concerns about the region's security grew less intense, as did its worries about the spread of communism in Asia. This was partly because China's policy was changing. For many years since its establishment after the 1949 revolution, Mao's regime had supported revolutionaries across the region, and especially in Southeast Asia. As China emerged from the disastrous internal experiment of the cultural revolution (which set back its own educational and economic advancement), China's approach became more pragmatic. This did not mean instant peace and harmony, however. Vietnam's invasion of Cambodia, where it dislodged the Khmer Rouge, led China to launch a brief war against its traditional rival in 1979. But rather than teaching Vietnam a lesson – the explanation often given by Beijing for its decision to use force – China learned a good deal about its own military ineffectiveness. As Xiaoming Zhang (2005, p. 851) has suggested: 'It was China, not Vietnam, which actually received the lesson.'

It is probable that this experience provided a negative reason for China to adopt a more pragmatic approach to foreign relations in Asia. The largest and most populous country in Asia was too weak to have the influence on regional affairs that it may have been seeking. This reduced the incentives for China to use force to challenge anyone in the region, let alone the still-mighty United States. But there was a positive reason why it was wise for China to step away from dreams of revising the system of states in Asia: coming to terms with that status quo actually fitted better with the country's changing domestic agenda. By the late 1970s a new leadership under Deng Xiaoping was in control, with the express purpose of strengthening China by opening the country up to the global (capitalist) economy. This may have been the biggest and most important transformation of all. It drew China into a position where it realised that its own efforts to destabilise the regional security environment could set back its own plans to strengthen itself through economic development.

Looking back on this important time of transition for Asia's security, it is tempting to argue that from this moment forward economic prosperity trumped military prowess as the index of power in the region. Many

Asian countries seemed to want to follow the Japanese model of export-led economic growth without the encumbrance of very large military spending and the distraction of fighting costly wars. If security is about the absence of threats to things that are valued, and economic growth is a top priority in a country's value system, then the security conditions for that growth become obvious. These include the absence of wars that destabilise the region, preoccupy governments, drain the public purse, harm trade and scare off investors. As will become clear in Chapter 4, this logic is not the last word on the relationship between economic conditions and Asia's security. But for our present purposes, there is already something inadequate in this economics–security nexus as an explanation for what has changed in the region. War between states may well have largely disappeared as a leading part of the military dimensions of Asia's security and insecurity. But the armed forces of the major players did not suddenly all go away at the same time. What explains this fascinating mixture of a region that has become relatively peaceful but has by no means been disarmed?

Military power: balance or imbalance?

The United States offers the most important piece of evidence to consider here. Since the late 1960s a series of administrations in Washington have been determined to avoid fighting another long and costly ground war in mainland Asia. The Guam Doctrine, which was announced by President Nixon in 1969 and signalled America's reduced willingness to commit land forces to Asia, has remained a staple of America's approach. But while the costly Vietnam experience cast a very long shadow over US policy it did not remove American power from Asia. The mainstay of America's military presence in the region had long been at sea rather than on the land, continuing the pattern of Western maritime supremacy in Asia. That pattern had been largely unbroken since the Portuguese explorer Vasco da Gama sailed into Asian waters at the end of the fifteenth century, beginning what Coral Bell (2007, p. 1) described as '500 years of ascendancy of the West over Asia'. The US Seventh Fleet was always the supreme part of America's presence in the region, because it represented a unique post-war capacity to project armed force, including via America's aircraft-carrier battle groups. This was a presence that only the Soviet Union at its height might have challenged. This dominant position was facilitated in part by bases in US territory in the Pacific Ocean, including Hawaii (where its Pacific Command was headquartered) and Guam. But it was also made possible by the stationing of US forces in Japan, at ports such as Yokosuka and on the island of Okinawa; in South Korea; and, for a time, at Subic Bay in the Philippines.

America's continuing military presence in North Asia and Southeast Asia, which for many years included the stationing of tactical nuclear weapons in South Korea, carried a strong signal regarding Washington's commitment to the security of its regional allies. In North Asia, these allies were Japan and South Korea and, until the United States recognised the People's Republic of China, also Taiwan. In Southeast Asia, there were two American allies: the Philippines and Thailand. And further south, Australia and New Zealand were connected to the United States through the 1951 ANZUS Treaty. More will be said about these arrangements and their contemporary relevance in Chapter 11, but for now it is important to note that these alliances functioned separately from each other, and therefore were unlike NATO in Europe, which connected allies in that region into a single group. Nonetheless, in Asia there was a significant number of countries with a firm interest in the continuing military presence of the United States as their security guarantor.

Neither the Soviet Union nor China had any comparable set of strategic relationships in Asia. Moscow's close links with a formally non-aligned and inward looking India, for example, had little if any outward implications. There was no China–Soviet Union–Vietnam triangular alliance because by the late 1960s China had developed very significant differences with both of these important neighbours. On its own, China had practically no ability to project military power beyond the close confines of its part of the Asian mainland. Faced with its own weakness, China's best option was to tolerate America's maritime dominance in the region. Moreover, Japan's strong alliance with the United States meant that China's main Northeast Asian rival had no need for massive armed forces, nor its own nuclear weapons. This meant that China benefited indirectly from at least one aspect of the US alliance system. Writing in the late 1990s, Thomas Christensen (1999, p. 58) recorded the widespread opinion among Chinese security analysts that 'by reassuring Japan and providing for Japanese security on the cheap, the United States fosters a political climate in which the Japanese public remains opposed to military buildups and the more hawkish elements of the Japanese elite are kept at bay'. Moreover, tensions between the United States and the Soviet Union, when they resumed in the second half of the 1970s, were not focused on Asia. Instead, much of their renewed competition was played out in Africa and Latin America, though it was the Soviet Union's invasion of Afghanistan in 1979 that brought the curtain down on the transitory relaxation of superpower tensions known as détente.

America's strategic dominance in Asia remained largely uncontested after the end of the Vietnam War, which was ironic inasmuch as many took this moment to be a sure sign of US weakness and withdrawal. But the ongoing American connection to the region was quietly welcomed by a number of regional governments who were free riders on the formal alliances their neighbours had established with the United States. This even included those who preached the importance of non-alignment, including Indonesia. For its part, the United States would come to see the Suharto regime in Jakarta as a bulwark of stability. This meant turning a blind eye to Indonesia's controversial annexation of East Timor in 1975 for the sake of geopolitical stability, though Brad Simpson (2005, p. 282) has argued that the United States and a number of its allies were also concerned that East Timor 'was too small and too primitive to merit self-determination'.

It is difficult to be sure how much Asia's peace from the mid-1970s relied directly on American power in the region. But with American power acting as an insurance policy against significant aggression, a breathing space was made possible for large parts of the region. This was in effect a continuation of a longer trend. Singapore's founding prime minister, Lee Kuan Yew (who died in March 2015), on a number of occasions argued that America's otherwise unsuccessful war in Vietnam had bought Southeast Asia's newly independent states precious time for early development. As Ang Cheng Guan (2009) showed, these sentiments were developed against a backdrop of Singapore's concerns about the collapse of British power in Asia, which made the continuing presence of US forces all the more important.

It is just possible that it was an *imbalance* of power in Asia, with the distribution skewed strongly in Washington's favour, that allowed for a reliable and generally positive external security environment in the region, and most especially in East Asia. Threats of force were not necessarily absent. In fact, the deterrence of war depends on the credibility of threats to use force when peace is in doubt. But the peace held. And with security from external dangers reasonably assured, at least in terms of the unlikelihood of direct military attack, the states of East Asia could focus their energies on national development, and in particular on the building up of their economies. The latter feature was possible not least because export-led growth required safe sea lines of communication, which America's maritime predominance did a great deal to protect and promote.

If this seems a one-dimensional argument that depends too heavily on the actions of one major state actor, that perception is certainly an accurate one. Even the strongest actors in international politics rely on the cooperation, or at least the tolerance, of other large states. But it is still possible to detect a remarkable, though somewhat accidental, conjunction of interests between the United States as the external balancer, and China as the region's largest and most important resident. This conjunction derived from a philosophical point. As soon as Deng Xiaoping declared to the Chinese people that it was good to be rich, and as soon as the pragmatic quest for economic development trumped Mao's ideological and revolutionary fantasies of the past, China had the ultimate reason not to challenge US regional dominance in a direct or violent manner. China needed a secure external environment in which its economic advancement might be possible. It could not create those favourable conditions itself, and by challenging the USA it would only make its security environment worse. The best strategy for China was to do what most of the rest of East Asia was doing: to free ride on US primacy in the region.

Internal explanations

One of the big questions for Asia's future, which will be explored in Chapter 3, is whether China is changing its approach and feels that it has a greater role to play as a provider of regional security as its own sources of power develop. But it is undeniable that the changing style of the leadership in charge (including from Mao to Deng) had a big impact on China's foreign policy, including its attitude towards the use of force. In fact, many of the external security problems that have already been considered in this chapter have significant origins in domestic politics, including the tensions between China and India, and China and the Soviet Union. This raises an important question for broader consideration: to what extent are the sources of Asia's security and insecurity attributable to domestic political developments as opposed to the external distribution of power in regional interstate relations?

This is no minor issue. Asia's last experience of total war (the Second World War, which began in the 1930s for parts of China, and in the early 1940s elsewhere) would not have been possible without the existence a profoundly militaristic Japanese government. Fortunately, the like of that regime has not been seen again in the region, at least not in control of one of the major powers of Asia. One of the gravest threats to the security of any region is a large state that regards war not only as a potential means to political ends, but as a necessary ingredient to its own personality. This was certainly evident beyond Asia as the middle years of the twentieth century approached. At the same time as Japan's imperialism was reaching its high point in Asia, Nazi Germany was waging war in Europe in the name of a racist ideology. The resulting war between Hitler's Germany and Stalin's Soviet Union was a contest between totalitarian political systems that concentrated absolute power in the hands of ruthless and paranoid dictators. The consequences of that contest for millions of people caught in between, as Timothy Snyder (2010) has recorded, were gruesome.

The effects of Japan's militarism in Asia, including in China, could also be drastic for local populations. The memories of these atrocities continue to haunt Sino-Japanese and Korean–Japanese relations today. Jennifer Lind (2009, p. 132) has observed that:

> Japan's neighbors still keep a wary watch over the country that brutalized them in the early part of the twentieth century. Tokyo's official apologies for its past aggression and atrocities are dismissed as too little, too late. Worse, they often trigger denials and calls of revisionism in Japan, which anger and alarm the country's former victims.

But one of the most positive factors in the evolution of Asia's security environment in the post-war years was the dramatic change in Japan's political personality away from an attachment to militarism. It took some time for others in the wider region to recognise what was happening. One of the main reasons Australia and New Zealand sought an alliance with the United States in the early 1950s was their concern over the possible remilitarisation of Japan. But as the post-war era continued, Japan grew into its new role as a liberal democracy constitutionally committed to the avoidance of aggression. Criticisms of Japan still come from China and South Korea, and are likely to continue as Japan becomes more active in security affairs, but the civilian governments elected in Tokyo are a world away from the regime of the 1930s and early 1940s.

How much of a difference domestic political change has made is brought into sharp relief by a contrast with North Korea. This self-absorbed dynasty (now into its third generation of leaders from the Kim family) diverts its meagre resources, which might have been used to avoid recurring famine, into an oversized military that includes a small but expanding arsenal of crude nuclear weapons. North Korea's bellicosity continues to give the widespread impression of an irrational regime that might do anything, but as Victor Cha and David Kang (2013) recently suggested, these colourful interpretations often hide more than they reveal. The judgement made over two decades ago by Denny Roy (1994, p. 313) remains apt: North Korea's actions do not betray madness, but are instead part of a deliberate 'propensity for ruthless but carefully calculated violence, combined with grandiose threats and brinkmanship'. This behaviour provides a constant challenge for North Korea's neighbours. These include China, which only puts up with Kim Jong-Un, the present leader, because of fears of the alternatives, including the existence of an American ally (in the form of a united Korea) right up to the border with China.

Yet the North Korean example is an exception to the contemporary rule in Asia. Very few regimes continue to justify their existence by fostering a sense of perpetual external danger, at least not to the same degree.

Indeed, technically speaking, the Korean peninsula has remained in a state of war since the 1953 armistice. But there are perhaps even more important links between regime type and security in Asia, and these occur in the context of internal conflicts rather than external tensions. It seems almost self-evident to think that if domestic politics can be a major driver of conflict (as well as of peace), it is wise to look at how this relationship plays out in terms of intrastate or domestic security in the region.

One very sad example of this argument comes from Cambodia, which, as the Khmer Rouge took power in 1975, became for a short time Kampuchea. Mobilised by an especially virulent totalitarian ideology which saw intellectuals and professionals as enemies of the state, and which sought to erase all memories of earlier social arrangements, the Khmer Rouge leadership was responsible for the mass murder of close to two million Cambodian people. One might expect this to have generated an open and shut case for a military response from outside. One response eventually came at the end of 1978, from Vietnam, but this step was unlikely to have been taken for altruistic purposes. One scholar presented the view of many observers in raising important questions about both Vietnam's invasion of Cambodia and India's violent 1971 intervention in what was then East Pakistan and would soon become Bangladesh: 'it was clear in both cases,' wrote Mohammed Ayoob (2002, p. 86), 'that Indian and Vietnamese strategic interests were involved in these interventions because of existing hostile relations between India and Pakistan and between Vietnam and the Cambodian regime, respectively'.

Indeed, Vietnam took the opportunity to retain a significant military position in Cambodia long after the infamous killing fields had done their worst. Geopolitical considerations were also in play for a number of other countries situated close to Cambodia, but who decided against intervention. These included China, which had supported the Khmer Rouge out of a common opposition to Vietnam. They also included several Southeast Asian countries that were not alone in seeming to be more worried about an expanding Vietnam, and in maintaining the principle of non-intervention, than in the unfolding human tragedy in Cambodia.

Independence and Asia's postcolonial world

These concerns about Vietnam were a unifying factor for the Association of Southeast Asian Nations (ASEAN), which had been was established in 1967 with the express purpose of creating a diplomatic community whose members did not interfere in one another's domestic affairs. While neither Vietnam nor Cambodia would become members for many years, the former's invasion of the latter violated that code of behaviour. 'For the next twelve years,' wrote Shaun Narine (1998, p. 204), 'removing Vietnam from Cambodia became the central focus of ASEAN's international diplomacy and internal activities, and the most important test of its ability to manage its regional security environment'.

This group of Southeast Asian states was quite a small one in its early years. ASEAN originally had just five members: Indonesia, Singapore, Malaysia, the Philippines and Thailand. The first three of these had been involved in a messy and undeclared conflict which demonstrates the mixture of postcolonial independence challenges, territorial competition, and concerns about subversion and insurgency that dominated the security concerns of many Asian countries in the sixth and seventh decades of the twentieth century. As British power in Southeast Asia was waning, moves were being made for a federated independent state of Malaysia which included Singapore as well as Sabah and Sarawak on the eastern side of the island of Borneo, the remainder of which was Indonesian territory. This was, in the words of John Subritzky (2000, p. 209), Britain's fleeting attempt to 'preserve its influence in the region in an era when empire was no longer tenable'. Opposed to this amalgamation, Indonesia's opportunistic leader Sukarno launched an undeclared campaign of cross-border infiltration, precipitating the war between Indonesia and Malaysia which came to be known as *konfrontasi* (confrontation) and which continued from 1962 until 1966.

While relations between Malaysia and Indonesia have not always been smooth since that time, there has been no repeat of this scale of conflict. But the plan for a fully federated Malaysia was still not delivered because of Singapore's decision to seek independent statehood, a status gained by the small city-state in 1965. This laid the basis for years of tension and mistrust in the Singapore–Malaysia relationship. Yet very soon Singapore, Malaysia and Indonesia were to be partners in ASEAN, linked by a common interest in restraining their potential for destabilising interference in one another's affairs. But it seems doubtful that this would have occurred without the rise to power in Jakarta of Suharto, who replaced Sukarno in a coup d'etat after a violent process of political change that cost the lives of as many as half a million Indonesians.

This is one further piece of evidence for the proposition that domestic changes often have more impact on Asia's security environment than changes to the external balance of power. Once in power, Suharto (who was pragmatic but neither liberal nor democratic) quickly ended Sukarno's confrontation policy that had isolated Indonesia diplomatically and drained it economically. ASEAN was only possible because of the determination of Indonesia's new leadership to turn their country's regional reputation around. As Dewi Fortuna Anwar has argued (1994, p. 45): 'Regional cooperation could be seen as further proof that the Indonesian

Government would devote its energies and resources to internal development rather than the pursuit of an adventurous foreign policy.' In that sense, Suharto was following for Indonesia a similar logic to the changes that were to be be implemented by Deng in China: external peace as a prerequisite for the main task of economic development.

But these changes were not just about the countries of Southeast Asia themselves. External players were also relevant to these difficult years. On the one hand, Suharto's coup was partly a response to domestic concerns that Indonesia's communist party had become too beholden to Mao's revolution-minded China. On the other hand, *konfrontasi* had precipitated British military intervention to support Malaysia against Indonesia. Joining that conflict were fellow Commonwealth members, Australia and New Zealand, who would also become parties to the 1971 Five Power Defence Arrangements (FPDA). This exchange of letters was formally designed for the external defence of Singapore and Malaysia (initially against the possibility of a return of Indonesian aggression).

These arrangements would form the foundation for ongoing military cooperation between the five parties that continues to this day. Crucially, at a time when local relationships were still fragile and lacking in trust, this would be the only venue at which Singapore and Malaysia could engage in a semblance of military cooperation. But, as Damon Bristow (2005, p. 5) explained, the FPDA were quite unlike the Anglo-Malayan Defence Agreement that preceded them, and the US alliance commitments to nearby Thailand and the Philippines. Nothing so much as a glimmer of a British security guarantee was on offer. This was as much as anything a reflection of the fact that the United Kingdom was now a smaller factor in East Asia.

This important point provides another external explanation for regional security problems: the increasing weakness of the Western European countries that once held large Asian empires created an environment in which these disputes could emerge. To a large extent this was just as important as changing Cold War dynamics in Asia. It was partly about a lack of economic resources and political determination in the West. Changing public attitudes towards the continuing subordination of what were once imperial possessions also had an impact. But a vital part of this story is the resistance of the Asian colonies to their European masters, which became as inexorable and unstoppable in Asia as it had in Africa and the Middle East.

That resistance explains the earlier communist emergency in what was then Malaya, to which Britain responded with a successful counterinsurgency campaign from 1948. It most definitely offers a significant explanation for the civil war in Vietnam against France, and the resistance to American counterinsurgency efforts there. It explains the violence in

Southeast Asia's largest country as the independence movement in Indonesia fought the Dutch in the second half of the 1940s. But elsewhere in Asia, the pattern could be quite different. Significant violence occurred in South Asia but most of it occurred *after* the establishment of India and Pakistan as independent states. The uneasy partition of contested areas displaced up to two million people and resulted in a war over contested boundaries between the two new sovereign states.

Elsewhere in the borderlands of Asia, there were examples of a later and much more peaceful transition to independent statehood. Most of the colonies in the South Pacific, including the Australian colony of Papua New Guinea, made a largely peaceful transition to sovereign statehood in the late 1960s and early 1970s. This was not an even process, especially in the eastern part of the Pacific Islands area. Gardner and Waters (2013, p. 114) recently argued that, 'While Britain and Australia encouraged self-government, France baulked at it, and Indonesia has perceived its right to all former Dutch territory as the justification for its claim to West Papua.' Moreover, as Hegarty and Powles (2006, p. 258) noted, events in the 1980s, including a rebellion in Vanuatu and military-led coups in Fiji, helped to swing opinion away from the view that the South Pacific would always live up to the second half of its name. But the Pacific pathway to independence, which remains incomplete, was less violence-ridden than had been the case in Southeast and South Asia. By the time it did come, the decline of European colonial power in Asia itself was largely complete, though exceptions remained, including Britain's control of Hong Kong (which would continue until the handover to China in 1997).

A different form of empire unravelled with barely a shockwave in the late 1980s and early 1990s. This was the Soviet Union, whose collapse led to the establishment of a number of newly independent Central Asian Republics: Kazakhstan, Kyrgyzstan, Tajikistan, Uzbekistan and Turkmenistan. This was a much more peaceful transition than might otherwise have been expected, and international concern about the consequences of this important political moment was fleeting and very selective. The main worry for a time, as an article by William Walker (1992) demonstrated, was the destiny of the nuclear weapons that had been part of the vast and scattered Soviet arsenal. In Central Asia, the most important of these, including long-range nuclear-armed intercontinental ballistic missiles, stood to be inherited by the now independent Kazakhstan. The successful return of these weapons to central control by what was now a geographically smaller Russia represents a rare example of complete nuclear disarmament.

But Central Asia's deeper political destiny, and the security implications of the changes in this part of Asia for the wider region, was always bound to be a much more gradual and complex process. A good deal of this is once again because of the nature of the domestic political arrangements within each country, and as Kathleen Collins (2002, p. 138) argued over a decade ago, these directions appeared initially to be divergent: 'Kyrgyzstan experienced democratization. Kazakhstan, Uzbekistan, and Turkmenistan saw a renewal of authoritarianism. Tajikistan slid into failed statehood and a bloody civil war.' Within a few years, she argued, 'all five new Central Asian states had settled down to one shade or another of authoritarianism in which informal, clan-based networks dominated political life'. Political systems of this nature cannot but have significant implications for the management of security issues and for wider regional security relationships, including the challenge for potential security partners who have more formal systems of political authority.

Ambiguity rather than clarity was often a product of these different processes of regional political change. As conglomerations of nations and empires of one sort or another broke down across Asia, a number of grey areas remained. In a number of cases the initial boundaries of the newly sovereign Asian states were hardly fixed or final. In 1975, Portugal's rule in East Timor ended as abruptly as it did in other parts of the world (including in Angola and Mozambique). Taking advantage of this vacuum, Indonesia seized East Timor, which did not become an independent state until a further 24 years later, once the Suharto regime had collapsed, and not before significant violence had been visited by pro-Indonesian militias in 1999. Sometimes the new state boundaries were not at all conducive to internal stability. The Papua New Guinean province of Bougainville, for example, had close cultural affinities with neighbouring Solomon Islands, and a controversy over the control of a copper mine led to a nasty civil war in the 1990s as an independence movement on Bougainville fought unsuccessfully to secede and the Papua New Guinea Defence Force fought to prevent it. (This example will be referred to again in Chapter 4.)

It is difficult to overstate the importance to Asia's security of the processes of state formation (the way that independence was initially gained) and state consolidation (as newly independent countries established the rules of domestic politics). The maintenance of sovereign independence is a value that most states hold very dear: defending sovereign territory is often the first task that is set for a defence force. As a consequence, the achievement of statehood can often be seen as an inherently positive factor for national security. But in the meantime the pathways to independence and the consolidation of statehood can involve messy and sometimes violent processes that can rumble on for years.

These difficult adjustments clearly accounted for a very significant proportion of Asia's wars in the second half of the twentieth century.

But their eventual settling down in many cases (as state consolidation occurred) is an important factor in the relative absence of internal conflict in good parts of Asia. The leading theorist of Asian security, Muthiah Alagappa (2011), has argued that Asia's peace since the 1970s cannot be explained adequately by interstate factors, including the sustained American military presence that underscored a long period of US strategic primacy in large parts of the region. These are just the explanations presented by scholars such as Hugh White (2012), for whom regional security is intimately connected to the external balance of regional power. But for Alagappa, Asia's violent past and more peaceful recent decades can be attributed far more directly to different stages in the messy state-building processes which so many countries in the region have gone through. As the analysis in this chapter has demonstrated, there is extensive evidence to support both sides of this important debate about the causes of Asia's relative peace in recent times.

Asia's interstate–intrastate nexus

Along with Africa and the Middle East, the Asian region made a leading contribution to the numerical expansion of the international system of states in the twentieth century. Because so many Asian countries were caught up in independence struggles and state-building challenges, otherwise separate internal political processes in the region could still be grouped together as part of a wider trend. This gave a clear impression of flux, fragility and insecurity across Asia. In other words, rather than the ripple effects that one might see in a series of closely interdependent events, the fact that so much of Asia was going through these changes was enough to talk about these various political processes as constituting a single regional security event with widespread effects.

But many of these individual processes of state formation and consolidation in Asia *were* in fact connected, giving rise to a more genuine system effect in security terms. In other words, ripples were occurring that affected interstate as well as intrastate (or internal) political relations of Asia. For a start, as is demonstrated in the case of India and Pakistan, a common political ancestry as parts of the same European empire (in this case the British one) gave some pairs of new states in Asia an instant and ongoing connection. This also applied to the former French colonies commonly known together as Indochina and which included today's Cambodia, Vietnam and Laos. Second, as has already been suggested, the division of larger political components in Asia into sovereign independent states with their own territorial boundaries created as much confusion and contestation as it did clarity and stability. To name but a few examples, Malaysia and the Philippines, Indonesia and Malaysia, and Pakistan and India all struggled at times to retain peaceful and cordial relations with one another over their contested boundaries. And no fewer than six East Asian countries regarded themselves as claimants to overlapping parts of the South China Sea. As Chapter 6 will make clear, the claims by Brunei, China, Malaysia, the Philippines, Taiwan and Vietnam are becoming more of an issue as China has grown stronger and begun to assert its own territorial claims more demonstrably.

There is a third and very significant reason why it is impossible to separate the internal dimensions of Asia's post-war security record from the international aspects. Cold War superpower relations on the one hand, and the independence struggles and state formation processes in Asia on the other, often became closely intertwined. Rather than seeing these as competing explanations for Asia's security and insecurity it is often better to see them as interacting forces. As would also often happen in Africa, for example, the main Cold War rivals were constantly on the watch for regional partners in Asia as participants in their global struggle.

The domestic political alignment of these various countries would often shape their participation in this broader competition. A series of right-wing military regimes in Asia, for example, including in South Korea, Taiwan and the Philippines, were by no means models of democracy. But their strong anti-communist credentials made them helpful partners for the United States, whose support for them unfortunately condoned (and perhaps exacerbated) the sometimes savage internal crackdowns such governments often turn to in the name of national security. Indeed, this is one reason why it is sometimes claimed that the United States was a spoiler of Asia's security rather than its protector. The corruption and incompetence of some of these governments also became a liability, both in terms of their reliability as American allies and for domestic opinion within the United States.

By a similar logic, as some independence movements found communist forms of revolutionary organisation an effective pathway to independence, their appeal as potential partners of the Soviet Union and/or China was almost inevitable. North Vietnam's Ho Chi Minh appealed for this very reason, though he proved far less malleable than his external communist supporters would have wished. When it did come, external sponsorship (especially if it increased the military resources available to revolutionary movements) could also intensify security challenges within these countries. It was not hard for even small signs of outside support to leave the governments of newly independent states especially nervous. This was one reason for the strong emphasis on non-interference among the members of ASEAN. Moreover, as the nuclear-armed superpowers were unwilling to fight each other directly (and wisely so), they would sometimes treat Asia's smaller conflicts, which often had a significant internal dimension, as proxies for that bigger and much more costly clash. Writing in 1967, while the United States was still at war in Vietnam, the soon-to-be president, Richard Nixon (1967, p. 122) accused the Soviet Union and China of this practice. But the United States was also doing this, leading to some devastating critiques, including by a young Noam Chomsky (1969), that the superpower which was meant to be a bastion of liberty was sustaining conflicts that were responsible for tremendous suffering and political damage in Southeast Asia.

A combination of internal fragility and external competition helps to explain the appearance of conflict and insecurity in Asia up to the end of the 1960s. But is there a corresponding mixture that can help to account for Asia's relative peace since then? The answer to this question is an affirmative one, but it must be a conditional response. It is yes because the pressure was taken off internally and externally at about the same time. By the 1970s, many of Asia's recently independent countries might have had only a relatively short experience of sovereign statehood, but they were already becoming increasingly confident about their internal cohesiveness. Most of Southeast Asia's maritime countries no longer feared internal challenges in the way they once had. Their fears of insurgency, including from externally supported communist movements, had largely receded. And after the establishment of Bangladesh, the contours of South Asia as it is known today has essentially been formed. This did not remove the tension between India and Pakistan over Kashmir, but it did mean that the violent struggles associated with the appearance of newly sovereign states in that part of the world were far less likely to reappear. Indeed, as Sisson and Rose (1990) have shown, the eventual appearance of Bangladesh as a newly sovereign state was the result of the surprising mixture of two interlinked wars. One was an internal conflict as the main political parties from West and East Pakistan were unable to accommodate each other. This helped to precipitate the second, an interstate war that began when India sent its forces into East Pakistan in November 1971.

Other, external, factors also gave Asia breathing space. One of these was the temporary break from severe superpower competition that the US–Soviet détente in the late 1960s and early 1970s had made possible. When this was coupled with the Sino-US rapprochement and the cessation of most of China's support for Asian insurgencies, a much more positive interaction between internal and external security features became possible. The primary political objective of most of Asia's new political units had been satisfied in the form of national survival. With this assured, much more attention could be devoted in many parts of Asia to the second objective – national development, including the growth of stronger economies. This in turn worked best in the presence of a stable external environment, which it was in the interests of most regional states, including China itself, to promote. Rather than a negative and vicious circle between external competition and domestic fragility, much of Asia became characterised by a much more virtuous circle between external peace and domestic development.

This positive answer must be conditional because internal conflicts, and the wresting of independence from colonial powers, were a major feature of life for some parts of Asia but not for others in the second half of the twentieth century. There was no single, consistent mix of internal and external factors across the region that applied in all cases. In some parts of the region, interstate security dynamics seemed to rule the roost. For example, as a very close ally of the United States, and with its own boundary dispute with the Soviet Union in the form of a group of northern islands, Japan was clearly part and parcel of the Cold War competition. So too were both portions of the divided Korean peninsula. Taiwan was also to some extent another tug-of-war. China regarded Taiwan as a renegade province and dedicated itself to building up armed forces (including missiles) to achieve unification by intimidation but without the invasion that China was incapable of mounting. As soon as it recognised the People's Republic of China, the United States could no longer treat Taiwan as an allied or sovereign state, but it continued to provide it with military support, a point of continuing contention with Beijing even today.

All of these examples come from North Asia, where America's forward military presence in the region was so obviously present. This is also where the nuclear dimension of the US–Soviet standoff in Asia was concentrated. An abiding mission for US forces was to keep Soviet submarines, which carried intercontinental ballistic missiles, trapped within the first island chain north of Japan. And an abiding mission for the Soviet forces was to break out of that very same chain. The effect of domestic political transformations on security was often hidden by a number of these external considerations.

Conclusion

For North Asia at the very least, then, there appears to be something to the argument that American military power was a central contributor to the absence of armed conflict in Asia's more recent past. But this was not the only factor at work. Changing internal dynamics in a number of Asian countries also help to account for the more violent first period and the less violent second period that characterised Asia's security after the

Second World War. At the same time, a number of Asia's internal security problems were exacerbated by the strategic competition that was going on between the United States, the Soviet Union and China. But across Asia this was by no means a universal phenomenon, a finding that should not be surprising. After all it would be very odd if such a diverse region were to have its security shaped by a single universal variable.

More factors that impinge on Asia's security in different and often uneven ways will become apparent as this book continues. But before moving to these factors in later chapters, one assumption that has been underpinning the current chapter needs to be acknowledged and questioned. This is the very widespread notion that the best way of evaluating Asia's security is to measure the occurrence (and absence) of armed conflict in the region. In short, this logic suggests that a secure Asia is one where wars are not taking place and are unlikely to do so. An insecure Asia is a region where wars are prevalent and possible. For reasons that were considered in Chapter 1, this is by no means a silly way to approach the subject. There is probably nothing that can threaten the various things that are held dear (acquired values) than an armed attack. Territory, lives, finances, and even forms of government, can all be put at risk by the initiation of hostilities. Yet it is still important to whether an Asia that is largely peaceful can still be a region affected by serious patterns of insecurity.

A peaceful but insecure Asia might be one where governments in the region feel under siege whatever the circumstances, including whether or not armed violence is actually taking place or is completely absent. There is evidence for this possibility in at least some parts of the region. Pakistan, for example, has an ongoing sense of insecurity because its geography does not allow it the strategic depth that India, its much larger neighbour and competitor, possesses. The Chinese leadership worries about the potential for internal division and disorder, and will continue to do so whether major incidents appear or are avoided. Neighbouring North Korea gives every impression of being inherently oversensitive about the possibility of an attack, and in turn routinely threatens South Korea, and sometimes Japan, with the vilest of rhetoric. Some of the countries that are most reassured by the strong presence of American forces in the region are also those who are most nervous about the possibility (small though it may be) of their departure. And the more that the United States reassures its allies that it will not leave, and will instead make Asia more of a focus in the future, the more China tends to regard itself as a target.

This latter interaction is especially significant. Perhaps the best example of an insecure peace occurred at the height of the Cold War when the United States and the Soviet Union came close on more than one occasion to nuclear war. Some have now come to argue that Asia is heading towards a new Cold War between the United States and China. The sheer size and influence of these two countries means that any such development would be certain to have significant ripple effects across the region. But is that where the relationship between these giants is heading? And is theirs the only really important security game in town? These are the types of questions that will be considered in the next chapter.

Chapter 3

Power: Is It All About the USA and China?

Explaining the improvement in Asia's security environment after the violent middle decades of the twentieth century is likely to remain a matter of some debate. That is as it should be. But when it comes to looking at Asia's current and future security environment, there is significant consensus about the one factor that really matters: the relationship between the USA and China.

The interactions between China and the United States have effects at the regional, or system-wide, level in ways that few other relationships in Asia can replicate. Accordingly, many of the other countries in the Asian region keep a close eye on the positions that the United States and China are taking on regional security issues. It would be a trap for any Asian country to set its own security policies based entirely on what they believed the United States and China were doing. But it would also be the height of folly to ignore Washington and Beijing, who are central to any sense of Asia as an interdependent system of interacting components.

Those seeking a quick measure of Asia's security might do far worse than to look at the quality of the relationship between the two giants. And for those seeking a sense of how Asia's security is changing, it might be best to look at how the United States is responding to the increase in China's power. These facts of regional life illustrate a further aspect of their central importance as the two leading powers in Asia and its borderlands. No careful observer of Asian regional security can escape the fact that a multitude of actors, issues, problems, relationships and perceptions are involved. The chapters in this volume testify to this multiplicity. But it is often tempting to seek an escape from this complexity. Reducing Asia's security to the quality of Sino-US strategic relations is appealing because it is analytically so simple.

This chapter reflects something of the struggle with this understandable but problematic tendency to focus on the United States and China. It is important, as later sections will suggest, to go beyond this analytical comfort zone where all that matters to Asia's security revolves around these two sizeable actors and their bilateral strategic relationship. But it is also very important first of all to see why these two players are so significant. Moreover, if China and the United States are not nearly as important as many assume, then a great deal of security policy-making around the region is taking place on the basis of a false and very hazardous premise.

Great powers and Asia's security

A shorthand way of expressing the importance of China and the United States to Asia's security is to refer to them as the region's great powers. But this is only the case if it is clear what a great power is, and if it is also obvious that these two large countries qualify for this lofty status. It may also be necessary to accept the argument that a great power framework still applies to Asia's security, if it really ever has. A contemporary study by Nick Bisley (2012) is one example of the opposition that has built up to the notion that such a framework, which often seems best suited to earlier periods of rivalry between Europe's great and imperial powers, is indeed appropriate for Asia in the twenty-first century.

As should be obvious from the terminology, to refer to the United States and China as *great powers* is to assume that it is their unusually significant connection to *power* that sets them apart. In turn, if power is regarded as an asset held by a given actor, it might seem logical to regard great powers as those countries that have much more of this capacity than others. Indeed, a standard way of measuring national power is to think in terms of the resources available to a country. Traditionally, great powers are expected to have formidable armed forces, strong economies and significant populations. The United States would appear to qualify on all of these fronts, with China still lagging in some but easily making the grade in others, as will be discussed in some detail below. While sheer size matters, quality does too, and this is a harder criterion to assess with any confidence. But it is in their relation to one another, and to other countries, that the real measure of a great power needs to be made.

This is especially so for relationships involving security, including the extent to which various powers can provide for their own security regardless of what others may be able to throw at them. Great powers have sometimes been seen as those countries strong enough so that they do not normally have to rely on the assistance of others to meet their immediate security requirements. A great power might be thought of as a country with enough military capacity to hold its own in a major conflict. This view might seem rather quaint, and more suited to the days of Napoleon and the European balance over 200 years ago. But in an important essay on the changing distribution of power in Asia in the 1970s, Hedley Bull (1971, p. 675) argued that Japan could not be regarded as a great power in the contemporary world if it lacked nuclear weapons (and thus the ability to deter nuclear attack through its own reciprocal capabilities).

According to this sort of logic, great powers might be more likely to survive an armed attack, whereas smaller and medium-sized powers might not be so lucky. This would make it wise for smaller actors to seek the protection of a great power as an ally or partner, often out of fear about what other great powers might be planning. But this already represents a change in thinking. Here, power is no longer simply a resource that a great power possesses in great quantity, hoarded or stockpiled somewhere. It is relational. As Robert Dahl (1957, pp. 202–3) argued in a famous article, actor A has power to the extent that it is able to 'get B to do something that B would not otherwise do'. Hence a power is great when it has a level of *influence* lacked by lesser powers through which it can shape the behaviour of others in genuinely significant ways. Great powers have a very rare capacity to shape the overall system of which they are a part, hence the close attention that so much of Asia pays to what China and the United States are doing.

This means that an analysis of power should not comprise simple comparisons of the things that various actors possess: the size and quality of their armed forces, the size and vibrancy of their economies and so on. It needs to examine the relationships of *influence* between them. Here it is useful to follow the Chinese example, pointed out by Xuetong Yan (2014, p. 164), where there are 'two separate words which basically cover the whole range of meanings of the concept of "power" in English'. One of these words, *shili*, equates to strength referring to capabilities or resources. The other, *quanli*, denotes power as influence. And it is the latter that is reflected in many of the ripple effects (that is, the system-wide security factors) in Asia. These effects are often the result of the conscious use of power by political actors as they seek to influence the behaviour of others, persuading one another to do things they might otherwise not choose to do. Great powers should be expected to have rather more of this influence in the region than others: their decisions are more likely to result in ripple effects than the decisions of others. And this is not just because they are very large in an objective sense. It is that other countries are conscious, in a subjective sense, of the size and weight of the great powers in the strategic interactions that combine to make up Asia's security environment. These other, smaller, powers will be adjusting their own approaches on the basis of what they believe the great powers are doing.

This also means that great power status is not only about material resources. Finances and military hardware do matter, but a significant portion of a country's influence may come from the ideas and approaches it brings to regional security, and whether these principles are taken up by other actors. This is reflected in the idea of 'soft power', introduced by Joseph Nye (2004) to draw a distinction with the forms of power where influence comes from exploiting material resources of some kind or another. In contrast to the wielding of big military instruments or the employment of economic inducements and punishments, soft power is meant to be about the ability to set the agenda through the appeal of one's ideas, values and society. It is no surprise that such thinking came from a prominent author from the United States, a country that scores favourably in these areas. Nor should it be surprising that the potential for soft power is often mentioned when doubts are being raised about the extent of American military or economic capacities.

But soft power on its own is not sufficient for great power status. For example, it is not easy to see how a small country which has a good reputation but lacks significant material power can persuade others to follow its lead when they do not wish to do so. Inspiration is one thing. Leverage is another. This is New Zealand's challenge in Asia: being well respected only buys it a certain amount of influence. Yet, by the same token, a country that has large material sources of power may not qualify for great power status if it is neither willing nor able to wield these resources to influence others. Giants with big pockets and strong shields but who also have credibility problems can sometimes be ignored. As will become clear later, that may well be what has been holding Japan back from playing a more significant role in Asian security relationships. And this is precisely what a newer crop of Japanese leaders (including Shinzo Abe) have been seeking to change.

The USA as a great power

All of this begs the obvious question: what is it about the United States and China that allows them to make the 'great power' grade? America is the obvious starting point here. It is widely recognised as not just the leading power in Asia and its borderlands but also as the one genuine great power on the global stage. These two features are intimately connected. Because of Asia's rising importance, any global great power worth its salt must have a significant profile in this region. The United States most certainly has both, but this combination can also produce challenges for Washington in terms of where it needs to concentrate its influence. Even global great powers have their limitations, and even the United States cannot be powerful everywhere all of the time.

As the focus of this study is on Asia's security, it seems natural to start with America's military power. According to frequently cited estimates by the Stockholm International Peace Research Institute, American military spending is at least as large as the combined spending of the other countries in the global top ten, including China, Japan and India (Perlo-Freeman and Solmirano, 2014, p. 2). The United States is one of only two countries (the other being Russia) with large and comprehensive nuclear forces, which can hold almost any Asian city hostage should an American administration wish to do so. And, unlike Russia, the conventional forces of the United States are both modernised and eminently usable. The United States certainly has global security responsibilities which extend to the Middle East and Europe. But the significant portion of America's armed forces that remains devoted to Asia, especially several of its air carrier battle groups, easily overshadows the military capabilities that any Asian countries can offer. The United States is also the world's leader in both military doctrine (the way armed forces fight) and the application of advanced technology for military purposes, including the cyberspace dimension.

Thanks in large part to these unique military capabilities, America's military influence in Asia is palpable. The United States Navy is the most able of any of the world's fleets to support (or restrict) the flow of goods along Asia's sea-lanes. Given the reliance of China, Japan, South Korea and other East Asian countries on energy supplied by sea, this means that they are all dependent to some extent on American naval power, even if, as Marc Lanteigne (2008) has explained, this is a dependence that China would rather avoid. The United States is regarded by several of its Asian allies, including Japan and Korea, as their main security guarantor. No other country is as obliged to come to the assistance of such a range of Asian countries should they be attacked, and no other country has the ability to do so, including via forces that are permanently and rotationally stationed in the region. For these reasons, any country seeking to change the overall shape of the security environment in Asia would need to think first about Washington's response.

But while military power (again in terms of both resources and influence) may be a necessary condition for great power status, it is by no means a sufficient one, and even if it was, armed forces and military relationships are so expensive to maintain that they become a declining and unaffordable asset in the absence of a strong and robust economy. Britain's retreat from great power status in the twentieth century was not just about the loss of an empire. It was exacerbated by economic decline. Indeed, Britain's withdrawal as a significant military actor in East Asia (involving its withdrawal from areas east of the Suez Canal) was precipitated and accelerated by a major financial shortfall that became especially apparent in the 1960s. While Karl Hack (2013, p. 22) has acknowledged that British perceptions of the change from a colonial to a postcolonial Asia were also involved, it was the dire straits of the mid-1960s financial crisis that really forced London's hand. Until then, a range of regional countries had welcomed Britain's regional military presence in Asia as a stabilising factor, especially as it applied in Southeast Asia. But after that rapid withdrawal, Malaysia, Singapore, Australia and New Zealand were all rather more on their own than they had been previously.

Despite the Nixon administration's limitations on future American commitments to Asia's ground wars, a parallel regional withdrawal did not come from the United States. Yet, in a celebrated study published in the 1980s, Paul Kennedy (1987) attributed a similar problem of over-commitment and decline to the United States to what had afflicted Britain earlier. While that diagnosis also proved premature, economic power remains central to America's prospects of retaining its great power status. Growing debt levels have caused a renewal of concerns about US economic strength, but as a country of over 400 million people and with the world's largest and most vibrant economy, it is perilous to write off American power. For most countries in Asia, the United States remains a vital export market and source of investment, and the US dollar remains the leading international currency. America has been a safe refuge for surpluses generated in the region, (including by China) and many of these funds are invested in bonds issued by the US Department of the Treasury.

More generally, the global capitalist system in which Asian countries make their living, and the organisations that support it (including the International Monetary Fund) continue to reflect the influence of American ideas. This may offer something of a counterpoint to the argument presented by Amitav Acharya (2014) that regardless of whether the United States itself is or is not in decline as a global power, the era when the international system was dominated by American norms and ideas is coming to an end. Few Asian countries seem to be acting as if the United States is a spent force economically in their region, and they know that one way to keep Washington engaged is to remind it of US economic interests in the region. Those who are less welcoming of America's regional role must also acknowledge that if Washington takes a severe dislike to their behaviour it has a track record of using its economic muscle for coercive purposes through the imposition of sanctions as a form of great power diplomacy.

The latter point is important because great powers also need to be major and influential *political* actors in international affairs if they are to live up to their lofty title. Here there are few material resources that are available to measure. For example, the sheer size of a country's diplomatic service is not always the best guide to this dimension of power. Even more so than military or economic power, this is a story about influence rather than resources, and here Washington has certain advantages over the field. Among allies, partners and even potential adversaries, a relationship with the United States normally ranks as either the first or second most important diplomatic link of all. Audiences with the president at the White House are among the most significant meetings any Asian leader can hope to have. Given America's political importance, influencing Washington is among the most important diplomatic games that any regional country can play. This even applies to those whose political systems and foreign policies seem to be diametrically opposed to America's. One of the main prizes North Korea wants to achieve from its on-again/off-again threats, where it brandishes its nuclear weapons programme to coerce others, is a normalised diplomatic relationship with the United States. And there is no more important bilateral relationship for China than the one it has with Washington. Beijing's emphasis on a new era in great power relations, which, as Lampton (2013) confirms, is closely connected to Xi Jinping's ascension as China's paramount leader, is not a generic commentary on world affairs. It is instead a remarkable sign of the judgement that, if China wants to keep on rising, it needs a workable relationship with that other very large power.

There are also organisational markers of Washington's political importance in Asia. The United States remains the most important member state of the United Nations, whose headquarters are based in New York. With the veto vote in the UN Security Council that only the five permanent members hold, the outcome of the Council's deliberations on any conflict or crisis in Asia depends on Washington's position. As the Cold War came to an end, the market economies in Asia were determined to make sure that the United States remained regionally connected in an institutional sense. 'A trans-Pacific regional economic organization,' argued John Ravenhill (2001, p. 62), 'had the potential, its proponents believed, not only to help defuse trade conflicts between the United States and East Asia but also to sustain US strategic interests in the region'. This new group was APEC (Asia-Pacific Economic Cooperation). Similarly, as China has itself become more powerful in Asian affairs, most regional governments have seen America's increasing participation in regional security forums such as the East Asia Summit (about which more will be said in Chapter 10) as crucial to the maintenance of a favourable distribution of power in the region. In short, they regard a strong America as the most important ballast they can have as China grows.

The latter point requires close consideration. Perhaps the most important way of explaining why the United States is a great power in terms of Asia's security is that most other countries regard it as the only country that can reassure them of their concerns about China's increasing influence. To date at least, the United States has itself seen this unique role as an essential part of its foreign policy mission. In the late 1990s, Christopher Layne (1997, p. 88) recalled a decades-long pattern in American strategy with a consistent focus on 'preventing the emergence of rival great powers in Europe and East Asia'. In the Asian context, this has specifically included rivals from the Asian mainland who are able to project significant power into the region's oceans where the United States has long enjoyed strategic superiority. And for just as long, the United States has recognised that while it may have the support of allies and partners in achieving this objective, it will generally be left to do most of the 'heavy lifting'. This means that, among other things, the United States sees the retention of its own great power status, including in Asia, as being essential to international stability. In Washington's eyes, a secure region requires a strong America. A region where only China is strong is a recipe, in America's mind, for insecurity. This is also a widespread view in Asia, even among countries who are unwilling to say so publicly.

Great power China?

Hence much of the discussion, including the various assessments of America's role and influence in Asia, seems ultimately to be all about China. To the extent that great powers are defined by their unique ability to shape the system of which they are part, China is giving every impression of being able to meet this criterion. Put simply, if the United States is commonly regarded as the one and only country that can retain the regional system largely as it is, China is the one country that seems most likely to be able to change it. Whether it will do so is as much about Beijing's power and preferences as it is about Washington's ability and willingness to counteract that change.

It became clear in the previous section that the foundation of great power status is not military power but rather economic strength, from which strong armed forces are then an outgrowth. This most certainly applies to China, whose main claim to great power status is its return to economic vitality on a truly massive scale since the 1980s. With a population of over one billion, of whom literally hundreds of millions are becoming the world's largest middle class, China's economic trajectory has been like no other country on historical record anywhere in human history. But this is also a *return*, because 300 years ago, as Europe was experiencing an industrial revolution that China's emperors were dismissing, to their eventual cost, China was still the world's largest economy. After a significant pause, it is set to become that once again within a decade or so. This is the main reason why the twenty-first century is likely to be China's period of renewed greatness. And this marks China out as a different challenge to American leadership than was presented by the Soviet Union in the Cold War years. As G. John Ikenberry (2008, p. 26) has argued, 'whereas the Soviet Union rivaled the United States as a military competitor only, China is emerging as both a military and an economic rival'.

The wider ripple effects of China's return as the world's largest economic power are likely to be more striking and systematic the second time around. One reason for this assessment is the level of China's integration in, and importance to, the international economy. No fewer than fourteen countries know China as a direct geographical neighbour. Many have long histories of coping with China as a sometimes strong and sometimes weak state, waxing and waning between expansion and influence, and shrinkage and division. But many more of the states of Asia now also have China as their leading trading partner. These include countries on China's borders but also many further away in Asia and its borderlands. If Japan was the economic engine driving Asia's prosperity in the 1970s and 1980s, that role has increasingly been taken on by China. This network of close economic linkages in which China is the common feature is the strongest argument that Asia and its borderlands comprises a system of interconnected parts. By the middle of the first decade of the twenty-first century, Thomas Christensen (2006, p. 91) was already concluding that 'Without the central role China has played, there would not be the truly impressive economic interdependence that one sees in the region.'

China's rise is in many ways the defining feature of today's Asia. Likewise, China's economic importance to so many of its fellow states in Asia is the main source of Beijing's influence in the region. This is a widely acknowledged reality across Asia, and China's leaders are fully aware of the way other states in the region depend on China so extensively for their own prosperity. As Mark Beeson and Fujian Li have noted, (2012, pp. 36–7) 'basic economic interdependence is forcing other states to recalibrate their relations with China whether they wish to or not'. To complete that circle, China's regional partners know that China is aware of the situation and the leverage this may offer. For the most part, Beijing has not tended to use this influence in a damaging or overtly threatening way, partly because it depends on the prosperity that these connections allow. None the less, a very intriguing situation has developed in much of Asia. While the United States remains the most important security partner for many regional countries, and for its allies is their security guarantor, China has become the most significant economic partner of these very same countries, notwithstanding the very close links some of them have traditionally had with the United States.

These security linkages favouring the United States and economic connections favouring China are both forms of ongoing influence. When considered in combination, they generate some intriguing dilemmas for the smaller states of the region: to what extent, on an important issue, do they vote with their security viewpoint and draw closer to the United States, and to what extent do they vote with their economic viewpoint and side with China? Even if some states regard a stronger China as a challenge, it will not necessarily mean that they will take Washington's view. Of course, the answer from most regional countries is that they want the best of both worlds. But as China's power is growing faster than America's, and as China becomes a more important security actor (especially as its military power grows) it may become harder for this combination to work. What is especially noticeable about this equation is that, for the most part, only China and the United States are considered as the leading actors in the drama. The dominance of this duopoly is evident, for example, in the argument from Hugh White (2012) that the United States should forsake its leadership ambitions and make a deliberate choice to share power with China. While there are continuing debates about the fine line that regional countries have to walk between China and the United States, there is really very little discussion of choices that regional countries might need to make involving other powers. Japan and India are largely off the radar in this respect. In short, this reinforces the common view that when it comes to the crunch, only China and the United States really do matter.

The idea that there may be choices to make between the United States and China also feeds the argument that they are natural competitors. Historically at least, that is what great powers have often done. Hence John Mearsheimer (2001) argues that war between the two giants is almost inevitable, regardless of the good intentions that either (or both) may have to avoid such a conflict. This is not a view that one finds reflected in the way that United States governments talk about their relationship with China. They do not deny that competition exists, but they also emphasise the common interests that unite the two big players in Asia. Many opinion-makers in China go even further than this, depicting their vast country as an historically unusual great power that bucks two negative trends. The first is the tendency for a rising great power to engage in armed conflict with established great powers. Bijian Zheng (2005, p. 22) has suggested that China is determined instead to 'transcend the traditional ways for great powers to engage'. The second is the tendency for great powers to build large and threatening armed forces that are ready to fight these conflicts. China, so the argument goes, has not put nearly as much effort into its military expansion, concentrating instead on economic development. Some Chinese writers continue to argue that China remains a relatively weak military actor, that the People's Liberation Army lacks the ability to project armed force at great distances, and that China's historical record is profoundly defensive rather than aggressive.

Whether or not this broader thesis holds, it is true that China could easily have devoted more of its growing wealth into building its military capabilities. Indeed, for a country widely seen as rising quickly, and which is spending 10 per cent more each year on its armed forces, China still faces significant military limitations. For many decades, China's continental strategy favoured a vast army to defend the People's Republic, an approach that severely hampered its ability to project military power beyond the Chinese mainland, including against Taiwan. But this approach has been changing. China now boasts an outward-looking maritime strategy designed to impart influence over the sea and air beyond the immediate coastline. China's growing influence, and willingness to use its armed forces to send signals in maritime Asia is a leading factor in the region's territorial competition, will be considered in more detail in Chapter 6. But this transition is still a work in progress. It requires China to devote decades of significant investment into advanced military equipment and training. In this qualitative regard, China will continue to lag for some time behind some of its neighbours, including Japan and the Republic of Korea, to say nothing of the advantage that is held by the United States. On the latter, Robert Ross (2009, p. 26) concluded that while 'China is buying and building a better maritime capability ... the net effect of China's naval advances on U.S. maritime superiority is negligible.'

But these qualitative gaps do not mean that China can be excluded from great power standing on military grounds. There are several reasons why it would be unwise to do this. First, unlike every other East Asian country apart from North Korea, China possesses a nuclear arsenal, and, unlike North Korea, China undoubtedly has effective systems with which to deliver those weapons. China has certainly struggled to master the launching of nuclear-armed missiles from the sea, a particular area of disadvantage *vis-à-vis* the United States, but it has for some time possessed land-based missiles with long ranges. Not all analysts will agree with the argument laid out by Michael Tkacik (2014, p. 162), that the numerical and qualitative improvements in China's arsenal signal 'a goal to create an arsenal that is useful for purposeful coercion, early use, and possible parity with the US and Russia'. But China is definitely enhancing its nuclear forces as part of a long-term modernisation effort.

Second, China has embarked on a concerted conventional military building programme. The numbers of Chinese submarines under construction, for example, dwarfs programmes elsewhere in the region. Third, with submarines, multiple missiles systems, and a strong commitment to information warfare, China is raising the costs for other powers, in particular the United States, to deploy forces close to the Asian mainland. China cannot *control* the sea (no state can do that in reality) and it has a long way to go before it can project substantial military power at a significant distance across the oceans. But it has certainly made it more risky for the United States to operate freely in China's neighbourhood. During the most recent major Taiwan Strait crisis in 1996, the United States deployed aircraft carriers close to China in response to what it saw as China's intimidation of Taiwan. It would be much more risky for the United States to do that now, because of China's increased ability to detect and threaten America's naval forces. America's main allies in that part of the region, especially Japan, know that this also has implications for what the United States might be able to do in a crisis that affects them. As Aaron Friedberg (2014, p. 45) has argued, the improvement in what are referred to as China's 'expanding anti-access/ area denial (A2/AD) capabilities will make it far more difficult and dangerous for the United States to intervene in a conflict on behalf of its regional friends and allies, and could conceivably deter it from doing so'.

Moreover, China's military modernisation is having distinct ripple effects, making it a wider regional security factor. Japan, Taiwan, India, Vietnam, the Philippines, and the United States are all considering how they can develop their own forces as a consequence of China's military development. It also feeds into the increased willingness of some regional countries, including Singapore, to host American forces on rotational deployments. A stronger People's Liberation Army (PLA) has also been associated with China's increasing confidence in asserting territorial claims, including in the East China Sea (in competition with Japan) and the South China Sea (in competition with Vietnam, the Philippines and other Southeast Asian claimants). If push came to shove, and in particular if the United States became involved, the PLA would still be caught short. At least in a technical sense, the United States is likely to be able to dominate China militarily at each level of any maritime conflict. But over time, China's capacity to enforce its claims is growing, and it probably has a greater political commitment to its regional interests in Asia than does the United States. As James Dobbins (2012) has argued, China's ability to escalate a conflict in its favour, and to America's disadvantage, is also likely to increase.

This raises a further question about China's great power status. Tensions with a range of neighbours and with other claimants to the South China Sea, for example, are certainly a mark of China's significance. But these problems denote an ability to influence the perspectives of other countries in a rather negative way. This may not always work to China's long-term advantage: the country may be able to shape the regional security environment as great powers are meant to, but this ability may not work to its benefit if the main result is to increase concern about Beijing's growing capabilities and future intentions. This may suggest that in the political-diplomatic dimension of great power status, China still has a fair bit of work to do. It may even eat away at China's perception of itself as a new and different kind of great power. Benjamin Schreer and Brendan Taylor (2011, p. 19) have argued that despite its growing material power, 'China still lacks a convincing regional leadership model to underpin its grand-strategic design.' In other words, China has a problem in converting *shili* into *quanli*.

None of this means that Beijing is an isolated great power, however. China does have some regional partners and allies, but most of these relationships come with built-in limitations. China's partnership with Russia, while seemingly impressive, is rather more a marriage of convenience between two geographically large neighbours that have historically not got on well. They share an antagonism towards American claims for global leadership, but generally are more effective at negating Washington's preferences than at imposing their own. China's closest Asian partners are a motley crew. Neighbouring North Korea, with whom China has perhaps its only formal treaty-based alliance, is a pariah state which diverts scarce resources to a bloated military system, including an internationally unpopular nuclear weapons programme. While North Korea is a foreign policy liability for China, it remains territorially important, and as Gregory Moore (2008) observes, the nasty status quo remains preferable to a collapsing North Korea that could bring all sorts of challenges to China. As a consequence, China's leaders put up with Pyongyang, though they are losing patience.

Pakistan, another close friend of China, is also a questionable asset as an ally. Assistance to Pakistan has allowed China to keep India back on its heels. But an unevenly governed and self-focused Pakistan has very little to offer to China's claims for regional leadership. Among the Southeast Asian countries, Cambodia is sometimes seen as most likely to do China's bidding. But neither it nor Laos or Myanmar are regional leaders. In short, there is no effective grouping of disaffected states under China's leadership that can compete with the array of mainly maritime democracies that constitute America's leading allies and partners in Asia.

Yet China has made more significant headway in gaining broader diplomatic influence in Asia through its involvement in the region's growing array of interstate groups that seek to foster security, and diplomatic and economic cooperation (these are discussed in more detail in Chapter 10). China's leaders were once shy of multilateral groupings of almost any stripe, seeing them as traps that would require a diminution of their country's hard-won sovereignty. But by the late 1990s, as Cheng-Chwee Kuik (2005, p. 107) has noted, China was becoming aware that 'multilateral forums ... might not necessarily be harmful to its national security', and that 'multilateral cooperation could be used as a diplomatic platform to promote its own foreign policy agenda'. Since that time, an increasingly confident China has seen that, as a large power, it can shape the agenda and even the membership of regional bodies. That the members of ASEAN often see the multilateral groupings they have championed in Asia as a way of engaging both the United States and China (and possibly as a way of moderating the competition between the two giants) is a sign of China's importance as one of the two obvious great powers in regional diplomacy. Moreover, the importance of China's burgeoning market to the countries around it automatically puts China at the centre of many of the groups that seek to build closer trade and investment relationships in the region. This includes the free trade arrangement that exists between China and the ASEAN countries, which reflects the growing importance of economic exchange within Asia.

The perils and limits of great power status

China's growing importance, even in economic terms, will always be a double-edged sword. Just as there is nothing the United States can do to completely reassure regional countries that it is paying them enough attention, there also is nothing China can do to completely reassure the region that its rise is not something to be feared as well as welcomed. But these concerns are also signs of China's importance. Indeed, one of the criteria for great power status is certainly not the requirement to be liked, though being liked does help when it comes to considerations of soft power. As Mingjiang Li (2008, p. 305) has observed, some scholars within China have argued that their country faces a soft power deficit because of the challenges of 'elucidating a set of rules that unites the Chinese population domestically and is convincing, appealing and attractive externally'. On the latter point, China seems especially vulnerable. Because the country is sometimes feared, its chances of being the one and only great power in Asia are seriously diminished, if indeed this is what its leaders are eventually seeking. The more that China's neighbours are aware of the country as a re-emerging great power, the more that many of them want the United States to stick around as the only great power that can guarantee a regional equilibrium.

Ironically, though, China's great power standing may be limited less by what others do and more by what happens inside the country itself. China's leaders remain very concerned about the internal cohesion of their vast and diverse country, not least because of major gaps between the rich, who are concentrated in the eastern coastal areas, and the poor, who often live further inland. This divide has been growing because of the unevenness of China's impressive economic development. But it is just one of several challenges facing the leadership and ongoing legitimacy of the Communist Party of China, which, according to David Shambaugh (2008, p. 4) include 'increasing social stratification and inequality, widespread corruption, pervasive unemployment, rising crime, and rural unrest'.

Internal problems are not new for this vast country. Over its very long history, China has experienced periods of growth and confidence (something that is happening now) as well as destructive eras of violent fragmentation and disorder (something that can never be ruled out). As Chapter 7 will argue, a weak China might in fact be more troubling for its neighbours than a strong one. Internal weakness could well reduce the ability of China's leaders to influence other countries in Asia deliberately, if for example, economic difficulties resulted in a reduction in external leverage. But if for some reason China were to fall apart, the rest of the region would surely know about and experience that phenomenon. There would instead be some different and more involuntary transmission belts that would allow effects to ripple out across the region.

This suggests that China is simply too big and too important to Asia and its borderlands for the countries of the region not to see it as a great power. China's weaknesses do not necessarily change that picture, because all great powers have their limitations. This certainly applies to the region's other main player: as a global great power, the United States has many calls on its time and resources, and even with the best of intentions it will sometimes find it challenging to focus as much of its attention on Asia's security as it might like. As it sought to free itself from a costly era of fighting long and complex wars in the Middle East and Central Asia (Iraq and Afghanistan), the Obama administration announced that it would pivot its efforts towards Asia. One of the architects of that policy, Kurt Campbell, has argued in a prominent article with Ely Ratner (2014, pp. 107–8) that this was not really a matter of choice: 'whether Washington wants to or not, Asia will command more attention and resources from the United States, thanks to the region's growing prosperity and influence – and the enormous challenges the region poses'.

In theory at least, one advantage of this change was the greater opportunity it would give a less distracted Washington to respond to China's growing influence in Asia. Indeed, some of America's friends in the region felt that the USA had not been paying nearly enough attention to a rapidly changing Asia. But that redirection of effort was always going to be challenged by developments elsewhere. The consequences of the Arab Spring for political order in Libya, Syria and Iraq, and the crisis between Russia and Ukraine (and thus between Russia and NATO) are examples of factors that might easily call for a further reprioritisation of American energies. This doesn't bring into automatic disputation the importance of Asia's security for Washington's policy but it does reduce its ability to command the stage. Hence the counterargument from Barno et al. (2012, p. 175) that 'Iran's quest for nuclear weapons, Israel's precarious neighborhood, and Iraq's turbulent sectarian infighting all require U.S. policymakers' attention today, not tomorrow ... Hedging against turmoil in the Middle East and inevitable shocks across the globe is just as important as pivoting toward the Asia-Pacific.' The logic of that hedging has only grown with the commitment (mainly of airpower) by the United States to the fight against Islamic State in Iraq and Syria.

Domestic factors may also conspire to reduce America's commitments as a great power in Asia. A significant part of this is about domestic political opinion within the United States. There is no guarantee that future presidents and Congresses will continue the tradition of a forward commitment of US military forces into Asia, or that they will sustain it at levels that work for America's leading regional allies. Unlike China, the United States is not a resident East Asian power located in the middle of the region. American leaders and officials refer to their country as a 'Pacific power', and the state of Hawaii and the territories in Micronesia and Guam (an increasingly important island military base) attest to this fact. But that is quite different from China's position, where all its population and territory are in Asia itself. America's primary sphere of influence has always been the Western hemisphere, as reflected in the Monroe Doctrine, which laid out America's determination to obtain strategic primacy in Central and Latin America. As Chengxin Pan (2014, p. 457) has noted, a number of American thinkers, and in particular those who foresee a more dangerous period of Sino-US strategic relations, fully expect China to copy this geopolitical example and develop a Monroe Doctrine of its own. To the extent that this is already occurring in terms of China's attitude towards its maritime territorial claims, that doctrine would lie squarely in the heart of Asia.

Economic constraints, which all great powers face at some time, may also restrict America's role in Asia. While it continues to outspend all comers on defence, the United States has had little choice but to reduce its overall military budget to help rein in the debt hangover from the 2007–8 Global Financial Crisis which began with a severe downturn in the US real estate market. A sustained reduction in defence spending would still leave the United States with the world's strongest armed forces. But it would almost inevitably accelerate the closing of the gap between the American presence in Asia and China's more concentrated deployment of military assets in the same region. As Michael Evans (2013, p. 174) wryly observed, 'The paradox of U.S. reinforcement in the Asia-Pacific while undertaking large-scale defence cuts numbering billions of dollars has not escaped notice' in the region, including in Southeast

Asia, where a number of countries have been strengthening their strategic relations with Washington. Even so, while both the United States and China face limits to the resources and influence they can generate, they can still be considered as the two genuine great powers that will continue to have dominant effects on Asia's regional security system.

What about Russia, Japan and India?

The obvious question arising from the preceding analysis is whether any other country in Asia counts nearly as much as the United States and China clearly do. An immediate response comes from the region's recent history. This is the unmistakable fact that during the Cold War the Soviet Union had a substantial impact on Asia's regional security. But since the collapse of the Soviet Union over two decades ago, Russia, which is now smaller both geographically and demographically, has not enjoyed the same influence and strategic weight in Asia. This applies in particular to East Asia, where the main Cold War contest in the region between the United States and the Soviet Union took place. Readers of the leading study of the development of Asia-Pacific international politics will find that, while Russia has a place among the great powers listed in the Cold War period, there is no corresponding listing of Russia in what Michael Yahuda (2011) has termed 'The era of American pre-eminence'.

Even so, Russia is no small fry in international politics today, and its annexation of Crimea from Ukraine in 2014 showed that Moscow could still attract world attention and Western concern in a way that few other countries can. This act also demonstrated Russia's ability to test global norms, including the importance of sovereign independence, norms which its sometime partner, China, takes very seriously. Russia is also something of an energy great power, courtesy of its enormous oil and gas reserves, even if, as Rutland (2008) argues, it is not always that simple to turn energy resources into influence. That said, the Putin government has shown a willingness to exploit the dependence of many European countries on Russian gas. But the crisis over Ukraine, which also led to serious tensions between Moscow and NATO, is further confirmation that Russia's main influence is directed more to Europe (to the west) and to the Central Asian borderlands (to the south). Of course, the latter include a number of independent states that were once republics of the Soviet Union; and alongside China, Russia has joined with these Central Asian countries in a regional grouping called the Shanghai Cooperation Organisation. But Russia's relationships in Central Asia (which are not altogether smooth) do not give it automatic purchase in wider Asia, including the very important East Asian parts of the region. Moreover, while Russia and India have retained an active partnership, it is not nearly as strong as it was during parts of the Cold War period.

At times in East Asia's security relations, Russia seems to take on the role of a spectator rather than an active participant. But this does not make Russia irrelevant to East Asian power considerations. By virtue of its extensive nuclear arsenal, and territory that extends to its Pacific coastline, Russia is still a factor in the strategic calculations of China and the United States. No account of the overall military distribution of power in Asia can leave Russia out entirely. And, in more recent times, flights by Russian long-range bomber aircraft, so often seen in the Cold War, have become more apparent, including near Japan, over the Korean peninsula and not far from Alaska.

But Russia is no longer a shaper of regional security in Asia in the way it once was. Its diplomatic presence is also less evident. It is more of a bystander than a leading participant in the East Asian regional forums of which it is a member. If because of its economic and political shortcomings the Soviet Union was in the words of Paul Dibb (1986) an 'incomplete superpower', Russia today is an incomplete great power as far as the Asian region is concerned. That may, of course, change, but Russia has a great deal of catching up to do. In terms of consistent patterns of influence, few countries in Asia find it necessary to keep Russia uppermost in their thinking in the way they find themselves considering the intentions of China and the United States. And while China often finds Russia a suitable partner with which to jointly resist American international leadership, this is still an option for Beijing rather than a strategic necessity.

In addition to Russia there are two other main contenders for great power status in Asia. The first of these is Japan, which by the 1930s was undisputedly a great military and economic power in the region. This fact was painfully brought home to a number of regional countries in many years of war and conquest. But after its complete surrender in 1945 and the establishment of a peacetime constitution that foreswore the projection of military force, Japan embarked on a novel political experiment. Could almost complete great power status be achieved by a large and fast-growing economy, and the diplomatic weight these economic resources enabled? This included a substantial overseas aid programme that worked in Tokyo's favour in parts of the region including in Southeast Asia. Writing of this policy as its stood in the late 1980s, Dennis Yasutomo (1989, pp. 495, 497) referred to a growing impression at that time that 'Japan will apparently greet the twenty-first century as an "aid great power"', having rejected 'the military road to great power status'.

For many years it looked as though this experiment was working. While acknowledging Japan's continuing dependence on America's protection for its security, and thus questioning the idea that economic sources of power had trumped military ones, Joseph Nye (1990, p. 180) posed the possibility more than 20 years ago that 'perhaps we are in a "Japanese period" in world politics'. Moreover, while Japan's defence spending was limited to a maximum of 1 per cent of gross national product (GNP), this was a small proportion of a very large economy, enough to support a regionally very significant Japanese military. That these capable armed forces could only be used in very limited circumstances because of constitutional restraints did not seem to be a liability when so much of the region relied on Japan's economy as the main engine of regional growth. Indeed, as a country without nuclear weapons and the extensive overseas military commitments that could easily strain an economy, the rapidly growing Japan became the example for what an apparently overburdened United States might to do if it wanted to rescue its own great power status. But Japan's example was not to last. Two decades of economic stagnation followed, and with them came a loss in political confidence. As China's more comprehensive great power rise became more evident, Japan's concerns multiplied regarding its own position, including the possibility that it was relying too heavily on the United States for military protection. With a declining population (something it has in common with Russia), it might be thought that time is running out on Japan's chances at modern great power status.

Yet Japan has retained significant influence on its regional neighbours. As China's most important neighbour and Asian competitor, and as the leading regional ally of the United States, Japan is uniquely situated to influence the great power picture, even if it does not reach that status for itself. In a study of how order might be produced in Asia, Evelyn Goh (2013) insists that the crucial question for regional countries will remain the way in which they see the United States and China relating to one another, but in whatever way this works out, Japan is the important third player in any future bargaining. This means that any new era in great power relations between the United States and China can only work with Japan's tacit support, but this is something that seems unlikely, because Japan's leaders have no interest in being forgotten by Washington or ignored by Beijing. A certain amount of tension between the 'big two' suits Japan much better than anything that resembles an agreement to run the region together.

Japan also has potential military options that it is impossible to leave out of the picture. Among the many countries in the world with civilian nuclear power facilities, Japan would be able to generate nuclear weapons in the fastest time. This move, it must be admitted, is politically unlikely. There is a strong aversion to nuclear weapons in Japan, and a broader opposition to nuclear issues following the release of radioactivity from a civilian nuclear reactor near Fukushima that was unable to withstand the effects of the devastating 2011 earthquake and tsunami. But a nuclear-armed Japan remains technically feasible, and it is something that China's and America's leaders cannot rule out as a possibility. On the conventional side, Japan's investment in advanced aircraft, submarines, surface warships, command and control systems, and its cooperation with the United States in the development of missile defence systems, are a reminder that the so-called Japan Self-Defence Force (JSDF) is one of the most capable military instruments in the region. The historical restrictions on its use are now under significant pressure as a new generation of conservative politicians want Japan to be able to play a more active role in military affairs.

These changes have been far from immediate. Christopher Hughes (2009, p. 27) refers to Japan's 'propensity to shift incrementally towards remilitarisation'. But they were accelerated after the election of Shinzo Abe as Japan's prime minister for the second time in 2012. The changes include a greater freedom for the JSDF to work with other regional armed forces. Yet military normalisation will not put genuine great power status within Japan's easy reach. Tokyo may have some friends in the wider Asian region, including India, Australia and some of the Southeast Asian countries who are most worried about China's approach to maritime territorial disputes in the South China Sea. But the staunch opposition of both China and the Republic of Korea to a growing security role for Japan, including its ability to come to the assistance of its security partners, indicates that Tokyo continues to be vulnerable on questions of regional legitimacy. And its continued dependence on America's military presence in Asia indicates that Japan cannot put itself forward as a strategically self-reliant great power in the way that Washington can, and that increasingly Beijing will be able to do.

The second contender for great power status in Asia alongside Russia is India, which gives the initial impression of moving in the opposite direction to Japan. While China's power is increasingly overshadowing Japan's, India seems to be catching up as the only other country in Asia and the world with a population of more than one billion people. Like China, with whom it shares long and sometimes contentious land boundaries, the modern state of India represents a venerable Asian civilisation. On this point, Stephen Cohen (2002, p. 8) has argued that India's 'distinct civilisational identity' allows this large country 'to mobilize its own people around a unique set of values, images and ideas'. But this remarkable history can be a liability as well as an asset. After gaining independence from Britain in 1947, the sense that India was in some way unique and special, coupled with a postcolonial desire for autonomy, fed into a quest among its leaders to remain separate from the world economy. This held the advancement of the country back for decades. India's preference for foreign policy non-alignment also held it back from taking a wider regional role in a political context. The shock of defeat in the 1962 border war with China exposed India's weakness, and encouraged greater military expenditure and modernisation. But, as Ganguly and Pardesi (2009) have recounted, policies of economic isolation meant that India remained a marginal player in global politics, and as an Asian power, India's influence was restricted largely to its own immediate South Asian neighbourhood.

India's big break came at the end of the Cold War, which made greater room for rising powers as the bipolar dominance of world politics ended. But it was a serious programme of economic liberalisation, an urgent response to a major financial crisis for India, that turned the corner by ushering in years of significant growth, and the global economy was seen increasingly as India's friend. The generation of one of the world's largest middle classes has meant that an enormous population is now part of the recipe for India's rise. And unlike China, the best years of the relationship between demography and economic growth still lie ahead for India. According to David Bloom et al. (2010), the ratio between China's working-age to non-working-age population peaked in 2010. In contrast, this ratio for India, which is still climbing steadily, is not expected to peak until 2035. This increases the chances that once China becomes the world's largest economy, it may not be too many years before it is displaced from its position by India.

Growing military capabilities suggest that India is preparing to meet at least one of the other prerequisites for great power status. By making the existence of its nuclear weapons obvious in a series of tests in 1998, India also announced itself as a de facto member of that important nuclear club, as also did Pakistan in the same year with tests of its own in response. And while India has often been locked in local struggles, once again including skirmishes with Pakistan, successive governments in New Delhi have shown an increased desire to make more of the Indian Ocean's connections to East Asia. This has helped to drive a larger and more able Indian Navy, which in general has been welcomed by many East Asian countries that are also rather worried about China's rise. Ironically, as David Brewster (2011, p. 845) has suggested, India's previous weakness in projecting military power beyond South Asia may make it more acceptable as a security partner in the wider region including Southeast Asia, where it is often perceived 'as essentially a benign power and not a would-be hegemon'. Even so, the idea that a materially stronger India might provide more ballast in the regional distribution of power has been one reason why it has been welcomed into many of Asia's multilateral forums. It has also been the central reason for America's attraction to the idea of India as a strategic partner. As a consequence, the United States has been willing to overlook India's non-membership of the Treaty on the Non-Proliferation of Nuclear Weapons (NPT) and agreed to cooperate on civilian nuclear matters.

But India's achievement of great power status in the twenty-first century is still a distant prospect. There is no doubting India's pre-eminence in South Asia. But in the wider Asia-Pacific region, the unevenness of India's economic modernisation (combining great advances with massive poverty), the challenge of governing such a diverse democracy, the tendency for strategic affairs closer to India to win out over wider Asian concerns, and a diplomatic service that is tiny in proportion to India's interests, all combine to dilute its regional influence. Unlike China, India is not positioned at the core of East Asia, where the region's major strategic relationships come together, and India does not play a consistent or prominent role in the multilateral processes that the ASEAN countries have established. Even the most articulate proponent of a growing Indian role in East Asia, C. Raja Mohan (2013, p. 89), admits that: 'Any suggestion that India might influence the balance of power in the western Pacific is met with skepticism from the community of diplomatic practitioners as well as regional experts in East Asia.' Yet India's own rise is real. But this will not happen quickly and for the time being very few analysts dare to place India in the same league as China, let alone America.

The importance of Asia's not so great powers

Even while India's time as a twenty-first century great power has not yet come, and while doubts remain about the extent of Japan's future influence, both of these Asian countries remain important to the overall picture of Asia's security. This is the case even when the focus of attention remains squarely on the United States and China, because India and Japan matter in the security outlooks of both Washington and Beijing. India is big enough to make a difference, depending on how it positions itself in its relations with the United States and China. For this reason, some American policy-makers are keen to see India as a partner in jointly managing the rise of China. Fully aware of these expectations, India has generally sought to retain autonomy. India may see China as its long-term strategic competitor in Asia but this has not stopped its leaders from seeking tolerably good relations with Beijing. While India's leaders have sometimes regarded China in threatening terms, Pratap Bhanu Mehta (2009, p. 221) asserts that this is largely because of the territorial disagreement between the two Asian giants (on land rather than at sea) and that there is 'relatively little desire to balance the Chinese elsewhere'. To the extent that India does more to balance China it will want do so on its own terms and not as an annex to American leadership.

Japan is in a different situation. The formal security alliance it has with the United States has been the mainstay of its foreign policy for decades, and the central factor that keeps the United States and Japan close at the time of writing is their common concern about China. But if Japan for some reason concluded that the United States has ceased to be a dependable ally, two stark choices will appear on the horizon. First, Japan might try to become self-sufficient as a great power in its own right. However, as Hugh White (2012, p. 88) surmises, this could well require the development of a nuclear arsenal, something that would change the strategic order in Asia. Second, Japan could simply accept China's growing weight on its doorstep and subordinate itself to this large neighbour. That would be a very unusual thing for Japan to do, at least in any comprehensive fashion. Indeed, Mike Mochizuki (2007, p. 769) has observed that 'Although Japan may be bandwagoning with China's economic growth, it is not bandwagoning with Chinese military power.' Yet whatever path Tokyo eventually takes – the nuclear option or the accommodation of China, or something in between – Japan's importance to Asia's regional security is abundantly clear, even if the system of which it is a part is dominated by China and America.

This logic also means that many smaller countries of the wider Asia region can in theory also remain pertinent in a region dominated by just two great powers. One such set of examples comes from the Korean peninsula, which is important to Asia's security not simply because it is among the most heavily armed and dangerous locations in the region (and in the world). If war were to break out in Asia, the Korean peninsula is one place where it might well happen. But even in the absence of such a dramatic development, the political future and international alignment of the two Koreas is crucial to what earlier parts of this chapter referred to as the distribution of influence in the region. In the event that North Korea's dynastic regime collapses, China could well be faced with an unpalatable future of a unified Korea allied to the United States running right up to its border. This prospect may be even more worrying to the leadership in Beijing than the flood of refugees that might appear if that regime collapsed. By the same token, should a unified Korea distance itself from the United States and embrace China, Washington will have lost one of its main points of influence in Asia, though, as Robert Art (2010, p. 377) has argued, this might also pave the way for the United States finally to withdraw its forces from the peninsula.

Such a move would probably involve further change in the distribution of regional power. As a pair, the two Koreas have the capacity to change Asia's geopolitics. Neither China nor the United States can afford to take their eyes off either of them. North Korea has claims to more direct forms of influence through its nuclear weapons programme, which will be referred to in Chapter 5. But this is one of Pyongyang's only cards to play, and its dependence on military threats as a form of foreign policy seriously limits any chance of broader influence. However, those threats are enough to ensure that a good deal of South Korea's focus remains firmly on events to the north of its border and make it hard for Seoul to make a claim for wider regional leadership.

Further south, the countries of Southeast Asia are aware that they are also important to the distribution of influence in the region. They know this partly because ASEAN provides a series of forums where the leaders of the great powers meet regularly, and where both China and the United States make a point of emphasising the importance of these interactions. Many of the Southeast Asian countries also know this because they are being courted by both China and the United States. Partly because of their size, Vietnam and Indonesia are particularly important in this context. Neither of them is a formal ally of the United States (a point that will be returned to in Chapter 11), but Washington increasingly regards them both as important security partners in Asia. That is quite a turnaround, given the very difficult (and violent) history of US–Vietnam relations, and to a lesser extent because of Indonesia's tradition of non-alignment.

There is at least some degree of reciprocity here. As China has been growing, the majority of the Southeast Asian countries have been keen to ensure that a strong American presence remains in the region. In an assessment that has since proved accurate, Ian Storey (2007, p. 11) once argued that 'America's role in the eyes of Southeast Asian elites is likely to become more important, not less.' But that does not mean many Southeast Asian countries have made a decision to choose the United States ahead of China. Vietnam, for example, has a long history of enmity with China, and this has been displayed in tense standoffs over maritime boundaries in the South China Sea. But Vietnam has also had to come to terms with the importance of its growing neighbour. Like almost all of the Southeast Asian countries for whom economic development is a central priority, China simply matters too much.

That creates a delicate balancing act for Southeast Asia. For example, Singapore has been careful not to make too much out of the fact that it offers port facilities to visiting United States Navy's ships, knowing that it also has to give significant priority to the engagement of China. As Evan Laksmana (2011, p. 31) has written, Indonesia has similarly pursued a deliberate policy of encouraging the simultaneous engagement of as many of the larger powers as possible, in the hope of creating and sustaining what it calls a 'dynamic equilibrium'. Other powers are part of this mix, including India and Japan. Nonetheless, it is quite clear that China and the United States are the most important contributors to any such regional equation as far as many of the members of ASEAN are concerned.

Further away from the core of Asia, a number of countries in the wider region also matter to this distribution of influence. Australia, which is among the closest allies the United States has anywhere in the world, hosts joint facilities that play an important role in US intelligence networks, and has one of the most able militaries for its size. In the event of a conflict in Asia (including with China), the United States would probably call on Australia very early to assist, and as Benjamin Schreer (2013) has indicated, Australia is likely to have a role to play even if such a war became especially heated. But China would not need to do much to remind Australia of the importance of the connections that so much of Australia's modern wealth relies on, including the massive Chinese investment in Australian natural gas exploration and Chinese demand for Australian minerals. This could create some interesting dilemmas for Canberra in a crisis.

China has also lavished attention on New Zealand, Australia's smaller neighbour. That New Zealand was the first developed country to sign a free trade agreement with China may have something to do with the former's non-threatening size. But New Zealand may also be attractive as a partner because of its relative autonomy from Washington after the breakdown of their formal alliance relationship in the 1980s. This appeal appears to have remained, even though New Zealand's security relationship with the United States has become noticeably closer, a process explored elsewhere by this author (Ayson, 2012). The small Pacific Island countries, some of whom with populations only in the tens of thousands, receive more great power attention than one might expect. China, for example, has a significant diplomatic presence in the South Pacific. This does not mean that Beijing has ambitions to build a string of bases on the islands. It is more likely that, because the world's smallest sovereign states have a vote at the United Nations, this is enough to attract attention, especially for a rising great power, which as Jian Yang (2011, p. 141) noted, sees itself as 'a leader of the developing world'.

Do the great powers matter that much?

The assumption running through the analysis in the previous section is a precarious one: that other countries in Asia matter, but that they matter precisely because of what they mean for the relationship between the United States and China. As further chapters of this book will demonstrate, it is eminently possible for these smaller countries to be important to Asia's security in ways that are far less connected to the great power picture. An even wider view can be taken by understanding the importance of non-state actors to Asia's security. Indeed, some non-state groups are likely to have a greater chance of creating ripple effects across Asia than some of the less consequential states in the region. As will become clear in Chapter 8, there have already been clear instances where groups who resort to terrorism, for example, have had a big impact on perceptions of security and insecurity in important parts of the region, especially in South Asia and maritime Southeast Asia.

But it is the contention of this book that great powers deserve especially close attention because, unlike the other residents of the region, the United States and increasingly China are influential enough to figure in the security calculations of just about every other security actor in Asia and its borderlands. This shaping effect can occur in different ways. A great power can extend security protection to other states in the region to increase the latters' sense of safety, to compensate for their vulnerabilities and reduce their need to provide for themselves. This, for the most part, has been the reason that so many states in Asia have welcomed a strong American military presence in the region. It is one of the main reasons why a number of Asian countries still enjoy strong formal alliance relationships with the United States. A number of these allies, especially Japan and South Korea, also sit under the American nuclear umbrella: Washington has extended nuclear deterrence to them partly so that they do not feel too tempted to go nuclear themselves. These are varieties of what are commonly referred to as positive security assurances which, in the nuclear context, are commitments by 'nuclear weapon states to come to the aid of a non-nuclear weapon state threatened by nuclear weapons, or against which nuclear weapons have been used' (Simpson, 1994, p. 25). It is not altogether clear how far positive security assurances (of a nuclear or non-nuclear variety) are available to any of China's allies, though a country deciding to take on North Korea, or perhaps Pakistan, would need to factor in the prospect of Beijing's potential reaction. On the whole, China is yet to act as a significant security guarantor in the way the United States has been for decades in Asia. But, somewhat ironically and unintentionally, it is China's growing power that has reinforced the importance of America's role in this regard.

A great power can also promise, explicitly by a formal statement, or implicitly by its actions, not to use force against other states in the region. This is a negative security assurance: a sign of what a great power will *not* do. In theory at least, the knowledge of this commitment to restraint is important, because great powers normally possess the ability to cause significant destruction to others should they decide to do so. One of the reasons that Asia has seemed peaceful in recent decades is that few countries fear that the great powers have designs of this sort on them. But this is not a universal feeling and there are at least two associated challenges. First, as Victor Cha (2009 pp. 119–20) has pointed out, if the United States thought that North Korea would be satisfied with a negative security assurance, then it needed to think again, because Pyongyang has wanted much more. Second, the very same military might that the United States uses to offer some means of protection to its allies can itself become a challenge to the security perceptions of others. China has to some extent been reassured by Washington's promise to Tokyo because this has reduced the chances of a more militarily active Japan. But China also senses that it is regarded by Washington and Tokyo as the likely antagonist. Moreover, while many Asian countries will continue to welcome China's economic prosperity as a continuing boost to their own success, they will also worry about Beijing's strategic intentions towards them. As China becomes a great power in a military sense, and even as it uses its economic weight to boost its diplomatic clout, such concerns will grow unless those neighbours can be convinced that China's intentions are modest and peaceful.

This brings another reason why the great powers matter so much. If China's neighbours grow increasingly concerned about Beijing's long-term designs and short term assertiveness (including in the South China Sea), there is only so much they themselves can do to influence Beijing's behaviour. The fact that China is a great power and they are not makes a real difference here. Historically, the countries of Southeast Asia were vulnerable to the European great powers for a not dissimilar reason: they had very little ability to contest the military, economic and political might of these external colonial powers. During the Cold War, several Southeast Asian countries sought to neutralise themselves from the contest between the United States and the Soviet Union, partly because they were aware of their own weakness at the time as newly independent states, often with domestic insurgencies to tackle. Hence the declaration in the early 1970s of a Zone of Peace, Freedom and Neutrality (ZOPFAN), which, in the elegant words of Kusuma Snitwongse (1998, pp. 185–6) 'conveyed ASEAN's intention to manage its own regional order'. This seems a much more difficult strategy when one of the great powers is as close to the heart of Asia as is China, and when that re-emerging great power is so central to the prosperity of its neighbours. But as China can intimidate as well as enrich, there is a strong preference in the region for the presence of another great power to balance the picture. Hence the main appeal of the United States.

The consequence of this logic is that security in Asia depends to a significant degree on a balance of power between the United States and

China. A balance of power is a slippery and variable concept: in one famous essay, Ernst B. Haas (1953, p. 458) was probably underestimating the variety in depicting 'eight more or less distinct meanings and connotations'. For the purposes of this analysis it is sensible to consider the balance as a relatively even distribution of power between the main actors so that none of them is able to dominate the system. But there is an immediate problem in applying this thinking in twenty-first-century Asia. For a number of years, many of the countries in Asia were reassured not by a roughly even balance of power, but by a gross imbalance in favour of the United States, which came into sharp relief once the Soviet Union had left the stage. And far from being a threatening condition, as Evelyn Goh (2008) has suggested, this American preponderance was welcomed by many in Southeast Asia. The United States was not a local great power growing within the Asian neighbourhood. It was often seen as an external provider of security whose lack of close proximity suited local interests. Moreover, at least for several decades, there was a widespread understanding that China was unwilling and unable to challenge America's primacy. But no particular distribution of power, balanced or imbalanced, lasts for ever. For this reason, the changing great power relationship is likely to be the first point of reference when the future of Asia's security environment is being considered.

Conclusion

This point becomes more significant as the gap between the United States and China decreases. First, those countries that have relied on America's ability to project military power into the region for their security will be less reassured as China raises the costs and risks for American action. From their perspective, a genuine balance of power where China and the United States are closer to being equals would mean less rather than more security. Second, while the countries of the region know what it was, and is, like to live with a very strong United States in the late twentieth and early twenty-first centuries, it is not clear quite what China will do with the extra influence it gains as its sources of power grow. This particular variety of uncertainty is not always going to provide an environment for good decisions. But it will increase the importance of other actors (including India) that might be independent-minded and strong enough to keep the region from descending into a relentless game of competition between the two great powers.

A zero-sum outcome, where America's loss is China's gain, and vice versa, would be much worse than the situation that has been characterising the relationship between these two great powers up to this point.

One occasionally hears predictions of a new Cold War in Asia between a communist China and a democratic America, but Narushige and van der Hoest (2013) are right to question this comparison. Certainly, a direct replica of the original version seems unlikely. China has no desire to be the leader of an international communist movement that no longer exists. Nor has China imposed itself physically and military on Asia in the way that the Soviet Union did on Eastern Europe at the end of the Second World War, even though some of its neighbours do worry about Beijing's claims in the East China and South China Seas, and even though China continues to coerce Taiwan. Moreover, there is widespread consensus that the US policy of containment, which was used to keep the Soviet Union in check and eventually to allow it to disintegrate, simply has no real chance against China. Unlike the United States and the Soviet Union during the Cold War period, which were part of separate economic systems, China and the United States are participants in the same international economy. A full policy of containment, described in detail by John Lewis Gaddis (1982), would, if directed against China, require the United States to cut off its nose to spite its own face.

Indeed, it can be argued in quite a different direction: that the increasing economic interdependence between these great powers is insurance against a hot war as well as a cold one. A major armed conflict between the US and China could be the most serious occurrence imaginable for Asia's security. Even if the two great powers could restrain themselves from using their nuclear weapons on each other (and there is no guarantee that they could) a war between them could easily be extremely costly in terms of human lives, regional confidence and regional prosperity. It would be a threat to almost everyone's acquired values in Asia. This would be the ultimate in ripple effects, the possibility of which (however remote) is a further reminder of why the relationship between these two great powers is so significant to Asia's security. Whether the common bonds of economic advancement that China, the United States, and much of the rest of the region share are enough to prevent deep tensions and even a hot war remains a hotly debated question. This is the question to which Chapter 4 will turn.

Chapter 4

Money: Can Economic Interdependence Keep Asia Safe?

It is debatable whether Asia is becoming more or less secure, especially as the great powers jostle for influence. But it is undeniable that Asia is becoming more prosperous. Despite problems with poverty that continue to confront a number of countries in the region, Asia is better known for being home to many of the economic success stories of the modern world. Several Asian countries have transformed themselves into advanced developed economies in little more than a generation, and the concentration of global wealth is moving increasingly in Asia's direction and away from the Western world. According to one study commissioned by the Asian Development Bank (Kohli et al., 2011), Asia is likely to account for over half of the global economy by 2050, up from under 20 per cent as recently as 1980. Given the scale of this economic transformation, the distribution of global power is also shifting. It is only sensible to think that this trend must carry some security implications with it.

It seems uncontroversial to argue that throughout Asia and its borderlands, prosperity is highly sought after as a national objective by most of the region's countries. It seems logical to surmise that this generates a common interest among states across the region in avoiding those factors and problems which may pose a threat to good economic conditions. A serious breakdown in regional security, including the outbreak of a major war in Asia, would seem to rank among the most serious of these potential threats to regional prosperity. As a result of this chain of logic it might be concluded that there is a strong common interest across Asia in the avoidance of war because of the threat such activity would pose to mutual economic interests. It might then be proposed that, as Asia becomes wealthier, it is also likely to become more secure, because armed conflict has become too costly to be permitted.

This means that the first task of this chapter is to address the argument that, as Asia's economies become more intertwined through trade and investment links in the common search for prosperity, these same ties have become a major and effective obstacle to the outbreak of war. Since most, if not all, countries in Asia and its borderlands are connected economically to a number of others, and since many of these connections are becoming tighter, this factor may well be a virtuous ripple effect with a positive effects for the Asian security system.

But the security implications of economic interdependence, and the economic implications of conflict, are not quite so straightforward. The second task of this chapter is therefore to take a broader view of the regional security implications of Asia's economic conditions and interactions. It is necessary to consider whether some of Asia's economic connections can be a cause for concern as well as reassurance. Interdependence may be uneven, opening up opportunities for influence that may harm the security prospects of those countries that are on the receiving end. The uneven experience of economic prosperity in parts of Asia may also have important implications for domestic security (and insecurity) in some of the region's countries. By the same token, as quickly growing economies in the region have growing resource needs, it is also necessary to consider the possibility of dangerous competition for energy supplies as one of the unintended consequences of a wealthier Asia.

Asia's economic interdependence

It is certainly logical to argue that the richer a country becomes, the more it has to lose from a crisis or armed conflict that would harm or threaten its wealth. The basic logic of risk aversion, as presented by Kenneth Arrow (1971), might then be used to suggest that these gains would make countries more cautious about entering into armed conflict. But it is not just the growing wealth of Asian countries that might reinforce the region's peace. Instead, the way in which Asian countries have been growing and developing together, and specifically the economic linkages that have grown up between them, might be regarded as especially pacifying. These intraregional connections are by no means automatic. The export-led growth strategies that Japan pioneered and many Asian countries (including China) have followed have relied heavily on a wider global demand for regional goods and services. But as Asia's growing economies have increasingly become markets for each other, and as Japan and then China have become major centres of regional demand and supply, an increasing proportion of this exchange has been happening within the region.

This economic integration includes the role that many of Asia's economies have played in global production networks or supply chains, where various components from different origins are brought together. This includes the many components needed for the manufacture of computer hardware, home electronics and cars. China's central position in a number of these supply and manufacturing chains, as described by Sturgeon and Memdovic (2011, p. 19) has been part of the gravitational pull of world trade in the direction of Asia, benefiting a number of other East Asian economies in particular. While the rapid acceleration in Asian trade integration that occurred after the mid-1980s has slowed, over half of Asia's trade now takes place within the region. According to the Asian Development Bank (2013), intra-Asian trade as a percentage of total trade conducted by the countries of the region grew from 45 per cent to 55 per cent between 1990 and 2012. Regional disparities are also apparent: trade integration is more intense (in terms of both volume and percentage) among the countries of East Asia than it is in Central Asia or South Asia. But if economic linkages do indeed have a pacifying effect, there is probably no more important part of the region for them do so than among some of the main East Asian economies. This is after all the part of Asia where the strategic competition between the larger regional powers is so often concentrated, a point confirmed by the analysis in Chapter 3.

These regional economic exchanges are about more than the trade in goods. They are also about investment and financial interactions. Some of the evidence for Asia's increasing financial integration comes from a negative example: the 1997 Asian financial crisis, which began when international money markets put severe downward pressure on Thailand's currency, the baht. This crisis quickly spread to other parts of East Asia, and in some cases the withdrawal of external investment from many of these countries left a severe dent in economic growth rates. In at least one example, this crisis had significant political and security implications. As an assessment by Hal Hill (1998) suggests, the regional economic downturn accelerated the collapse of the already tottering Suharto regime in Indonesia. Looking back, one can see that Indonesia managed a successful transition towards a more democratic political system which has since allowed for peaceful transfers of power, a hallmark of domestic political stability. But at the time, no such mechanisms existed and it seemed entirely possible that Indonesia might descend into a violent domestic struggle for power (remembering that Suharto had come to power himself only after hundreds of thousands of Indonesians had lost their lives in political reprisals).

The resolution of this particular financial crisis did not owe a great deal to international efforts from outside the region to restore confidence to the situation. Probably the most effective response that slowed down the crisis and its broader effects came from within the region. This was China's decision not to devalue its own currency, a factor that had positive ripple effects simply because of China's growing economic importance to its East Asian neighbours. As might be expected, this decision also generated favourable views within the region about the stabilising role that China could play. Writing not long after the crisis was at its worst, Marta Dassú (1998, p. 35) had already concluded that 'there can be no doubt that the crisis has already clearly favoured China's image and external prestige'. These benefits for Beijing lived on and served to balance and dampen some of the concerns in Asia about the security implications of China's rise. They are just one sign of the way that China's growth can translate into regional confidence rather than regional alarm.

Similarly, while China and the United States may be strategic rivals in some areas, they are also close economic partners who benefit from one another's increasing prosperity. The generation and intensification of these connections since China opened up to the international economy has been nothing short of spectacular. As Morrison (2014, p. 2) has observed, total China–US trade rose from just US$2 billion in 1979 to US$562 billion in 2013. This has made China 'the second-largest U.S. trading partner (after Canada), the third-largest U.S. export market (after Canada and Mexico), and the largest source of U.S. imports'. China also has very strong economic interconnections with India and Japan, who are also frequently seen as being among its main strategic competitors. This is further evidence for the proposition that China is an asset in economic terms even for those who see it as a challenge from a security point of view. As a result, most countries in the region have a vested interest in China becoming a stronger power, even if this same development also makes them nervous strategically.

For the United States, still the region's strongest actor, interdependence with China goes much further than merely trade. China's purchases of American treasury bonds have helped to fund the debt of American consumers who have been spending beyond their means. This consumption includes goods imported from China, thus doubling the latter's vested interest in the debt connection. Were China's purchases or American consumer spending to stop, both countries would suffer significantly. Those writing about nuclear strategy in the early Cold War spoke of a 'balance of terror', where both the United States and the Soviet Union were deterred from using nuclear weapons because of the mutual catastrophe that would ensue. A financial variation on this theme appears to have developed as the mutual reliance of the United States and China grew rapidly, and in particular once China became the world's largest holder of foreign reserves. Mark Beeson (2009, p. 733) records that: 'In what was described by former US Treasury secretary Larry Summers as the "balance of financial terror", China and Japan in particular have been willing to invest huge sums in US securities to underwrite the US budget and support overall consumption patterns'. If that balance of financial terror works, at least some reassurance should result: neither side seems to have an incentive to upset the applecart; indeed, they should operate cautiously and sensibly in the fear that this might happen.

Lest it be thought that mutual dependence only characterises the relations between China and the other major regional powers, there are many examples of mutually beneficial ties that cross the region in different directions and involve states of all shapes and sizes. Australia, for example, largely escaped the effects of the global financial crisis, which began in 2007 when banking failures in the United States quickly spread to destabilise the global economy. Australia was insulated from these effects, which put the American and European economies into recession, primarily because of the continuing demand in Asia for its mineral and energy resources. And while a good deal of the demand historically had come from Japan, it was China's massive appetite for iron ore, coal and natural gas that was widely seen as being responsible for Australia's continuing prosperity. Of course, there is always a cloud to every silver lining, hence the observation from Saul Eslake (2009, p. 237) that for the same reason Australia 'could be more seriously affected than other countries by any significant downturn in the Chinese economy'.

The economic costs of Asian conflict

Much of the rest of the world would have settled for this risk, but there is no guarantee that things will continue in the same fashion. China in particular faces challenges in sustaining the high levels of growth it has enjoyed since its economic reforms of the 1970s. These include the possibility of China being stuck in what economists call a middle-income trap, the structural problems that can afflict China's vast state-owned enterprises, and the challenge in shifting the economy in the direction of stronger domestic demand. China's multiple environmental problems, affecting water, soil and air, are also likely to reduce long-term rates of growth. As noted in Chapter 3, and as Banister et al. (2012) confirm, China's demographic challenges mean that its working age population is set to shrink. While India's demographic factors are more favourable, the largest country in South Asia has nonetheless found it difficult to continue the high growth rates that followed economic reform in the early 1990s. The stagnation that afflicted Japan for two decades is a reminder of the hazards of extrapolating high growth rates from a favourable period into what is always an uncertain future.

Moreover, while regional economic shocks may also be brought about by problems with the business cycle in Asia, exogenous (or external) problems can also feature, and here the connection with Asia's security environment becomes particularly clear. As it has been several decades since Asia experienced a major interstate war, it is not easy to determine what the economic consequences of such an unwanted development would be. This is not least because armed conflicts tend to take on a life of their own and can be very difficult to predict or control. The savage interaction between belligerents is one of the most inflammable interdependencies that can occur. In the early nineteenth century, Carl von Clausewitz (1976, p. 77) wrote:

> war is an act of force, and there is no logical limit to the application of that force. Each side, therefore, compels its opponent to follow suit; a reciprocal action is started which must lead, in theory, to extremes.

While those theoretical extremes are rarely reached, the ripple effects of any war can still be very damaging. And given the nature of Asia's increasing integration, it seems only logical to expect that among the earliest of these effects will be some very significant economic costs for regional countries. Even the mere prospect of a war might be enough to cause significant economic implications. International markets are very sensitive to tensions and crises, because forward prices need to reflect the changing elements of risk. It is clear from the Asian financial crisis of the 1990s that problems in one part of the region can radiate outwards courtesy of tight linkages in both currency and stock markets. If investors fear that the security environment is under threat, a run on currencies and a withdrawal of short- and long-term credit could be one of the first developments as a crisis unfolded.

Some of these effects might even be noticeable if an *internal* conflict was thought to be likely and if markets connected to the country concerned decided to discount prices ahead of such an event. One assessment of the impact of an earlier period of the civil war between the Sri Lankan government and the Tamil Tigers indicated that the costs of 12 years of war from 1984 to 1996 equated to two full years of the country's total economic output (Arunatilake et al., 2001). But these calculations only measured the domestic economic impacts within a small economy in Asia. It is difficult to imagine the full economic ripple effects throughout Asia if a large and heavily interconnected country in the region, such as China, was faced with a crippling civil conflict. But these are likely to be very significant.

Because *interstate* wars by their very nature cross state boundaries, ripple effects are built into the process and may occur especially early. In the tensions that might signal the coming of an interstate conflict, for example, restrictions on trade might well appeal as a way of putting pressure on an adversary. More severe action including the blockading of ports might also be employed in the lead-up to armed conflict: in the past,

Taiwan's government has feared that, short of full-scale war, China might one day express its unhappiness by seeking to close access by sea to its ports, thus curtailing the exchange of goods that is essential to its economic prosperity. In a systematic treatment of this possibility undertaken over a decade ago, Michael A. Glosny (2004) concluded that it would be harder for China to impose costs on Taiwan than many had assumed. However, as a measure short of full conflict, and given the mutual costs that might result from the more serious eventuality this action might still appeal to the leadership in Beijing.

There is a much bigger picture here, because Taiwan is not alone in the region in its heavy dependence on maritime trade. For many Asian economies, the vast majority of the goods they import, including the bulk of their energy supplies, arrive by sea. Disruption to ports or shipping routes and massive increases in insurance premiums would all have a negative economic impact. War, or perhaps just the threat of war, could disrupt the supply chains that connect many Asian economies and the manufacturing and assembly processes on which their prosperity relies. The sensitivity of this interlinked system to even relatively small and localised security events should not be underestimated. In a study of what a single terrorist attack might mean for the flow of sea-borne container traffic, Earnest et al. (2012, pp. 588–9) found reason to believe that even the 'disruption of a small port' in Asia, could 'induce a costly cascading disruption', with effects felt keenly across the other side of the region in the United States. In some cases, production might cease if the exchange of components were to become prohibitively expensive or logistically infeasible. The resulting loss of employment in many sectors – raw materials production, manufacturing, and service and transportation industries – would place a significant imposition on government spending at a time when their sources of revenue were already being depleted.

It might take a significant regional war to demonstrate quite how tightly interdependent the economies of Asia have become, but that may be an experiment that hardly any country in the region can afford to run. By studying data from a 127-year period up to 1997, Glick and Taylor (2010, p. 102) have found that the negative impacts of war on trade are not only grave, but also persistent: 'even after conflicts end, trade does not resume its prewar level for many years, exacerbating total costs'. And these effects are not confined to the warring parties: 'unlike the direct costs of war, which largely affect only the belligerents, commercial losses affect neutral parties as well, meaning that wars generate a large negative externality through trade destruction'.

In theory, this logic should offer one of the best promises for Asia's security. War is too costly to afford in the region even before consideration is given to its human costs in terms of casualties, or to the potential loss of territory and valued political relationships. The economic costs of conflict alone would appear to make war an irrational choice, and the greater Asia's economic interdependence is appreciated, the more irrational that choice becomes. As almost every Asian country depends on its regional economic partners for a significant part of its prosperity, it seems logical to expect that there is an especially strong and robust common interest among regional governments in avoiding conflict. Indeed, the limited economic interactions that have been possible between India and Pakistan, a result of their difficult bilateral security relationship, and a hindrance to economic development for both of them, may serve as a warning here. The absence of strong and open trading, and transportation connections between the two countries, hindered by years of tension and mistrust, is one of the main reasons for the relatively low level of economic integration in South Asia. Kochhar and Ghani (2013, p. 107) suggest that a free trade agreement between India and Pakistan might on its own realise a twenty-fold increase in the trade between them.

Even if the countries of Asia and its borderlands were completely consumed with their own interests, out of pure economic selfishness it might still make eminent sense for them to do what they can to avoid armed conflict. In other words, a positive security environment in Asia, where war is a rarity, does not necessarily require the countries of the region to like one another. It only requires them to understand that in combination, their selfish interests can in fact be a recipe for mutual restraint. They cannot, in this sense, achieve their competitive economic goals without some sense of cooperation in the security realm. This seems to be a rather undemanding recipe for regional peace, and for the prosperity that this peace can support.

Peace through interdependence?

For some of the reasons explored in Chapter 2, it is only in recent decades that the countries of Asia have been able to benefit from the levels of economic interdependence apparent today. In earlier periods, Asia's economic connections with the outside world, including with the European colonial powers, meant domination, extraction and insecurity rather than national confidence, re-investment and wealth. As a consequence, the argument that Asia ought to remain peaceful because of its economic interdependence remains a rather novel, and untested, proposition. To properly test this logic it is wise to look to other parts of the world, and to other eras of modern history.

One of the main origins for the argument that economic intercourse has pacifying effects comes from the mainly British school of liberalism.

In the nineteenth century, writers such as Richard Cobden argued that restraints on trade would encourage wider conflict, and that an open international market was therefore a precursor to peace. A similar judgement is part and parcel of the liberal international order that the United States has long sought to establish and maintain. As G. John Ikenberry (2011, p. 60) has commented, there was an important place for open economic relationships in Woodrow Wilson's plans at the end of the First World War for 'a "one world" vision of nation-states that would trade and interact in a multilateral system of laws'.

The difficulty in Wilson's proposition does not lie simply in the subsequent years as the world became more rather than less open economically, stumbling towards a second world war that would draw in countries of Asia as well as Europe. There were also problems with what immediately preceded his vision: a catastrophic First World War that had occurred when Europe already seemed so interconnected as to make conflict far too costly to contemplate. This seemed to be the conclusion drawn by, among others, Normal Angell (2010), whose book, *The Great Illusion,* was first published in 1909. As Robert Zoellick (1997, p. 29) has observed, Angell argued that 'the new, complex financial and commercial interdependence, forged over the previous 40 years, made the old logic of waging war futile and absurd'. That Zoellick's observation was included in an essay on the relationship between economics and security in Asia is no accident. Angell's logic seems eerily similar to the argument that Asia's trade- and investment-dependent economies are likely to be less prone to conflict because of the common interests they share in protecting the wealth that results from this exchange.

Quite whether the First World War is the correct model for Asia's contemporary dramas remains unclear. Erik Gartzke and Yonatan Lupu (2012, p. 130) have argued that while a number of Europe's western states, including Britain, Germany and France, were very interdependent economically, the Eastern powers were not. And it is in the latter states that the war began. As a result, according to this logic, there should be more concern about war starting in Asia between the relatively separate economies of India and Pakistan, or between North Korea and the United States, than between the closely connected Japan and China. Yet the economic interdependence between the Western European states did not prevent their involvement in this catastrophic war.

Indeed, part of the problem in 1914 was that the great powers of Europe were connected in ways that were often more provocative than pacifying for their relationships. On the eve of war they were divided into rigid alliances with all sorts of obligations making it almost impossible for any of them to stay out of the war once it had begun. Rather than a stable balance of power between two groups of countries which checked the temptation to war on both sides, these connections actually made war more likely. Once Germany had responded to Russia's mobilisation, which was itself a response to German and Austrian pressure on Russia's ally, Serbia, a more widespread conflagration ensued. This political spiral of conflict differs from the logic of peace through economic interdependence, which tends to assume a cool and rational calculation of costs and benefits before decisions to go to war are made. Decisions to embark on war are not always made calmly and with a clear understanding of the full consequences. One part of the tinderbox came from the strong currents of patriotic sentiment that left cooler judgements to one side. To return to Zoellick (1997, p. 49) this all meant for Angell the hard lesson that 'the greatest illusion can be that new conditions eliminate the need to protect against old causes of rivalry, conflict and war'. Whether a similar lesson can be anticipated by Asia's leaders today remains to be seen.

This does not necessarily indicate the irrelevance of economic considerations in decisions about war and peace, but their subordination to other, more powerful motivations. Governments may be fully aware that war will be costly to their economies, but this may be less important to them in their decision-making than other factors. Thucydides, the historian of wars between the Greek city states, famously argued that there were three motivations for the imperial policy of Athens: fear, honour and profit. The argument that trade, and the mutual prosperity it brings, is a strong antidote to war tends only to deal with the relationship between two of these variables – the desire to maintain economic profit is expected to generate a positive fear of war. But the remaining factor seems much more problematic. Honour is about status. It is a mercurial notion suggesting that wars might be fought out of envy, the desire to dominate, an exaggerated sense of one's own importance, and grievances over the lack of respect one has been paid.

This may sound rather old-fashioned, as if conflicts occur mainly between two crusty old gentlemen who are left trapped in their memories of ancient squabbles. Yet old and unresolved squabbles are part and parcel of Asia's security challenges. Michael Mastanduno (2014, p. 36) has observed that, for at least some of the realist thinkers who find Thucydides strikes a chord, Asia's 'unfortunate combination of states with strong nationalist sentiment and unresolved historical grievances' increases the chances of conflict in the region. As will become clear in Chapter 6, some of the emotions and passions that divide countries in Asia provide combustible fuel for security problems in spite of the economic linkages that are supposed to keep them from harming each other. In the meantime, it is enough to say that China and Japan, and Japan and Korea, are locked in a competition for status. This may not be good news for security, and may make economic interdependence a less significant factor than many realise if, as William Wohlforth (2009, p. 35) suggests, 'the search for status will cause people to behave in ways that directly contradict their material interest in security and/or prosperity'.

There are other constraints on the thesis that trade means peace. The first is the possibility that fears about the impact of an interstate war on trade may be exaggerated, undermining the pacifying effects of economic interdependence. In one widely cited study, Barbieri and Levy (1999, p. 475) conclude that, on the basis of a series of historical examples, 'There is no consistent, systematic, and substantial reduction in trade between belligerents during wartime, and trade between adversaries appears to recover quickly after the termination of war.' This does not mean that any future war will fail to have the negative impact on trade that is commonly expected. It does mean that it might and that it might not. Perhaps even more worrying is the suggestion from Dale C. Copeland (1996), that an economically interdependent country might still prefer war now if it fears it might eventually be shut out of lucrative trading relationships. In other words, in the event that Japan believed that China eventually wished to exclude it from future economic dealings, or if China felt that in some way a form of economic containment was being devised against it, the immediate economic costs of war might not seem as serious as one would wish. Indeed, interdependence can cut more than one way. Early signs of the economic impacts of a growing crisis, including trade restrictions and consumer boycotts, may be just as likely to deepen resentment as to be a clear warning of the greater costs on the horizon.

This does not mean that economic interdependence is unimportant in relation to Asia's security. Instead, it means that it is unwise to put too much reliance on it as insulation against worsening rivalry and war in the region. Analysts who argue that Asia's economic interdependence is a sound reason why war is unlikely may be offering a very restricted view of the interests and other factors that motivate government decisions. The governments of Asia are political animals whose decision-making cannot be expected to be consistently rational or rationally consistent. War may still be unlikely under these conditions, but it is by no means impossible.

The political uses of economic relationships

A further limitation to the argument that trade encourages peace is the tendency to treat economic interdependence in the Asian region as an impartial factor that benefits all the players in the region in relatively equal measure. But the benefits of interdependence are not evenly spread throughout Asia. Moreover, rather than globalisation being a force that shapes the actions of political actors, the reverse can easily apply: governments in the region can use economic connections as a form of influence for more competitive purposes. This does not always make the region safer, however.

A particularly graphic example of this asymmetry can be found in the economic relationship between China and Taiwan, with the latter becoming so connected with the mainland that it is partly integrated into the much larger economy of China. This includes the approximately two million Taiwanese people who are working on the Chinese mainland. Any rupture to the economic relationship across the Taiwan Strait would harm Taiwan rather more than China. As this imbalance grows, it has clear implications for China's plans for the relationship, where Taiwan will one day come under its control in an act of unification. Beijing has never been able to achieve this objective directly by military means, as it lacks the ability to send an effective invasion force across the Taiwan Strait to physically control Taiwan. China does have the ability to coerce Taiwan militarily by building up its missile forces to deter any move towards formal independence, but the closer integration of the two economies seems most likely to bring the goal of unification closer. Nearly a decade ago, Kahler and Kastner (2006, p. 536) were already observing that, while the outcomes of China's efforts to gain politically from this engagement remained somewhat unclear, 'growing cross-Strait economic ties have given rise to a new constituency in Taiwan that has an economic stake in China and that tends to oppose foreign policy goals (like independence) that are starkly at odds with Beijing'.

At least one of Taiwan's neighbours is concerned that it may be party to a similarly unequal relationship where there is as much dependence on China as interdependence with it. Japan's economy was becoming sluggish in the 1980s, just as China's was expanding rapidly. As Japan's leading trade partner, China now accounts for about a fifth of Japan's exports and imports. China is also the leading financial supporter of Japan's public debt, but rather than the financial balance of terror alluded to in US–China relations, the pattern of influence seems to be moving much more strongly in China's direction. Japan's concern that China is becoming too rich and strong (including militarily) can add to feelings of resentment and entrapment, especially when it is realised that Japan's economic exchange with China is contributing to the latter's increasing girth. And as Evelyn Goh (2011, p. 899) has argued, Japan has also been keen to promote patterns of economic cooperation in the wider region that are alternatives to China-centred approaches to economic integration. In short, the more Japan's regional partners are also heavily dependent on China, the more Tokyo has a reason to feel politically insecure.

But these same economic connections can also provide opportunities for escalating tensions. Based on a study of repeated flare-ups with regard to their respective East China Sea claims, which will be evaluated further in Chapter 6, Min Gyo Koo (2009, p. 228) has argued that China and Japan have 'been deterred from pushing for a more definitive political showdown with respect to the island dispute in the interest of maintaining the lucrative trade and investment relations that both countries have enjoyed since 1972'. Yet some of these apparently pacifying economic connections have become *less* close in the wake of more recent tensions. Nationalist protestors in China have burned Japanese cars and called for a boycott on their purchase as a sign of their displeasure towards Tokyo. From time to time these actions are tolerated by the Chinese government, which can find them useful as a way of putting pressure on Japan. It is common to argue that this is a dangerous strategy because of the risk of these protests spinning out of control. But, at least in the case of the relationship with the United States, Jessica Chen Weiss (2013) has suggested that China's leaders exploit the knowledge of this destabilising possibility to signal their country's resolve. If she is right, this is a risky business, as the danger of a serious explosion remains.

The use of bilateral economic ties in strategic bargaining was completely impossible between the United States and the Soviet Union during the Cold War, primarily because they were in two quite separate economic blocs. But that is not the case today. Some analysts have argued that Washington might be wise to curtail its economic engagement with China on the basis that this only helps to build its power, which may then threaten American interests. For example, John Mearsheimer (2014) maintains that 'the United States has a profound interest in seeing Chinese economic growth slow considerably in the years ahead'. This is effectively a call for an economic containment of China, echoing the Cold War strategy adopted by the United States, where, by military, political and economic means it would resist the Soviet Union's expansion and let it collapse under the weight of its own contradictions. Even if such an approach were feasible it would be undesirable for America's own economy (a point Mearsheimer himself acknowledges) and could well prove disastrous to large parts of the business sector in the United States that relies heavily on trade with China. That sector could be counted on to lobby the president and the Congress to avoid any such policy.

Even so, the United States is faced with the conundrum that the stronger China becomes economically, the more that America's influence in Asia is likely to be challenged. This poses a problem in theoretical terms to liberal arguments that treat open trading relationships as win–win games. At the very least the wins are unlikely to be unequal, and yet in the current era almost every country in Asia (with the exception of North Korea) has been seeking closer and more open trading relationships in the region, and not least with a growing China. A patchwork of free trade agreements (FTAs) have been sprouting up all across Asia, designed in part to compensate for the very slow progress that has been made in recent years in global trade negotiations under the auspices of the World Trade Organization. These smaller agreements are sold on the basis that the removal of trade barriers allows members to maximise their competitive economic advantage. They are sometimes seen as building blocks towards an eventually region-wide and world-wide free trade environment.

Indeed, Jong-Wha Lee and Innwon Park (2005, p. 44) have suggested that, under the right conditions East Asian free trade agreements can certainly 'be a building block for a global FTA'. But this is far from being uncontested territory. Some observers, such as Richard Baldwin (2007, p. 6) argue that the 'noodle bowl' of Asia's FTAs contains preferential rather than truly open agreements, and that 'preference is just another word for discrimination'. It is not logically necessary that in combination a set of trade agreements within Asia will add up to a more open region let alone be an automatic boost to global trade.

Moreover, in some of the more recent trade negotiation processes it is possible to see signs of strategic competition. Having developed a quadrilateral FTA, the so-called P4 countries (New Zealand, Singapore, Brunei and Chile) were then joined by many large regional economies in negotiations for a Trans-Pacific Partnership (TPP) agreement. But once Washington agreed to participate in these talks in 2008, and later presented the TPP as a sign of its pivot towards Asia, concerns grew that this was simply part of an effort to counter and curtail China's increasing prominence in the region's economic architecture. For example, Lanxin Xiang (2012, p. 113) lists the TPP as a 'game-changing economic project' alongside Washington's new military commitments in the region as part of 'a comprehensive "containment" package in Asia'. This sense of wider competition has made it easy for the negotiation of a Regional Comprehensive Economic Partnership (RECP) centred on ASEAN and including China but not the United States, as a strategic competitor to the TPP. For small, trade-dependent economies in the region, such as New Zealand, both the TPP and RCEP are attractive as potential stepping-stones to a wider free trade area in the region. It is therefore not a choice of one or the other. But it is difficult to dispel the perception that the United States and China are approaching these arrangements as a vehicle for their strategic competition.

Economic coercion: sanctions in Asia

As a form of power, economic connections can also be withheld in competitive relationships so as to influence the behaviour of others. States may find economic instruments of intimidation useful when they are reluctant to use military force and where there is no real expectation that a full-blown armed conflict is around the corner. Economic pressure of this kind has frequently been favoured by the United States in its dealing with North Korea. For example, after it became clear in 2002 that North Korea had been proceeding with a secret programme to enrich nuclear materials, the United States eventually suspended the shipment of oil to the country, which had been designed to reward North Korea for not proceeding with a nuclear weapons programme. Yet, as Gary Samore (2003) wrote around that time, Washington's mix of less than forceful punishment measures and leaving open the possibility of a return to renewed cooperation as an incentive for North Korea, appeared to have little positive effect on Pyongyang's decision-making.

One of America's problems in this particular context is that its economic connections with North Korea have been so limited that they offer little in the way of real or potential leverage. This means that the cooperation of other partners in Asia is very important. From time to time, South Korea has suspended aid to North Korea to register its unhappiness with Pyongyang's threatening behaviour, but South Korea is alone in having to share a peninsula with its northern neighbour. This puts a limit on the degree of punishment Seoul is willing to apply. As a long-standing ally and close economic partner of (and provider for) North Korea, it is often hoped that China will be ready to apply pressure. Glaser and Billingsley (2012, p. 18) have noted that, in 2006, China did suspend oil supplies briefly, and that 'While not officially announced, the suspension was widely believed to be a sign of Beijing's disapproval of North Korea's nuclear ambitions.' But that disapproval is often trumped by China's interests in avoiding the collapse of the North Korean regime, and North Korea's continuing pursuit of its nuclear programme indicates the limits of even Beijing's influence.

The use of economic statecraft in this way to apply pressure on other governments is a form of coercion: efforts to influence the behaviour of another party through threats of harm and through limited acts of harm itself. Such a broad approach to coercion expands on the tendency to focus on threats that imply or use physical force. For example, Byman and Waxman (2002, p. 1) have defined coercion as 'the use of threatened force, and at times the limited use of actual force to back up the threat, to induce an adversary to change its behavior'. But there is no obvious reason to exclude economic varieties of intimidation from the concept, not least because, in the 2014 Ukraine–Russia crisis, Vladimir Putin's government seemed very keen to give the impression that it would interfere with gas supplies to Europe as a way of dissuading EU countries from taking measures against it. In today's Asia, where economic linkages provide the lifelines for regional countries and therefore also provide opportunities for applying pressure, the prospects for economic coercion are considerable. Indeed, one major study by Brendan Taylor (2010) considered the very real possibility that many major powers now view sanctions as an increasingly important part of their security policy towards the Asian region.

Like all forms of pressure, however, sanctions can sometimes generate unexpected and uneven impacts and may even be counterproductive. In 1998, when India and Pakistan tested nuclear weapons, the United States was required by law to impose sanctions on both states, and was joined in this by Japan and a range of other countries. These were lifted fairly quickly afterwards, leading some to wonder if the sanctions were little more than empty symbols. As analysis by Morrow and Carriere (1999) showed, because Pakistan's economy was already weakened, the withdrawal of the International Monetary Fund (IMF) assistance, which came with the sanctions caused very significant economic damage. Whether it was wise to increase the chances of political instability in Pakistan is open to debate. Moreover, while the stronger Indian economy was able to withstand the sanctions more easily, the memories of this punishment as a form of political embarrassment were not easily dispelled. As Kapur and Ganguly (2007, pp. 652–3) argued, 'Indian leaders' resentment over American non-proliferation policy broadly tainted the Indo-U.S. relationship, impeding cooperation in areas wholly unrelated to nuclear weapons.'

Energy interdependence and insecurity

Moreover, it would be unwise to regard economic pressure as a safer option, on the grounds that it necessarily forestalls the prospect of more violent action. Sanctions themselves can be regarded as hostile acts for which there is an important historical precedent in Asia. In 1941, the United States, which had yet to join the Second World War as an active combatant, responded to Japan's aggressive military expansion in Southeast Asia by imposing what amounted to an oil embargo. As Japan was heavily dependent on external energy supplies for its industrial capacity and for its war effort, and as the vast majority of its oil supplies came from the United States, this action had much broader consequences for Tokyo's decision-making. Scott Sagan (1988) indicates that after the embargo Japan's decision-makers were under intense pressure to escalate the war sooner rather than later, lest the shortage of supply made it impossible for them to hold any sort of advantage if they waited. America's embargo did not *force* Japan to launch the attack on Pearl Harbor that brought the United States into the war just a few months later, but it certainly increased the chances that their relationship would become a more violent one.

It might be thought that this kind of interaction belongs to an earlier era and is unthinkable in the twenty-first century. But it is unquestionable that many countries in Asia regard their access to the same sort of resources as an essential element of their national power. A deliberate restriction on that supply could therefore easily be regarded as a threat to their energy security as well as their economic security. China and Japan, and many other East Asian countries, are acutely aware of their heavy reliance on external energy supplies. Unlike the United States, which is becoming increasingly self-sufficient in oil, China is becoming more reliant on external energy supplies as its consumption increases. As well as being the world's largest energy consumer, China is especially dependent on external sources of supply for its oil needs, which are concentrated in its booming transportation sector. Figures prepared by Guy Leung (2011, p. 1331) indicate that while in 1991, China imported only 10 per cent of the oil it consumed, by 2004 over half of its oil consumption was already being imported. Especially in light of the boom in America's domestic production, China's position since 2013 as the world's largest net importer of oil is unlikely to be altered in the near future.

A number of consequences flow from this situation that could complicate the security picture in Asia. First, it might be thought to encourage territorial competition in areas of the region considered to be rich in energy supplies. The idea that under the rocks and shoals in the South China Sea might lie significant oil and gas fields is by no means the only or even the primary reason for competition among the claimant countries, including China's very extensive claim that will be examined in Chapter 6. Leszek Buszynski (2012, pp. 139–41) has noted that it is the way the South China Sea has been linked to broader rivalry between China and the United States that has really raised the stakes, but at the same time Vietnam's growing exploitation of oil reserves, which is strenuously opposed by China, is also part of the recipe for potential conflict. Recent years have witnessed some hazardous episodes between China and Vietnam on this score, including the sabotage of drilling platforms, the intimidation of commercial vessels, and stand-offs between maritime fleets.

Second, the major oil- and gas-importing countries in Asia have an abiding interest in the safety of the sea lanes through which their supplies of Middle Eastern oil are transported. This is one of the considerations that is encouraging China to develop a navy able to operate over longer distances. It is also a factor in India's efforts to increase its naval capabilities, and an encouragement to Indo-Japanese maritime cooperation. 'Securing energy supply routes,' says Madhuchanda Ghosh (2008, p. 287) 'is a core strategic interest' for India and Japan, who also converge to at least some degree over their concern about the increase in China's naval power.

Even so, a wider cooperative approach is not logically impossible. There is a shared interest among China, Japan, India and the United States for open sea lanes through such narrow points as the Strait of Malacca, partly because those lanes are about the movement not only of naval forces across the region but they are also vital in the safe and effective transport of supplies of goods and resources. These flows between the Indian and Pacific Oceans form part of the logic behind the idea of an emerging Indo-Pacific region. The notion that the focus of Asian power is moving eastwards is reflected in Robert D. Kaplan's arresting and popular book *Monsoon* (2010). However, in a more sombre analysis, Bisley and Phillips (2013) argue that as the Indian Ocean does not have the same importance as the Western Pacific, and similarly, as India's power is not close to China's, it makes little sense to collapse the two maritime areas into a single amalgamation, and that the continental aspects of Asia's security order should not be neglected.

Yet even critics of the Indo-Pacific idea cannot deny that energy flows between the Indian and Pacific Oceans, including their passage through the Malacca Strait in maritime Southeast Asia, are an essential part of Asia's increasing economic interdependence. And there is no guarantee that this will be a cause for common interest and cooperation. On top of the potential for great power rivalry spelled out in Chapter 3, China takes the controversial position that the movement of the military vessels of other countries through the exclusive economic zone around islands and features it claims is not provided for in international law. To the extent that India, China and Japan are building their longer-range naval and aerial capabilities in line with their expanding energy security interests, the result could just as easily be competitive and potentially dangerous. The potential for competition is increased by the facts that oil and gas supplies are finite, and demand for them across Asia is increasing.

This lack of certainty of long-term supply, which is a feature of energy insecurity, has also encouraged the quest for alternative pathways. As Suisheng Zhao (2008) showed, China has not been at all content to rely on international energy markets for supplies, but has adopted a strongly state-led approach in building a series of bilateral supply agreements. This has had a significant impact on China's political relationships.

For example, close links with Pakistan and Myanmar are valued partly because of the possibility of land-based pipelines that will help to supply China with some of its energy needs in ways that bypass the longer and more exposed maritime supply chains. China's quest for energy security has also encouraged it to value its controversially close relationship with Iran. That Iran is a valued source of oil for China has undoubtedly encouraged Beijing to veto attempts by the United States, the United Kingdom and France in the United Nations Security Council to sanction Iran for its nuclear programme.

These connections not only create stronger linkages across the wider Asian region and beyond, linking Central Asia and the Middle East to East Asia. They also generate concerns, including on the part of the United States, India and Japan, that China may be gaining too much influence over some of its main energy partners. That influence may even extend to countries that have long been close military allies of the United States. Australia, for example, is on its way to becoming the world's largest supplier of liquid natural gas (LNG). Much of this product is destined for China, which has made heavy investments in Australian gas facilities. This feeds into interesting conversations about how Australia manages its relationship with China, the most important direct source of its economic security, and its relationship with the United States, its traditional ally, a potential dilemma raised by Hugh White (2011). This also brings the argument back to the inequality that can be seen in many regional economic links that favour some countries more than others: even if all are gaining from economic exchange, some will be gaining more than others.

Uneven effects on domestic security in Asia

The unequal effect of Asia's economic expansion and interconnectedness for Asian security is not just a matter of relations between countries. It can also be felt *within* them. The time-span of the economic transformation occurring in parts of Asia is an intensely accelerated version of the much longer processes of development that have occurred elsewhere. Again, China is a prime example here. The uneven benefits of economic growth that have favoured the coastal fringe of the country at the expense of the inner provinces has helped to drive a massive eastwards population flow. The social consequences of tens of millions of unemployed and underemployed citizens are the stuff of nightmares for China's leaders. As Yongshun Cai (2002) explained, simultaneous lay-offs of workers by large state-owned and state-connected enterprises can sometimes be followed by collective action among former employees.

More generally in China, the knowledge in more inland regions that city dwellers on the eastern seaboard are doing so much better is not always conducive to social stability. An additional challenge for China is that the legitimacy of Communist Party rule depends partly on a trade-off between economic growth and political liberty. Indeed Minxin Pei (2006, p. 19) has argued that instead of generating incentives for greater political openness, China's experience of sustained economic growth has provided the political leadership with breathing space in which serious political liberalisation is far from an urgent priority. By the same token, however, a severe reduction in growth rates may just be enough to undermine this unusual social contract, with unpredictable implications for China's security picture.

China is by no means alone, despite the relative uniqueness of its political system in today's Asia. The dislocating effects of uneven economic growth are also a factor in India's future. In a comparison some years ago between the two Asian giants, Pranab Bardhan (2006, p. 14) argued that while India's social and educational inequality was more severe than China's, India's 'heterogeneity and pluralism have also provided the basis for a better ability to politically manage conflicts'. Yet, as India grows wealthier in aggregate terms, the stubborn divide between the very rich and the very poor can still be a rallying cry for local politicians opposed to India's economic liberalisation. It may be too simplistic to suggest that these economic gaps are the main cause of the internal violence that parts of India sometimes experience. But, as the words of Ahuja and Ganguly (2007) indicate, attacks by the Naxalites, the Maoist guerillas who operate in some of India's poorest states, are a sign that the rising total wealth of a country is no guarantee of domestic peace when the benefits are experienced unevenly.

Central government decisions to allow for economic development, including through the encouragement of foreign investment, can have complicating implications for security within some countries in the region. One such example comes from Papua New Guinea (PNG), whose government, in the quest for revenue and economic development, has allowed foreign mining companies to operate in a range of provinces. A leading source of PNG revenue for many years was a copper mine on the island province of Bougainville, an area more closely connected ethnically and historically to the neighbouring Pacific Island country of Solomon Islands, as was pointed out in Chapter 2. Unhappiness that the revenues from the mining activity were not available locally, and anger over the pollution being caused by the mining activity, added to the tensions in the province's relationship with the PNG government. This was a central cause of a civil war that cost the lives of thousands of Bougainvillean people in the 1990s, a leading example of the connection between resource development and what Ballard and Banks (2003, p. 289) refer to as 'low-level conflicts experienced in the Asia-Pacific region' during this period.

Yet this conflict did not have particularly significant ripple effects. While it complicated Papua New Guinea's relations with neighbouring Solomon Islands and Australia, it hardly appeared on the radar screens further north in the region. But there is one consequence of economic activity in the region and beyond that certainly does bring wider effects, and that is climate change. This issue will be covered in more detail in Chapter 8, but for now it is important to note two main points. First, that the processes of economic growth that are often regarded as so positive for Asia's security are also generating some of the world's most significant carbon emissions. And while they may claim to still be developing countries which should not be subject to the same restrictions as the long-established developed countries, India and China, which according to Ramanathan and Carmichael (2008, p. 226) already account for between a quarter and a third of global soot emissions, are crucial to any international agreement on climate change mitigation measures.

The second consideration is more significant. The climatic and broader environmental consequences of rising carbon emissions will hit parts of Asia especially hard. The region is home to large population concentrations in low-lying delta areas, such as in Bangladesh, and other vulnerable coastal and island communities. Some low-lying Pacific Island countries, including Tuvalu, may completely disappear should estimates of sea level rise be at all accurate. Moreover, food production and water availability in many parts of Asia, including in the already dry South Asian subcontinent that is home to more than a billion people, are already being affected negatively by climatic events widely believed to be early signs of the effects of global carbon levels. David Lobell et al. (2008) have pinpointed South Asia along with Southern Africa as two of the areas of the world where agricultural methods are in greatest need of revision because of the short-term implications of climate change for food production and scarcity. As a result, a curious cycle may in some cases be occurring where economic growth triggers changes that may in fact make future economic development ecologically unsustainable and materially infeasible. And significant population movements to some of the most vulnerable areas (including to coastal cities) in these regions may well exacerbate some of these problems.

As Chapter 8 will show, there is a fascinating debate to be had on how climate change fits into the picture of Asia's security, and whether it is wise to consider this in the first place. If climate change brings greater water scarcity, for example, might this become a major security problem between countries that are upstream and downstream along the same water system? But given the focus of the present chapter, it is important to be especially mindful of the economic implications of climate change. Some of the countries in the region most vulnerable to its effects are not in trouble simply because of where they are located, but because of the limited coping mechanisms some of them have to deal with the long-term impacts. Many are already challenged economically, socially and politically. Ironically, the quest for economic growth, which is a particularly strong feature across Asia, may have long-term implications that make future economic development in some parts of the region unsustainable.

Conclusion

It should be evident from this chapter that Asia's economics–security nexus is more complex than it might first seem. Of course, there may be questions about whether economic issues should be included in Asia's security picture, but it would be an odd decision to exclude them, given that the region's economic advancement is the main reason for the international attention it has been attracting. Economic advancement among many of Asia's larger countries is also the fundamental reason why so much attention is paid to the changing distribution of power in the region, and to Asia's place in global affairs.

It is possible to argue that economic security – the absence of threats to prosperity – warrants inclusion as an intrinsic part of understanding Asia's regional security. As Miles Kahler (2004) has suggested, approaches to economic security in Asia have increasingly needed to factor in questions of significant volatility as globalisation processes have made their impact. But it might be wondered if the performance of businesses, investment levels and national economic indicators should be viewed as intrinsic parts of the security situation. A more selective approach, as reflected in the current chapter, is to acknowledge that economic changes, while not always being security issues in their own right, can have security implications. This logic can apply to the even more conservative position that security is about the absence of *violent* threats, because even in such a restricted intellectual universe, the connections between regional security and economic conditions are multiple and work in different directions. As the preceding analysis has suggested, it is not feasible to say, for example, that the wealthier Asian countries become, the more secure they will be in their relations with one another. Economic growth can be both a security liability and an asset for Asia.

Similarly, linear arguments suggesting that Asia's greater economic interdependence has an automatically pacifying effect for the region also need to be put to the test. One reason to question this proposition is that interdependence is not just something that happens accidentally to regional countries as they trade with one another. It is a feature of their relationships that can be encouraged deliberately and manipulated by governments. In other words, interdependence can be a form of influence and even of pressure. Yet it does seem likely that governments believing that prosperity depends on a positive regional security environment, and thus the absence of significant armed violence in interstate relations, would think twice about using force as an instrument of policy in their relations with each other. But what remains uncertain is whether other considerations in the heat of the moment will trump the logic of the economists and the liberals. And if that occurs, the fact that Asia's increasingly wealthy states are also becoming rather heavily armed may not be an especially reassuring factor. Hence Chapter 5 will consider what the spread of new weapons systems in Asia means for regional security.

Chapter 5

Guns: Will Military Technology Lead to New Conflict in Asia?

There is a famous old Roman saying that those who wish for peace should prepare for war. According to this logic, Asia is in line for a significant amount of peacefulness with so many governments in the region making extensive military preparations. But if the Roman maxim is wrong, and if insecurity rather than peace comes from the development of military capabilities, Asia may be facing a difficult time. A major war in Asia could involve some of the world's largest and most advanced armed forces. For that reason alone it could be especially costly, including in the loss of human lives.

The other side of this coin is the reality that, if the world were ever to head genuinely towards disarmament, that process would be unlikely to start in Asia, where military power is regarded as being neither antiquated nor irrelevant. Indeed, there are concerns in quite the opposite direction: if there is an arms race going on anywhere in the world, it is most likely to be taking place in Asia. Yet, as this book has already shown, Asia's security environment is complex and diverse. Sweeping generalisations about singular trends are best left to one side. Hence this chapter will break down the arming of Asia into its component parts to gauge the real security implications.

Given that the interest here is in security developments that have broad regional effects, it seems only right to begin with the spread of nuclear weapons in Asia. If any form of military power could be expected to have ripple effects across the region, the various nuclear arsenals and programmes that exist in Asia must be leading candidates for that role. This means considering the extent to which Asia's security is influenced (and perhaps supported) by relationships of nuclear deterrence. And it will also be necessary to ascertain whether the use of these weapons in anger is remotely possible.

While nuclear weapons can grab regional security headlines for obvious reasons, this chapter will also pay attention to the development of advanced conventional forces in Asia. Many countries in the region once focused their military expenditure and training on the fighting of internal conflicts. But for some years now military programmes in many parts of

Asia and its borderlands have focused increasingly on the external projection of force. This is especially the case for the development of advanced maritime capabilities. This trend feeds into the important relationships involving the region's major powers, including the tensions that sometimes occur between the United States and China, and between China and Japan. Does this competition offer a recipe for the prevention of war, as the old Roman maxim would suggest? Or is it just as likely to provide the seeds for conflict? And if a serious conventional war were to break out, do the countries involved have the mechanisms they might need to prevent its escalation, including into a nuclear nightmare?

But perhaps today's bad security dreams are of a different nature when it comes to the military technologies that might raise the greatest concern for Asia. There are high-technology and low-technology versions of this argument. In terms of the former, the application of the very rapid developments in information technology to modern conflict is an important story that is playing itself out in Asia. This makes it necessary to ask whether tomorrow's regional security problems will be caused by the competition that is occurring in the cyber domain. Cyber insecurity could be just as prevalent as cyber-security.

In terms of the low-technology horizon, the spread and easy availability of small arms (including automatic rifles) in large parts of the region is a complicating factor for hopes for domestic peace in many countries. Nobody in Asia has lost his or her life to a new nuclear weapon attack since August 1945, because no such events have occurred since the end of the Second World War. But millions of people have died as a result of the use of small arms in the messy internal conflicts that have occurred over the same period. Perhaps this should be the main focus of arms control and disarmament efforts in Asia. Alternatively, it might be argued that the sheer scale of the potential death and destruction from larger weapons systems require greater attention, even if these larger and more catastrophic wars are less likely. Finally, it needs to be recognised that there are new weapons systems, such as drones, which combine high technology with a very focused impact. Is that where things are heading in Asia? Together these features creates a full agenda for this chapter.

The nuclear dimension of Asia's security

As soon as they were matched to long-range (intercontinental) missile systems in the intensifying Cold War, nuclear weapons appeared to make geography and distance far less important factors in modern security. As in the case of information technology, nuclear armaments also increased the tempo of military interaction. It took weeks for the forces of Russia and Germany to be mobilised and deployed in the First World War, and still a significant amount of time for forces to be assembled and put into action in the Second World War. But in the missile age, nuclear weapons could be launched and reach their target within minutes. And given its enormous capacity for destruction, a single nuclear weapon could cause levels of casualties and damage equivalent to waves of aircraft armed with conventional explosives. Because of their uniquely concentrated ability to cause harm and damage, nuclear weapons are the world's only genuine weapons of mass destruction. As Christian Enemark (2006) has argued convincingly, chemical weapons and attempts to turn diseases into biological weapons systems do not really come close.

With the exception of the American atomic bombs dropped on Hiroshima and Nagasaki in Japan in August 1945, a number of these features have been observed from the testing of nuclear weapons and computer modelling of their effects rather than from their actual use in war. Nuclear weapons take to an almost extreme level the idea that some forms of power derive their influence from what is promised (or threatened) rather than from what is used on the battlefield. In a unique representation of coercion rather than control, they offer an immense amount of what the American strategist Thomas Schelling (1966, p. 134) referred to as the 'power to hurt' in his suitably entitled *Arms and Influence*. If the logic of deterring rather than fighting wars has any credence, then nuclear weapons are the ultimate expression of this way of thinking.

Two factors made it highly likely that nuclear weapons would play a part in Asia's security almost as soon as they were developed by the United States and quickly thereafter by the Soviet Union. The first was that these two nuclear-armed superpowers were both interested in shaping Asia's security environment. Indeed, as the Cold War spread to Asia, the nuclear factor was not far behind. The United States in particular used the possibility of nuclear use (normally as a last resort) to back up its more readily usable conventional military power in Asia. And as time went on, the Soviet Union spread its nuclear forces to locations that brought Asia (including China) into range. Had the Soviet Union unwisely decided to join the wars in Korea and Vietnam, for example, it is hard to see how the two superpowers would have avoided threatening each other with nuclear annihilation as a means of coercion. That might well have ended up very badly indeed: the situation could easily have been far worse than the famous Cuban Missile Crisis of October 1962.

Hardly any part of the wider Asian region was completely spared the nuclear element of Cold War competition. Britain tested some of its early nuclear weapons in the Australian desert, and France and the United States both used South Pacific islands as locations for their testing. Before the ban on atmospheric nuclear testing, these detonations resulted in some noticeable environmental and health effects. American testing in Micronesia, for example, made some of the atolls of the Marshall Islands uninhabitable for many decades because of residual radiation levels. Ian Anderson et al. (2006, p. 1780) have noted that the United States' compensation agreement of the mid-1980s recognised no fewer than three dozen health problems: 'most commonly cancers of the lung, thyroid, breast, and ovary, and leukaemia and lymphoma'.

But for at least some Asian countries who felt especially vulnerable in dangerous parts of the regional neighbourhood, the possession of nuclear weapons appealed as a way to address their senses of strategic inferiority. At least some of that vulnerability and the further spread of nuclear weapons it inspired was linked to the initial round of proliferation: John Wilson Lewis and Xue Litai (1988) have argued that America's nuclear coercion in the 1950s was the leading reason why China's leaders felt that their country simply had to go nuclear. Indeed, by the mid-1960s China would become the fifth and final member of a special club of nuclear weapons states whose status would be recognised in the Nuclear Non-Proliferation Treaty (NPT) which came into operation in 1970. Under this widely signed treaty, the spread of nuclear weapons was meant to stop right there. Non-nuclear weapons states signed on to promise they would not develop nuclear weapons, a promise which most of them kept. The five nuclear weapons states were supposed to move towards the eventual removal of theirs, a promise that rather unsurprisingly none of them have held to. Indeed for China, with a still recent history of external subjugation, tense relations with the two nuclear-armed superpowers at various times, and with a realisation that nuclear possession was a clear indicator of great power status, nuclear weapons once gained would prove very difficult to relinquish.

One of China's large neighbours, India, did not sign up to the NPT, in part because it objected to the discrimination in the Treaty between nuclear weapon states and others, and largely because it was pursuing its own programme. Moreover India's calculations were also influenced by its border conflict with China in the early 1960s. In one of the rare instances of a direct chain-effect in nuclear proliferation (rare because the causes are often more indirect and complex), India's widely understood but covert programme was too much for Pakistan, another hold-out from the NPT, to ignore. The very public testing of nuclear weapons by both India and Pakistan in 1998 constituted one of two major contributions from Asia to the ending of the optimism about nuclear disarmament which was a feature of the immediate post-Cold War period. At the time, it encouraged warnings, including by Joseph Cirincione (2000), that a wider 'Asian nuclear reaction chain' was in process. This may have been a slight exaggeration, but there has been one more contribution from Asia to global nuclear proliferation in the form of North Korea's nuclear ambitions. Unlike Iran, which has long claimed that its nuclear programme was designed to produce nuclear energy for civilian use – a right under the NPT – North Korea has been far more brazen, openly claiming that it requires its own capacity for nuclear deterrence as a security equaliser.

The threatening way in which successive North Korean leaders have brandished their nuclear programme (which at most consists of a handful of crude devices) have led some to wonder just when Japan would follow suit and abandon its long-held prohibition on the development of its own nuclear weapons programme. But for decades Japan has been able to rely on the extended nuclear deterrence provided by its main ally, the United States. So too has South Korea, which like Japan has the advanced scientific and industrial base and civilian nuclear energy facilities that would make such a programme eminently possible. For now at least, the American nuclear umbrella has helped contain the further spread of nuclear weapons in East Asia. As China's military capabilities have strengthened (including a modernised but still not very large nuclear weapons arsenal) and as North Korea's nuclear threats have continued, both Japan and South Korea have sought new assurances from Washington on its extended deterrence commitment to them. Those assurances for the most part have been given. But it would still be a brave act to argue that nuclear proliferation has stopped in Asia. Indeed some experts, such as Andrew O'Neil (2007), write as if the further spread of nuclear weapons in the region is a distinct possibility.

In fact, it is hard to see how Asia could easily denuclearise even if there were wider international pressure for that to occur. Holding over 90 per cent of the nuclear weapons in the world, the United States and Russia have substantially reduced their arsenals since the late years of the Cold War. But China has consistently resisted the idea of turning this into a triangular process, arguing that its arsenal is as small as it can be to guarantee a minimum of nuclear deterrence. As David Santoro (2012, p. 141) has suggested, China has adopted a 'wait-and-see position' pending further American (and Russian) moves. And it is very difficult to envisage Indian nuclear reductions, let alone full disarmament, before China takes a major step in that direction. Pakistan is even less likely to do so, disadvantaged as it is by a major conventional military inferiority vis-à-vis India and lacking the strategic depth that India's much larger land territory offers. There also seems to be a very broad consensus that North Korea sees its nuclear weapons programme as one of the only cards it can play in international politics.

In short, not a single one of the Asian countries which possesses nuclear weapons seem likely to give them up. That said, there is something of a regional example of nuclear disarmament: when the Soviet

Union came to an end, the newly independent state of Kazakhstan stood to inherit the nuclear weapons on its territory, but by international agreement these were transferred to Russia. Ariel Levite (2002) is right to argue that rather more attention needs to be given to the decisions by states to reverse nuclear programmes that have been started (or in Kazakhstan's case inherited). But this hardly seems to be a precedent for the nuclear-armed states of Asia that have deliberately developed their own arsenals. Once the threshold of nuclear possession has been crossed, it is very difficult to walk back.

Could nuclear war really happen in Asia?

It is difficult to imagine an event in Asia that could cause more fundamental and far-reaching ripple effects than the detonation anywhere in the region of a nuclear weapon on or near any military or civilian target. Even the explosion of the smallest such weapon would have major international ramifications. The world has lived without such a development since the end of the Second World War. The breaking of what Nina Tannewald (2008) calls a taboo on the use of nuclear weapons would have consequences that are as uncertain as they are substantial. Many questions come to mind quickly. Would the perpetrators of the attack need to be punished by the use of nuclear weapons in return to avoid any future such acts? Would the second use of nuclear weapons in Asia (after their use against Japan in 1945) signal that nuclear weapons were no longer seen as the ultimate deterrent but as usable military systems? Would hopes for nuclear disarmament be galvanised or dismissed? It is almost impossible to know the answers to these questions in advance. But it is clear that the region would be dealing with unexplored territory.

Even a modest exchange of nuclear weapons in Asia could have very serious wider effects. Towards the end of the Cold War, western scientists began talking of a nuclear winter, where a major superpower nuclear war would cause a long-lasting blanket of smoke in the atmosphere that would shut out the sun, reduce temperatures and damage food production. These estimates eventually attracted a great deal of controversy and were largely forgotten as the Cold War era disappeared. But more recent modelling, including by Robock and Toon (2009), has suggested that an exchange of one hundred nuclear weapons between India and Pakistan could be enough to generate significant nuclear winter effects. Of course given the high population densities within missile range on either side of the often-contested India–Pakistan border, there would be immense human suffering in both countries. Millions of civilians would lose their lives. Millions more would be injured, many with long-term health effects. Entire cities could be destroyed. The functioning of India and Pakistan as modern states would be grievously affected, not least because large parts of their political leaderships would likely have been killed. Moreover, if the climate modelling is correct, tens or hundreds of millions of other people in other countries could find that they no longer lived in areas where there was enough light to support agricultural production. Massive migrations might be even likely, upsetting social stability in numerous places.

This is reason enough to ascertain whether the use of nuclear weapons in Asia is at all likely. The first point is that such an event is mathematically possible, at least for as long as these weapons exist. The sheer consequences of the event mean that its very low probability does not remove the need for concern. Nuclear war by design or accident can never be ruled out entirely even if these weapons have been acquired solely for the purposes of deterring war. But the second point is more reassuring. Nuclear weapons have not been used in anger in Asia, or anywhere else in the world, for many decades despite significant periods of stress and tension. This would seem to corroborate the judgement that for the most part, nuclear armaments are regarded by their possessors as weapons of the very last resort.

That makes it necessary to consider the extreme situation under which nuclear weapons might still be used. Most of the countries that possess them have a series of non-nuclear and even non-military options to select from in showing their displeasure with others before nuclear weapons even become a distant consideration. In fact the presence of nuclear weapons can set an undesirable upper limit of conflict which almost makes it safer to conduct a limited war. This seems to be one of the lessons of the brief Kargil conflict between India and Pakistan in 1999, although some prominent analysts, including Sumit Ganguly (2008), are not so sure about this conclusion.

Even so, it is hard to be sure that if a conventional conflict begins in Asia between countries with nuclear weapons, the countries directly involved know quite how to observe these limits. Some will have lower thresholds for nuclear use than others. Because of its conventional inferiority, Pakistan is likely to use nuclear weapons in a war before India does. Vipin Narang (2010, p.44) refers to Pakistan's approach as one of 'asymmetric escalation' which is 'geared for the rapid ... first use of nuclear weapons against conventional attacks to deter their outbreak'. This is not an illogical strategy, but it is an especially risky one, and Pakistan's development of shorter range or 'tactical' nuclear weapons could raise the risk of early nuclear use in a South Asian conflict. Moreover, it is simply impossible to know at what stage North Korea might try to use its nuclear weapons in any future war on the Korean peninsula.

Moreover, India and China do not have a strong history of effective military communication in times of crisis, a problem which extends into what Cunningham and Medcalf (2011) have identified as a stunted conversation on their nuclear relationship. This reduces confidence that these two neighbours and rivals know how to manage a quickly escalating situation without any thought being given to the use of weapons of mass destruction.

The first requirement for avoiding a nuclear war in Asia might then be to avoid conventional wars between the nuclear-armed states of the region which might get out of control. A second would be to ensure that there are adequate command and control systems in place to reduce the chances of a completely or largely inadvertent use of nuclear weapons in Asia. This also requires closer cooperation between the nuclear-armed states of the region to ensure that they are confident in one another's control systems. That would require a higher degree of transparency than has been evident to date, for instance, in the China–India relationship.

It may sound unusual, but the nuclear-armed states of Asia could learn a good deal from the Cold War superpowers. Despite their deep political differences, the United States and the Soviet Union found ways to reassure one another that an inadvertent war was not just around the corner. The problem is that, on the whole, the nuclear-armed states of Asia have been reluctant to agree among themselves on arrangements that might offer a similar level of reassurance. This reluctance may itself be partly a consequence of the unevenness in their capabilities: unlike the US–Soviet picture as it matured in the Cold War, the nuclear arsenals of the United States and China, to take the most obvious example, are highly uneven. But this asymmetry, as Avery Goldstein (2013) has argued, may put further pressure on what might still be only rudimentary levels of mutual understanding about what either side might do in an escalating conventional conflict.

The third requirement is a somewhat debatable one: that the avoidance of nuclear war in Asia is more likely in the absence of a further spread of these weapons. This is a debatable argument because in some cases the acquisition of nuclear weapons by one or more additional countries may bring a measure of stability, through additional deterrence, as well as understandable concern. Japan may offer the main future test case of this logic. A nuclear-armed Japan would likely frighten and alarm its neighbours, but as Hugh White (2012) has implied, it would also give Japan greater confidence in its ability to dissuade an attack regardless of whatever assistance the United States was able to provide. Yet it is not clear that this development would in overall terms generate an improvement for regional security. And there are few other cases of potential nuclear proliferation in Asia which might recommend themselves in this regard.

There is a fourth requirement as well and all the countries of the region would appear to have a common interest in its fulfilment: avoiding the acquisition of nuclear weapons in Asia by non-state groups. The regional security implications of activities by terrorist groups will be considered in Chapter 8. But for the purposes of considering Asia's nuclear challenges, there are some important judgements that can be made at this point. The first is a general one that it is extraordinarily difficult and almost impossible for non-state actors to build their own nuclear weapons. The second is that only the most reckless of states would pass on their own nuclear weapons to a terrorist group. The third is that few terrorist groups anywhere in the world have given signs of a systematic, let alone an effective, attempt to acquire nuclear weapons. In assessing the prospects of nuclear terrorism in Asia, S. Paul Kapur (2008) concludes that fairly much every active group in the region can be excluded from consideration and that even Al Qaeda's nuclear intentions are unclear. The fourth is that it is understandable but also not necessarily accurate to assume that in the very unlikely event that a terrorist group in Asia managed to obtain a nuclear weapon, they could be guaranteed to use them.

But against all of these considerations is the prospect that if a non-state actor were to gain access to one or more nuclear weapons anywhere in the world, then it is probably most likely to occur in Asia. Imagine if, in the competition for political power in an increasingly unsteady Pakistan, an extremist group were somehow to gain nuclear weapons. This is considerably less likely than some have assumed. But if it were to happen, and as this author has argued elsewhere (Ayson, 2010), the consequences could be greater than some assume. Just one nuclear weapon exploded somewhere in India could trigger a wider nuclear exchange between South Asia's two nuclear-armed countries in the panic and hostility that would easily result. Again the likelihoods are small, but they are not zero.

The competition for conventional military advantage in Asia

It is rather tempting to conclude that Asia is safe militarily if it can avoid the ultimate catastrophe: a war involving the use of nuclear weapons. But that is a temptation which should be resisted. There is no real non-nuclear equivalent of the NPT which seeks to control the spread of advanced conventional weapons systems. And while there is a tradition of not being the first country to use nuclear weapons in a crisis, the countries of Asia do not seem committed to a similar norm regarding the use of submarines, fighter aircraft, and non-nuclear-armed missiles.

This is a challenge for at least three reasons. First, the manufacturers of advanced weapons systems find that Asian countries are among their most lively market for sales anywhere in the world. Indeed Wezeman and Wezeman (2014) have reported that nearly one half of all the imports of arms around the world in the period from 2009 to 2013 were purchased by countries in Asia and Oceania, and that the top three importers were India, China, and Pakistan. Second, some states are very nervous about what their regional neighbours in Asia are procuring: the acquisition of conventional systems is certainly associated with rising regional tensions. Third, the appearance of these weapons systems is allowing some Asian states to contemplate more ambitious plans for how they might use their increasingly capable armed forces in a crisis. In the presence of the territorial disputes, historical animosities, and general mistrust that characterise a number of Asia's strategic relationships, this might not end very well.

Asia's central place in international military modernisation suggests that much of the region is following a pattern suggested nearly two decades ago by Eliot Cohen (1996, p.50) whereby 'a contemporary revolution in military affairs' was opening up 'tremendous opportunities to countries that can afford to acquire expensive weaponry and the skills to use it properly'. The modern conventional weapons systems that many Asian governments have been acquiring have a number of distinguishing features, most of which confirm that advances in information technology are at the heart of this apparent revolution.

The first of these features is enhanced accuracy: advanced missiles (which can be launched from aircraft, surface vessels, submarines and from the land) tend to be so precise that the destruction of one target often only requires one shot involving one warhead. These systems are often accurate over large distances so that effects on land, for example, can be achieved from quite a distance out at sea. The second feature is awareness: modern military platforms (ships, planes and tanks, for example) normally have advanced sensors that allow those commanding and controlling them to be aware of potential threats, including enemy radar systems that may be locking onto them as targets, and corresponding sensors allowing them to lock on to enemy forces in return.

The third feature is connectedness: thanks to advanced information technology, the weapons systems of the modern battlefield come with advanced communications abilities and the capacity to process significant amounts of data. This allows various components of a country's armed forces to work together, and it also allows co-ordination between allies so long as their armed forces have compatible communications systems. The fourth is stealth: modern submarines, for example, are extremely quiet and difficult to detect, as are the latest generation of fighter aircraft.

This creates uncertainty for an opponent who may not therefore have the freedom they once enjoyed to operate their forces in forward theatres although it may also allow stealthy platforms to evade defensive measures undetected.

With their combination of stealth and firepower, advanced conventional submarines appear to be at the top of the shopping lists for many regional countries. No less than a dozen regional navies – Australia, China, India, Indonesia, Japan, Malaysia, Pakistan, Singapore South Korea, Taiwan, Thailand, and Vietnam – have acquired or are acquiring and modernising submarine fleets. Advanced submarines have long been a staple of America's maritime military power in the region, and while much less able than it once was, Russia's navy has a long experience in this area. By comparison, China has until recently not been much of a player in the submarine business but it now boasts one of the largest ship-building programmes the world has seen. This has no doubt egged on Japan and Taiwan. For its part Australia sees its submarine programme as central to its attempts to maintain a military capability edge in the areas to its north and west. That means being stronger in this area than its Southeast Asian neighbours even though these are not normally regarded as potential adversaries. One can be certain, however, that Singapore, Malaysia and Indonesia keep a very close watch on what they each are developing. Indeed, one experienced analyst of Asia's maritime security, Sam Bateman (2011, p. 74), has argued that 'The submarine capability of a possible enemy requires a response and invariably this means a submarine fleet of one's own, particularly because a submarine is a most effective, anti-submarine weapon platform itself.'

It would be easy from this assessment to conclude that Asian countries develop military capabilities in response to one another. As Asia's militaries have grown, claims that a regional arms race is occurring have been frequent and persistent. There are in fact few historical examples of genuine arms races, defined by Hedley Bull (1961, p. 5) as 'intense military competitions between opposed powers or groups of powers, each trying to achieve an advantage in military power by increasing the quantity or improving the quality of its armaments or forces'. But there are some exceptions. Before the First World War, Britain and Germany were involved in a naval arms race to produce increasingly capable battleships. That a major conflict followed this example fed a widespread argument that arms races were a sure-fire recipe for war. Yet even in this case, the arms race logic is questionable. David Stevenson (1996, p. 421) has argued that while the armaments may have been 'the wheels and pistons of the locomotive of history' they were 'not the steam, and if considered in isolation they offer neither a sufficient nor an all-embracing explanation of the destruction of peace'.

The missing factor from most arguments that weapons cause war is the important political one, which more often than not drives decision-making over questions of war and peace. Yet the notion that the accumulation of advanced weapons system could drive global conflict helped inspire naval arms control in the interwar period, including an agreement between the United States, Britain and Japan to limit their competition in the construction of major battleships. Looking back it can be seen that this act of cooperation failed to prevent the onset of Second World War hostilities across so much of Asia. And in an ironic twist, the last major act of violence in that global conflict, the atomic bombing of two cities in Japan, helped set off the most famous arms race of all. This was the intense competition for nuclear military advantage between the United States and the Soviet Union. By the 1970s each had tens of thousands of nuclear warheads, accompanied by thousands of delivery vehicles (including long-range missiles).

There is no doubt that there are elements of competition in the weapons acquisition programmes of at least some Asian countries. But truly intense arms races in the region are not that easy to find. There is certainly not a single interconnected arms race across the region that explains the bulk of military expenditure. Instead there are some specific bilateral arms competitions which probably come closer. These include the competition between North and South Korea, in which the latter is well out ahead in terms of technological advancement, and the competition between India and Pakistan. Another such example is the way that Japan's decisions on armaments, including the acquisition of what looks very much like a small aircraft carrier, is a response in large part to the modernisation of China's armed forces and the increased capabilities these bestow. In one piece of analysis, Bjørn Elias Mikalsen Grønning (2014, pp.5–6) cites increased investments in submarine and anti-submarine warfare, amphibious warfare, ballistic missile defence, and fighter aircraft as part of Japan's deliberate response to a more powerful China.

But it is important to avoid the assumption that these interactions typify developments across the region. Competitive explanations for weapons programmes (including the arms race claim) assume that the countries of Asia are motivated by external considerations, where what one country produces today is the reason for what another country acquires in several years' time. By contrast, several Asian countries have a history of developing armaments for domestic rather than international purposes. Many of the armed forces of Asia have their origins in the strengthening of their government's ability to fight and deter internal wars. Many of the largest armed forces in Asia grew for that very reason. For decades the main purpose of Indonesia's armed forces were to preserve national unity after independence, and these forces were descendants of the military units that had fought for that very independence. A similar phenomenon accounts for the development of armed forces across large parts of Southeast Asia.

Moreover many of these armed forces have been so internally preoccupied that they have neglected more outwardly focused defence needs, including the need to defend the exclusive economic zones (EEZs) that exist courtesy of the United Nations Convention on the Law of the Sea. That deficiency still can be observed, for example, in the case of the Philippines, where the government continues to face a significant insurgency in its south but has neglected its ability to project military power beyond the land and in the air. In what may constitute something of an arms race in reverse, Cruz de Castro and Lohman (2012, p.3) have suggested that the Philippines Air Force, 'the most advanced Southeast Asian air force in the 1960s, had by 2001 become the weakest air force in the region'. Most of the Philippines' neighbours in Southeast Asia, however, have had greater freedom and willingness to develop these power projection capabilities, in part because their concerns about internal conflict have waned. But this may still mean that what seems to be a rapid arms build-up in parts of Asia may actually be the normalisation of some of the region's militaries, who are now looking beyond their national boundaries.

Growing power projection capabilities

In fact when a region-wide assessment is made, few countries in Asia actually possess the military capability to project significant power at a distance. In naval terms, some may now have navies that can operate reasonably effectively within their 200 nautical mile EEZs, but very few have genuine 'blue water' capabilities. Yet that picture has been changing, and this is not just because a number of Asia's governments are less concerned about domestic security challenges than they once were. One other factor is the increased national funds that are available to governments to purchase advanced military equipment as a result of tax revenues from their growing economies.

Of course this can still be regarded as a mainly domestic process where governments in the region are simply taking advantage of their more rosy fiscal situations rather than responding to every new weapon system which their neighbours are purchasing. Over two decades ago, Desmond Ball (1993) used the phrase 'Arms and Affluence', a play on the title of the influential book by Schelling (1966), to sum up that situation. But over the next twenty years Ball and other close watchers of Asia's military hardware came to see patterns of acquisition which more closely resembled the intense rivalry that characterises arms racing, including the competitive purchasing of cruise missiles, advanced submarines and fighter aircraft, and the advanced sensors that all of them require to operate effectively. Writing mid-way through that period, Hartfiel and Job (2007, p. 2) argued that there was 'a dramatic accumulation of potentially destabilizing weapons systems...by traditional rivals in East Asia'.

Yet the projection of military power whereby advanced forces can operate effectively far from their home ports and airfields remains the province of the great military powers of the region. And even here the field is restricted: the United States continues to have a huge lead over all comers. Some Asian navies, including those of China and India, have launched their own aircraft carriers, but to some extent this is for show. It does not mean they have mastered the art, nor does it mean that these capabilities would make a big difference in any battle. It continues to be useful to consider the reminder from Paul Dibb (1997, p. 111): 'While concerns over the acquisition of increasingly modern platforms and associated missiles are legitimate, there is a difference between acquisition and operational capabilities.' In particular, the United States remains the only country with multiple aircraft carrier battle groups (formations of ships and aircraft based around individual carriers) which offer an internationally unique concentration of firepower deployable over long distances.

The main representation in Asia of this American military advantage is the 7th fleet, involving tens of vessels, hundreds of aircraft and tens of thousands of marines. With its main forward location in Yokosuka, Japan, this fleet has been a reassuring sign for Japan and South Korea in particular of a distribution of military power in the region that works to their advantage. That this presence has also given Japan less reason to build its own forces (which are nonetheless substantial) has also been somewhat reassuring to China. But the development of China's own armed forces, especially for maritime operations, is changing this picture. China cannot reach out far from its own shores in the way that the United States is able to operate in Asia and it offers no noticeable challenge to America's dominance in the western hemisphere. Yet by developing its new submarines, missile systems, and information warfare systems China has significantly raised the risks for the United States in operating close to the Asian mainland. As Goldstein and Murray (2004, pp. 191–4) observed several years ago, America's aircraft battle groups will increasingly have to consider that quiet Chinese submarines may be present and offer a stealthy threat to their normal free access. For this reason, it would be much more risky today for the United States to repeat its response to the 1995–6 crisis with China and deploy aircraft carriers next to Taiwan.

The name now given to China's growing ability to negate easy maritime access by the United States is 'anti-access area denial'. It is *denial* because while China is unable to *control* the maritime area by dominating the seas, it is developing the ability to make it prohibitively costly for others to operate in the ways that they might choose. At one level this might be considered promising because China appears to be contributing to a new balance of military power in East Asia. But as has already been found in this book, most of China's neighbours, and indeed many countries throughout the region, have been quite content with a long era of American dominance. They have not generally seen America's big lead on all comers as a threat to their security. A more even balance in Asia therefore can seem oddly threatening, especially if there is concern that China may wish to go further and eventually establish itself as the dominant player, replacing the United States.

That day seems for now some way off. Richard Bitzinger and Barry Desker (2008, p.121) have argued that 'China's current military build-up is ambitious and far-reaching, but is indicative of evolutionary, steady-state and sustaining, rather than disruptive or revolutionary, innovation and change'. Situations such as these do not change overnight. Nonetheless China's increasing anti-access area denial strengths have been noticed by Washington, which attaches great importance to the full freedom of navigation for its maritime forces in Asia. Analysts like Aaron Friedberg (2014) believe that America has been rather too slow to respond to the challenge. But the United States has developed a concept called 'air–sea battle', which combines the firepower and mobility of its aircraft, surface and submarine vessels, and sensors and missile systems as an integrated fighting force. The Pentagon hopes that just as the United States used 'air–land battle' to devastate Saddam Hussein's armed forces in the 1990–1 Gulf War, so it will be able to employ 'air–sea battle' to negate, degrade and bypass the systems and strategies that China has been developing.

Unlike Iraq, whose army had become by the early 1990s a shadow of the force which attacked Iran in the mid-1980s, China is a rising military power in Asia. That brings with it a certain amount of confidence, a quality that has rarely been in short supply for America and its armed forces. This is not a reassuring mixture, but it goes to show that it is not the mere acquisition of advanced forces that is likely to spell trouble for Asia. The way that countries think about their use is also an essential part of the mix. Japan, for example, has long had one of the most advanced navies anywhere in the world. But it has been very restrained in using its armed forces for anything that cannot be explained as a mission of self-defence. That restraint appears to be lifting as Japan becomes a more normal military power, less restricted by post-war constitutional limitations.

There are corresponding signs from other important countries in Asia. Vietnam, whose claims in the South China Sea are contested by China, has long shown a willingness to use armed force (both coercively and in actual battle), and is now acquiring maritime capabilities that extend its ability to support its own claim to these disputed maritime areas. India, whose armed forces have traditionally been focused north against Pakistan, has been devoting increasing attention in recent years to the development of its naval forces and has indicated an interest in patrolling beyond the Indian Ocean into East Asia. As Walter C. Ladwig III (2009) notes, this wider effort is still in its formative stages, but India's ability to shape the distribution of regional maritime military power in is growing. The growing marriage of intent and capabilities in Asia means that this is one part of the world where it is difficult to pronounce the complete obsolescence of interstate warfare, even though the absurdly high costs of such activity may be universally acknowledged.

High-tech: Asia's cyber challenges

But as this study has already demonstrated, the states of Asia do not have to resort to armed hostilities to put pressure upon those whose behaviour they may wish to influence. They often seem keen to fulfil the dictum of the ancient Chinese strategist Sun Tzu (1994, p. 177) that 'attaining one hundred victories in one hundred battles is not the pinnacle of excellence. Subjugating the enemy's army without fighting is the true pinnacle of excellence'. Writing during the period when China was paralysed by conflict between feuding armed factions, Sun Tzu emphasised the role of deception and psychological ploys in getting the other side to do one's bidding without resorting to costly fighting. Over two thousand years on, most Asian countries would agree that weakening a potential adversary is much preferable to seeking an uncertain and potentially hollow victory on the battlefield. And many such non-kinetic options (non-violent in a physical sense) are making themselves available courtesy of the digital age.

Knowing someone's weaknesses is essential to effective strategic interaction. In this context, one of the main weaknesses of the many advanced countries in Asia is also one of their strengths: the extent to which their economies, societies, and governments rely on cyberspace. Some of the highest broadband speeds in the world are present in Asia, with Hong Kong, Japan, Singapore and South Korea leading the way. Without the almost instantaneous exchange of commercial information, and the electronic transfer of funds, the booming regional economies of Asia which were discussed in Chapter 4 would be shadows of themselves. Utilities such as electricity and water companies, and transportation systems, all of which are essential for economic activity and public safety, rely heavily on connected computer systems. Private and government organisations rely extensively on their websites and social media to communicate policies, opinions and warnings, including travel advisories and major announcements. Moreover, modern military weapons systems, especially those which rely on sensors for early warning, guidance, target acquisition and communication, are also heavily dependent on cyberspace.

The connections between the digital age and military power are extensive and have a long history. As Singer and Friedman (2014, p.16) have observed in a recent study, the internet has its origins in research commissioned by the United States government to enhance America's military effectiveness. Correspondingly, the ability to interfere with, disable or even control, the information technology systems within another country is becoming an increasingly significant part of the security debate. Thomas Rid (2012) has argued that claims of a looming 'cyber-war' are not especially credible, especially when they suggest that an armed conflict can be undertaken completely within the cyber domain. But his jeremiad has not stopped Gompert and Libicki (2014) from talking about elements of a cyber-war occurring between the United States and China in maritime Asia, given the importance of digital capabilities in the contest between A2AD and Air–Sea Battle which has already been discussed in this chapter.

Moreover, cyber-attacks of one sort or another are happening frequently globally and a good deal of this activity is occurring in and from parts of Asia. These do not need to reach the severity of cyber-war per se (whether such a thing exists) to be of concern. They include denial of service activities, where websites are bombarded and overcome by a multitude of bogus messages; exfiltration, where the firewalls of a system are penetrated to extract valuable data (including commercially and officially sensitive information) without the knowledge of the host organisation; and using the internet to damage or disable infrastructure by the introduction of malicious codes.

Because of the ubiquity of access to cyberspace, with over a billion internet users in Asia (and hundreds of millions more in the United States), the issue of correctly attributing the source of attack can present a major challenge. It is, for example, at the opposite end of the spectrum from an attack involving nuclear weapons, which could realistically only be mounted by one of a handful of countries. Intelligence agencies often have a good idea of the sources of major cyber-attacks, but their governments are often unwilling to reveal this information for fear of compromising their ongoing ability to keep track. Nonetheless, claims of malicious behaviour in cyberspace have strained some strategic relations in Asia. The American government has expressed its concerns that many attacks, including on the information systems used by banks and government agencies, have been coming from China. Especially telling was a prominent report published by Mandiant (2013), an American information security company, which suggested that cyber-hackers associated with the People's Liberation Army were likely to be responsible. There has also been concern that large Chinese companies which produce information devices and networks used overseas (including mobile phones) are too close to the government in Beijing. The use of exfiltration for commercial espionage is a particular concern.

China's response has been to suggest that it too is a victim of cyber-attacks and that many of the attacks emanating from Chinese internet protocol addresses have been the work of non-state groups. Beijing will be aware as well that the United States has the largest and most effective surveillance network, able to monitor phone calls, emails, social media and other electronic transmissions. The United States government is widely thought to have been involved with Israel in the creation of malicious code which was used in an attempt to shut down elements of the Iranian nuclear programme. But more damaging to the American position has been the release of classified information by Edward Snowden about the extent of the National Security Agency's digital surveillance efforts.

In Asia the main immediate effect of these revelations was, as Adam M. Segal (2014) points out, to reduce America's capacity to keep China's activities under the spotlight. The United States may still be ahead in the technological game, but it has taken a battering in the credibility stakes. Yet while the Snowden controversy placed severe pressure on America's partners in some parts of the world, including in Latin America and Europe, Asian allies have been slower to express their concerns about US activities. Leif-Eric Easley (2014, p.148) argues that Japan and South Korea may be preoccupied by their own domestic intelligence concerns, and that 'their geographical proximity to the main source of international cyber concerns' (i.e. to China) may temper any desire to criticise the United States. This is a further sign that security ripple effects do not happen autonomously, but can be limited (or accelerated) by the interests of political actors.

But this does not mean that US–China disagreements on cyber-security have attained a quiet stalemate. Finding common ground between the region's two most powerful countries on cyber-security questions will not be easy. China has proposed an international treaty to set global rules, although it has also proposed a less binding code of conduct. Here it has had the support of Russia, from where a significant proportion of malicious cyber-attacks emanate. But the United States has not been enthusiastic about this idea, partly because of its concern that the major objective for Russia and China is to legitimise the restraints they place on their own citizens' access to the internet. This has set the scene for significant differences of opinion on political values (a subject which Chapter 12 will address) to make cooperation on this aspect of security more difficult. Paul Meyer (2012, p. 18) puts this gap in perspectives aptly: 'What one state might view as a "disturbance" or "sabotage" in its cyberspace may be seen by another as a legitimate exercise of freedom of expression.'

Yet even the most authoritarian governments face limits in controlling how much their citizens can and cannot see. Because of the centrality of cyberspace to modern commerce, any government interested in a more prosperous future has to reconcile itself with the widespread private use of the internet. To the extent that the malicious manipulation of cyberspace can be seen as a weapon, and because there are vast numbers of internet users in Asia, this creates a fascinating comparison with the situation regarding the availability of military technology. Many if not most states, exert good levels of control within their territories over access to the major tools of armed violence. But it is quite clear that no government could ever expect to have a national monopoly on the use of cyberspace. That means that at the very least the private sector needs to be a partner in any policy that is developed for the purposes of cyber-security. And it also guarantees that while some attacks will come from government sources, the relative democracy of the internet also means that challenges will come from non-government sources.

Low-tech: the problem of small arms

The use of cyberspace as an instrument of potential harm isn't the only such area where a complete monopoly is well beyond the reach of Asia's governments. Non-state groups, whether they are guerrilla movements and armed factions (to be discussed in Chapter 7) or terrorist groups and transnational criminal organisations (to be discussed in Chapter 8) may seldom have ready access to the advanced conventional systems highlighted in an earlier portion of this chapter. And even among states, only a small handful have aircraft carriers and advanced fighter aircraft. But at the other end of the weapons spectrum, there is widespread access in many parts of Asia to small arms and light weapons. These can have significant implications for security in particular Asian countries and can also bring implications for the wider region.

The sources of these weapons, which include the ubiquitous Kalashnikov AK47 assault rifle, are numerous, complicating any attempts to control their spread. Some are remainders from earlier civil conflicts, including independence struggles. Some, such as weapons still in use in some South Pacific countries, e.g. in the highlands of Papua New Guinea, are even older, emanating from stockpiles of weapons used in the Second World War. Others are trafficked across porous borders for use in growing internal conflicts. Some have been raided from government armouries. Others have been supplied by external governments unhappy with the status quo in a particular conflict situation, with uncertain and unintended long-term consequences. For example, some of the weapons supplied by the United States to the mujahideen in Afghanistan during the Soviet invasion were in later years used by the Taliban against US forces. At times weapons have been sold, or leased, by underpaid soldiers to earn income. Preventing this leakage is not easy, especially when, as David Capie (2008) has argued with respect to some governments and armed forces in Southeast Asia, there is a reluctance to allow a more transparent accounting of the flows of arms which have occurred.

These armaments are hardly major factors in the overall distribution of military power between states in Asia. They are not generally useful in calculations of the capacity of one state to deter attack by another. But they are linked to serious problems within a number of these countries. The political and social context is important. The United States has a very high level of private gun ownership, (with perhaps as many as 300 million such weapons) and experiences high mortality levels from gun violence. But many of these fatalities are due to criminal behaviour (including between gangs and involving drugs) which does not threaten the political integrity and cohesion of the country.

By comparison even low numbers of light weapons in some societies, where governments may be lightly armed themselves and lacking in political authority, can be deeply troubling. The seizure of about five hundred rifles from Solomon Island police armouries by militants in 2000 helped change the distribution of power in that Pacific country of less than one million people. As Alpers, Muggah and Twyford (2004, p. 287) explain, some of these weapons 'were used to overthrow the elected government. Quickly diverted to criminals, the guns then enabled armed gangs to embark on a three-year spree of intimidation and violence.' Disarmament of light weapons was an early and urgent priority for the Australian-led stabilisation mission which was deployed to Solomon Islands in 2003 (and about which more will be said in Chapter 7).

Small arms are available in much vaster quantities in some regional countries including Afghanistan, Myanmar and Pakistan. Yet it is important to be wary of sweeping statements about their availability. The Central Asian states, for example, have often been regarded as hotbeds of small arms trafficking. Their proximity to Afghanistan (whose conflicts have been a major encouragement to the regional arms trade) and their former status as republics of the heavily armed Soviet Union, would suggest that small arms are ubiquitous. Their use certainly featured heavily in Tajikistan's civil war, which began in 1992 (shortly after the Soviet Union's collapse). Some years later, MacFarlane and Torjeson (2005) argued that nearby Kyrgyzstan had in fact been relatively unaffected by arms flows from both Tajikistan and Afghanistan and did not appear to have a significant domestic small arms problem (partly a consequence of the relative absence of civil conflict, a record that would be tested by the violence which would later affect Kyrgyzstan in 2010).

Moreover, while the use of small arms is part and parcel of the internal violence which has occurred in parts of Asia and its borderlands including in Afghanistan, Myanmar and Pakistan, they are unlikely to be a direct *cause* of war. As discussed earlier in this chapter, the argument that arms build-ups cause armed conflicts is shaky when applied to the larger and more destructive munitions; it is even less tenable in the case of small arms and light weapons. Yet that does not mean that the proliferation of the latter can be dismissed in security terms, not least because there is often a very low threshold to the use of light weapons (as opposed to more destructive larger armaments), especially in the messy internal conflicts which parts of Asia and its borderlands continue to experience. In a region where major war has become relatively rare, and has claimed few if any lives in Asia for a generation, the use of small arms in smaller wars remains a major contributor to deaths associated with armed conflict, among both combatants and civilians.

This means it is useful to consider *whose* security is at stake here. Local communities which are exposed to intimidation by an armed group may not feel at all secure in the knowledge that there is no threat to the territorial boundaries of the country that they may be living in. They may find it irrelevant that the government in the distant capital city is not in danger of collapsing. But, to adopt the definition of security from Wolfers which was employed in Chapter One, they will encounter and feel a direct threat to the thing they value most: their personal survival. And if this is a widespread phenomenon, it may be difficult to argue that the country in which these citizens live really is secure in a deeply meaningful sense. This is the province of the idea of human security, an approach which will considered in Chapter 7.

A proliferation of drones?

But what about weapons where the human connection is claimed to be absent, and which have been operating in dangerous parts of Asia? In the last several years the United States has been using aerial drones (which do not have a pilot or crew on board) to track down and then kill individuals who are considered to be responsible for terrorist acts in Pakistan in Afghanistan. The issue in Pakistan is a fascinating one. Unlike the case of Afghanistan, where American forces were operating on the ground in large numbers against Al Qaeda and the Taliban in the wake of the 9/11 terrorist attacks, the United States has not been present in nearly the same way in neighbouring Pakistan. At times it is not even clear that the United States has fully informed the government in Islamabad, which is meant to be a security partner, that it is about to launch a drone into Pakistan's airspace and conduct the attack. This has been an issue of some controversy in Pakistan. While the United States may feel more secure because of these efforts to dispatch terrorists, some voices in Pakistan have different views of the real aims and implications of these strikes. At the very least, as Brian Williams (2010) notes, these drone attacks, which began in 2004, have divided Pakistani opinion, in some cases feeding anti-American perceptions and complicating the security relationship with Washington.

The advantage of drones is that they can operate in dangerous areas without the need to put the lives of American personnel at risk. They can be controlled from a great distance from the battlefield: indeed their operators (or remote pilots) can sit at terminals in the continental United States, one further sign of the impact of information technology on modern conflict. Significant debate has been attached to the legality of the use of these systems, including the extent to which the use of a drone across national boundaries constitutes a violation of sovereignty and can amount to an act of war. In response to these concerns, Daniel Byman (2013, p. 34) has argued that 'although a drone strike may violate the local state's sovereignty, it does so to a lesser degree than would putting U.S. boots on the ground or conducting a large-scale air campaign'. Yet this is an argument that the violation could be worse, rather than a claim that it does not exist. Moreover, drones do not necessarily overturn all of the disadvantages of standard military operations. For one thing, while drones are able to deliver lethal force very accurately, there are still occasions in which the casualties from their missile attacks are bigger than anticipated. Secondly, drones can sometimes be used to target individuals who fit fairly broad categories of people thought likely to undertake terrorist acts. Indeed, they may well encourage a "shoot now and ask questions later" approach. And unlike soldiers on the ground, drones simply cannot exert any sort of physical control over a territorial area.

But drones offer enough advantages to be sought after by the militaries of a range of countries internationally, and they will come to play an increasing role in Asia's security as their ownership extends well beyond the United States. Kreps and Zenko (2014, p.72–3) have argued that within a few years, China's spending on drones will be as large as America's, and that 'Australia, Japan, and Singapore have developed unarmed surveillance drones that could be used for more military purposes – some of them in highly volatile regions.' If they chose to do so, the states of Asia would be unlikely to regard drones as being able to work completely in isolation but instead in conjunction with other weapons sytems. Likewise a war in Asia fought entirely by autonomous robots (a step up from drones whose actions are still decided in real time by their controllers) seems a farfetched idea. Rather similarly, the notion of a completely separate and non-kinetic war conducted entirely in cyberspace hardly stands up to scrutiny: armed forces are much more likely to be enhanced than replaced by advances in digital bandwidth.

Conclusion

The human element, the violent element, and in particular the political element, all remain important in all of this analysis. Indeed in a more general sense it can be concluded from this chapter that while advances in military technology have been able to affect and even change Asia's security environment, they will rarely, if ever, determine it. Modern warfare in Asia, whether it remains a distant possibility or a more pressing reality, requires modern weapons. But in every case the factor of political decision-making (including choices about whether to go to war in the first place) never disappears.

Even if the arms competition occurring in Asia continues to intensify, it will not remove all of the scope for governments to make the choice for peace rather than war. But because armaments affect perceptions, and are commonly viewed as a sign of broader intent, they can still help generate perceptions of security and insecurity. Indeed the two can mix in a rather volatile way. Most Asian countries regard growing armaments as recipes for increasing the ability to defend their territory and offshore interests. Increased military expenditure can therefore increase one's own sense of security. But that same expenditure can be viewed quite differently by a potential adversary (or by a country that mistakenly feels that it is being targeted). Indeed weapons systems that are developed for domestic reasons may also have an unanticipated external effect of this nature. The arming of Asia, where each step in the chain seems rational for the country procuring the given system, may produce a collective version of the situation where, in the words of Robert Jervis (1978, p. 169), 'many of the means by which a state tries to increase its security decrease the security of others'.

In other words, what scholars call the security dilemma may be lurking as a significant problem for Asia's security future. That seems to be a fairly good way of explaining some of the interactions between China and Japan, for example, and between India and Pakistan. That is certainly also one of the dangers in the contest between America's air–sea battle doctrine and China's A2AD capabilities. And efforts by one major country in the region to boost their cyber-security capabilities might easily be seen as threatening by another. So perhaps an armed Asia isn't quite so secure after all. Perhaps that old Roman adage should be turned on its head. If the countries of Asia really do wish for peace, it might not always be best to them to be giving the impression that they are preparing for war. Because they might just get what they say they don't really want.

Chapter 6

Rivalry: Will Territorial Competition and Nationalism Ruin Asia's Peace?

As the last two chapters have shown, both economic interdependence and the spread of weapons can certainly affect Asia's security conditions. While the first may offer some encouragement to the avoidance of war, the second may in some instances increase tensions and make war more likely. But neither of these factors is likely to be a final determining factor in the decisions that governments make about whether or not to maintain peace. Those decisions, as the old but still relevant arguments of Clausewitz continue to suggest, are primarily political ones.

This raises the question of what lies behind the political decision-making of governments in the region. If such decision-making is regarded as a cool and calm process, in which national leaders are consistently guided by the long-term interests of their respective countries, then there would seem to be a good chance for Asia and its borderlands to be peaceful and secure. That is, unless those long-term interests diverge so fundamentally that even a minimal level of mutual satisfaction is not possible. Yet, while not inconceivable, such a severe divergence of interests seems unlikely if regional political leaders agree about the basic desirability of a peaceful external environment. From what has been addressed in this book to date, there are reasons to expect that there is an informal agreement along these lines in large parts of the region.

That's the good news. The bad news is that it may not always be easy or straightforward for regional countries to act in accordance with this mutually beneficial logic. This difficulty is especially likely when political decision-making becomes a contested and heated process in which concerns about national reputation and inflated fears about the intentions of other countries in the region make themselves felt. In truth, politics is a mix of cool policy-making and decisions made in the heat of the moment, and Asia's security is shaped by the interaction between different national political systems in the region, each of which brings a special mix of cool-headed logic and emotional energy. As a result, while armed conflict is far from inevitable and is eminently avoidable, there can be no guarantee that Asia's international relations will produce a peaceful region.

Those who study Asia's security need to consider the possibility that emotions will spill over into wider regional relationships, threatening to break Asia's peace. One sign of this problem is apparent in the disputes over territory in important parts of Asia. Few issues seem to get the juices running in the region on a more regular basis. Some are established differences over land boundaries, including the ongoing competition between Pakistan and India over Kashmir. But, as this chapter will show, there are maritime territorial disputes, concentrated especially in East Asia, which are of increasing concern and prominence. These include the Sino-Japanese and Korean-Japanese contests over island groups in the East China Sea, and the competition between multiple claimants over parts of the South China Sea, an important body of water which stretches from the Malacca Strait in Southeast Asia to the Taiwan Strait in North Asia. In this case, the relations between China and Vietnam, and between China and the Philippines, all three of them claimants, have become especially tense.

It is possible to provide a cool-headed explanation for at least some of these disputes. The South China Sea, for example, contains valuable oil and gas reserves which some energy companies are already exploring. Although estimates of the scale and value of these reserves vary considerably, there is no doubting their increasing appeal in an era of rapidly growing energy demand in the region. Valuable fishing grounds are also to be found in these waters, and with some of Asia's fisheries being among the most overworked on the planet, this provides an additional economic incentive to South China Sea competition. But something more seems to be happening than the pursuit of commercial gain. There is a level of tension and even enmity that suggests that much deeper passions are at work. As Buszynski and Sazlan (2007), for example, argue, cooperation on energy exploration in the South China Sea, which could readily be of mutual regional benefit to the various claimants, seems held up by non-negotiable territorial claims by some of the parties.

Resource questions themselves are unlikely to be underlying cause of the problem. Indeed, as the second half of the chapter will indicate, contending nationalist sentiments between neighbours are fomenting some of Asia's most fractious and dangerous relationships. This is not to say that nationalism is necessarily a recipe for disorder in the region. The process of self-determination, while volatile, has been one of the foundation stones of modern Asia as a region of independent sovereign nation-states. Nationalism can be part of the glue that holds many of Asia's countries together, especially if they have histories of violent internal division. But when neighbouring countries see each other's national sentiments (and national histories) as a threat to their own and are unwilling or unable to restrain their own nationalist tendencies for the sake of regional peace, the outcomes for regional security may be especially worrying.

The importance of territory

It may seem rather old-fashioned to emphasise attachments to territory as a major ingredient in Asia's regional security (and insecurity) in this twenty-first century. The organisation of any region on a territorial basis hardly seems suited to an age where information technology has removed many of the traditional geographical obstacles to communication. And, as argued in Chapter 4, the growth of intra-regional trade suggests that a single integrated Asian economy is not too far off. An emphasis on territory would seem to take these processes backwards, and away from the more harmonious world which many have hoped for down the years. John Lennon's famous song, *Imagine*, for example, speaks of such a world where because separate countries no longer exist, a major reason for fighting has been removed. It might be thought that this point should already have been reached.

But territory (and geography more broadly) really does count in Asia, and it does so in the context of both interstate relationships and domestic politics. The reasons for this situation are many. As a general point, the sovereign control of a territorial space, delimited by boundaries, is a fundamental element in the definition of modern statehood. This matters internally and externally. It matters internally because if internal sovereignty is contested, including by a breakaway province whose leaders have declared independence, the integrity of the state in question is at stake. For example, the aim of the Moro Islamic Liberation Front to create a separate state that is independent from the Philippines, involves wresting a defined area of territory away from Manila's recognised authority. Independence for West Papua, which Indonesia is very keen to resist, would have similar political implications. The 26-year-long drive by the Tamil Tigers (LTTE) to create an independent Tamil-majority state in the northern part of Sir Lanka was a bloody and violent affair, of which territorial control was a central aspect. Indeed Neil DeVotta (2009: p. 1023) has noted that 'At its apogee between the mid-1990s and 2006, the LTTE controlled nearly one-quarter of Sri Lanka's territory.' But this control was contested forcefully by the Sri Lankan government, which eventually surrounded and defeated the Tigers with a significant loss of civilian life in 2009: a further sign of the seriousness of sovereign territorial contests in Asia.

Formal authority over territory is also important in external relationships among states. It contributes to one of the fundamental principles of a functioning system of states where each member of that system is expected to respect the sovereignty of the others. The Charter of the United Nations (1945, p. 3) stipulates that 'All Members shall refrain in their international relations from the threat or use of force against the territorial integrity or political independence of any state, or in any other manner inconsistent with the Purposes of the United Nations.' Respect for the territorial boundaries that separate one state from another is among the most fundamental contributors to regional order. Undertaking activities across those territorial boundaries, including the movement of armed forces, without the permission of the state concerned, is a potent sign that this expected standard of international behaviour is not being respected.

Governments normally regard the preservation of the territories for which they are constitutionally responsible as something worth fighting for. This is a general international principle but is an especially lively one in many parts of Asia where memories of territorial intrusions are still fresh. Attempts to seize the territory of another country remains one of the most direct causes of war imaginable. And the protection of sovereign territory remains one of the primary rationales for the raising of armed forces: many governments in Asia have justified some of the military build-ups which were discussed in Chapter 5 on this very basis. Increasingly these forces are developed not only to protect land boundaries, but also to protect the legal rights that states have within the maritime zones extending from their coastlines.

As the countries of Asia are members of such a system of sovereign states, all of them should in theory have a common interest in respecting the concomitant norms of international behaviour. But in Asia there are significant factors that complicate relations over territory. These factors operate in some of the most important parts of the region and involve some of its most powerful countries. The first of these is proximity, which is fundamental to any sense of Asia as a coherent collection of states. As was indicated in Chapter 1, a region is *more* than a group of countries who are located side by side. They must be connected with each other in ways that make sense of the region as a system of interacting parts. Yet by the same token it would be odd to think that a region might consist of countries which are connected (economically, culturally or politically) but which are scattered to the four horizons. Proximity, especially among countries who share land boundaries with one another, immediately creates a sense of interdependence that no other factors can completely erase.

In the most significant example of this phenomenon, no less than fourteen Asian countries share land boundaries with China: Russia, India, Kazakhstan, Mongolia, Pakistan, Myanmar, Afghanistan, Vietnam, Laos, Kyrgyzstan, Nepal, Tajikistan, North Korea, and Bhutan. The assessment that each one of these fourteen countries makes about the security of their immediate neighbourhood depends to at least some degree on the quality of the relationship they have with China. This is true whether or not their boundaries are agreed and settled. In fact, as M. Taylor Fravel (2008) shows us, for the most part China has settled its land boundaries with the exception of some of those it shares with Bhutan and India.

The India–China border, which totals more than 3000 kilometres, features two areas of contestation which the two countries briefly fought over in 1962. One area is claimed by India as part of its state of Arunachal Pradesh. This claim is based on an agreement between British India and Tibet, but Tibet has since been absorbed by China, which refuses to acknowledge that earlier agreement. Further to the west is the area of Ladakh which India claims as an extension of the state of Jammu and Kashmir, a point of view that China also contests. Of course since the Kashmir portion of this Indian state is itself contested with Pakistan, which just so happens to be an ally of China, further complications are almost inevitable. Indeed it is perhaps unsurprising that India refuses to accept the exchange of territory between China and Pakistan which formed the basis of their own border agreement in 1963.

These particular territorial disagreements are problematic. China and India have been unable to agree on the delineation and demilitarisation of a good deal of their long border. While strategic rivals, neither China nor India has an interest in the eruption of a serious bilateral border conflict. Even so, as Jonathan Holsag (2009, p. 824) explains:

> The maintenance of the balance of power in the border area remains prominent in both countries' strategising and is still nourished by frightening reports about small-scale but provocative troop deployments and the construction of new transportation arteries that facilitate swift mobilisation.

The complex array of intersecting boundary claims in this part of continental Asia is not easy for any of the parties concerned. At one and the same time, for example, India can be exchanging fire with Pakistan over their Line of Control in Kashmir and disputing with China the presence of PLA (People's Liberation Army) soldiers in Ladakh. As the connections between North Asia and South Asia become more intense (associated with the economic growth that both China and India have been experiencing), more challenges might well be expected.

Elsewhere on the Asian mainland, there are other territorial differences between neighbours which have led to at least small-scale military incidents. Thailand and Cambodia, for example, have held different interpretations of the status of an eleventh-century temple. As a venerable article by L.P Singh (1962, p.23) indicates, a French map from 1907 which includes the temple in Cambodian territory seems to contradict a marginally earlier and quite different Thai understanding. Just over a century later, several Thai soldiers, who were accused by Cambodia of violations of sovereignty, lost their lives in a small but troubling military exchange between the two neighbours. This was a significant development given the commitment of ASEAN's members to peaceful relations among themselves, Yet at least in terms of the publicity given to territorial claims in Asia, land boundary issues have tended to take a back seat. Maritime territorial disputes have instead been dominating many of the regional security headlines in recent years.

At one level this is something of a puzzle. The United Nations Convention on the Law of the Sea (UNCLOS), which has been in operation for over three decades since coming into force in 1994, recognises maritime interests in a formal fashion. States with coastlines have territorial rights for the first fourteen nautical miles out into the sea, and then special resource rights over an Exclusive Economic Zone (EEZ), which extends out to 200 nautical miles. This has been of considerable benefit to many countries in the wider region. New Zealand, for example, enjoys one of the world's largest EEZs, despite its small land area and population of little more than four million people. It is also responsible for the maritime zones of three Pacific Island groups, the Cook Islands, Niue and Tokelau. But New Zealand's advantage is its lack of proximity to main countries and trouble spots in the region. Australia, its close neighbour and friend, is still 1500 kilometres away. The north of Australia, by comparison, is as little as one tenth of that distance from Papua New Guinea, its former colony, with whom it came to an agreement on maritime boundaries in 1978. And it is similarly close to East Timor which upon becoming independent little more than a decade ago, returned to an old disagreement with Australia about the division of a lucrative offshore oil and gas deposit which lies between them.

But it is further north still where things start to get really interesting. The crowded geography in and around Southeast Asia, including parts of the South China Sea, which are depicted in Map 2.1, makes it physically impossible for countries in that part of the world to enjoy New Zealand's luxurious position. Here EEZs overlap readily. The internationally recognised principle of equidistance (where two countries essentially agree divide their claims at the half-way point between them) provides the most obvious answer to this common problem. But in practice the claimants to parts of the South China Sea (which include China, Taiwan, Malaysia, Vietnam, the Philippines, and Brunei) often cannot agree on quite who has sovereignty over small islands and other features. In other words, claims about maritime rights are in many ways extensions of disputed claims about rights on land, in which case even an agreement on how UNCLOS operates would not be sufficient to bring order to the situation if fundamental questions about the sovereignty of a particular feature remains contested. And even then, ambiguity would remain. At the time when the Convention was coming into force, Jonathan Charney (1995, p. 725) warned that 'The articles that directly address the delimitation of maritime boundaries are general and indeterminate.'

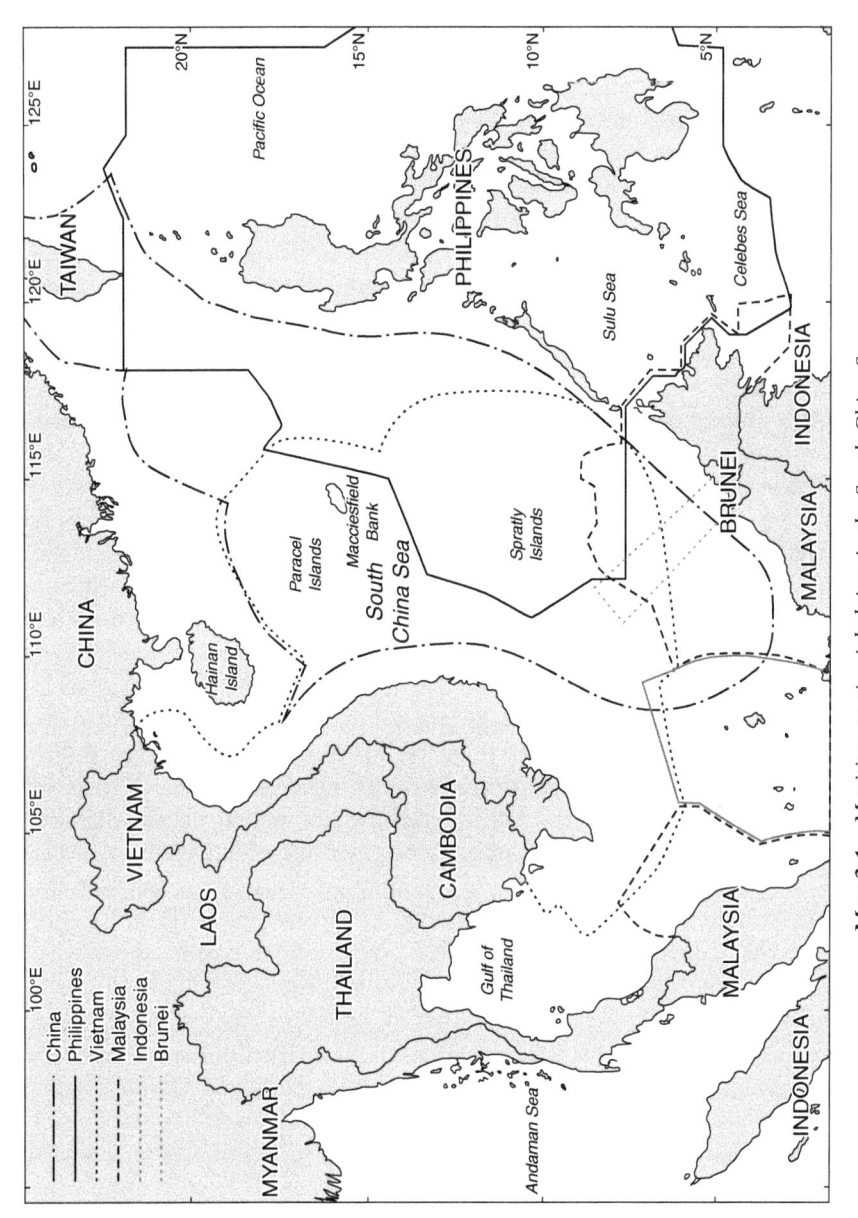

Map 2.1 *Maritime territorial claims in the South China Sea*

That situation is even trickier because some of the features in the South China Sea are barely more than rocks and offer a legally dubious basis for the drawing of maritime boundaries around them. But this is no impediment to such claims being made. And the most famous and expansive of these claims comes from China, the country whose military capacity to enforce its claims is growing faster than any other country in the region, and whose confidence to do so is also rising. This is the famous 'u-shaped' or 'nine-dashed line' claim which encompasses the vast majority of the South China Sea, including islands and other features that Vietnam and the Philippines regard as their own territories (and which on maps can look closer to them than they do to the mainland of China). Vietnam's disagreement extends to the naming of the sea itself, using the term 'East Sea' not 'South China Sea' and in the Philippines the term 'West Philippine Sea' has become more popular.

As to enforcing this claim, here China relies heavily on what others think it is willing to do –a reminder that security is as much about subjective perceptions as it is about objective reality. Lyle Goldstein (2011, pp. 323–4) rightly argues that one should not be swayed simply by what Chinese leaders say, because military 'actions (or investments) always speak louder than words', but nonetheless he concludes that in this maritime territorial context 'Beijing's build-up in the South China Sea region is significant'.

A further complication is that there are differences of view on what states who transit through these areas are obliged and free to do. China has aken the view that it is the sovereign power in the South China Sea, and as such its permission is needed for the transit of military vessels through the EEZ it claims, but which others dispute. This runs contrary to the principles of freedom of navigation under UNCLOS as they are understood by most of its signatories. But the United States, which insists most strongly on these principles, has not ratified the Convention due to a view in Senate that this would disadvantage American commercial and security interests. And while the United States government argues that it does not take a position on any of the boundary disputes, its insistence on freedom of navigation is interpreted by Beijing as a taking of sides. Moreover, China resents the USA's use of that freedom to conduct naval surveillance activities close to the Chinese mainland, including around Hainan Island where much of China's growing submarine fleet is housed. Even among analysts like Erik Franckx (2011) who assert that the two major powers might be able to find some way of settling these differences, there is a recognition that these gaps in interpretation are significant and potentially hazardous.

Historical complexities

Quite how settlement would arise is unclear. For example, China has been reluctant to seek international legal redress to the ongoing differences among the claimants over the South China Sea, and has rejected the move by the Philippines to seek such redress on the Scarborough Shoal, a groups of reefs lying about 200 kilometres from the Philippines. Legal redress should Beijing accept it, might well penalise China. This may be one reason why the Chinese government retains a degree of ambiguity about what its claims actually mean. This includes China's use of maps issued in the pre-communist era which include the famous nine-dashed line, and which Taiwan also supports. Exactly what is meant by these dashes is unclear in a legal sense, and as Zou Keyuan (2012) has noted, China has increased its efforts to enforce this u-shaped line without removing the ambiguity about what the line represents. In any case, as one might expect, a number of the other countries with claims to the South China Sea are prepared to come up with maps and records which are more sympathetic to their own claims.

Here the analysis has moved from proximity to history, a second factor which has in many cases also not been kind to the territorial element of Asia's security. Further north, for example, Japan and Russia continue to dispute the ownership of four of the Kuril islands which lie between Hokkaido (in northern Japan) and the southern tip of Russia's Kamchatka peninsula. Russia continues to claim that the Soviet Union, which seized the islands at the end of the Second World War, was recognised internationally as the rightful owner. Japan continues to dispute this assessment of the historical record. Similarly, Japan's claim to the Senkaku islands in the East China Sea is contested by China's claim to the same group, which it refers to as Daioyu. Again the historical factor intrudes. These islands were occupied by Japan in 1895, a period when its imperial expansion extended to the Korean peninsula and when it had defeated China in the first Sino-Japanese war. Japan's claim, which it argues was confirmed in the peace settlement following the Second World War, is increasingly being contested by China. Likewise, South Korea's administration and control of the Dokdo (in Korean) or Takeshima (in Japanese) islands is contested by Japan who had incorporated them as its power was growing in the first decade of the twentieth century. The overlapping nature of a number of these East China Sea and North Asian claims is depicted in Map 2.2.

These contentions are by no means restricted to the Asia's major powers. Despite their claims to amity and cooperation, (enshrined in a 1976 treaty by that name) several of the members of ASEAN are also involved in disagreements over the territorial boundaries between them. The Philippines and Malaysia, for example, have divergent views on what

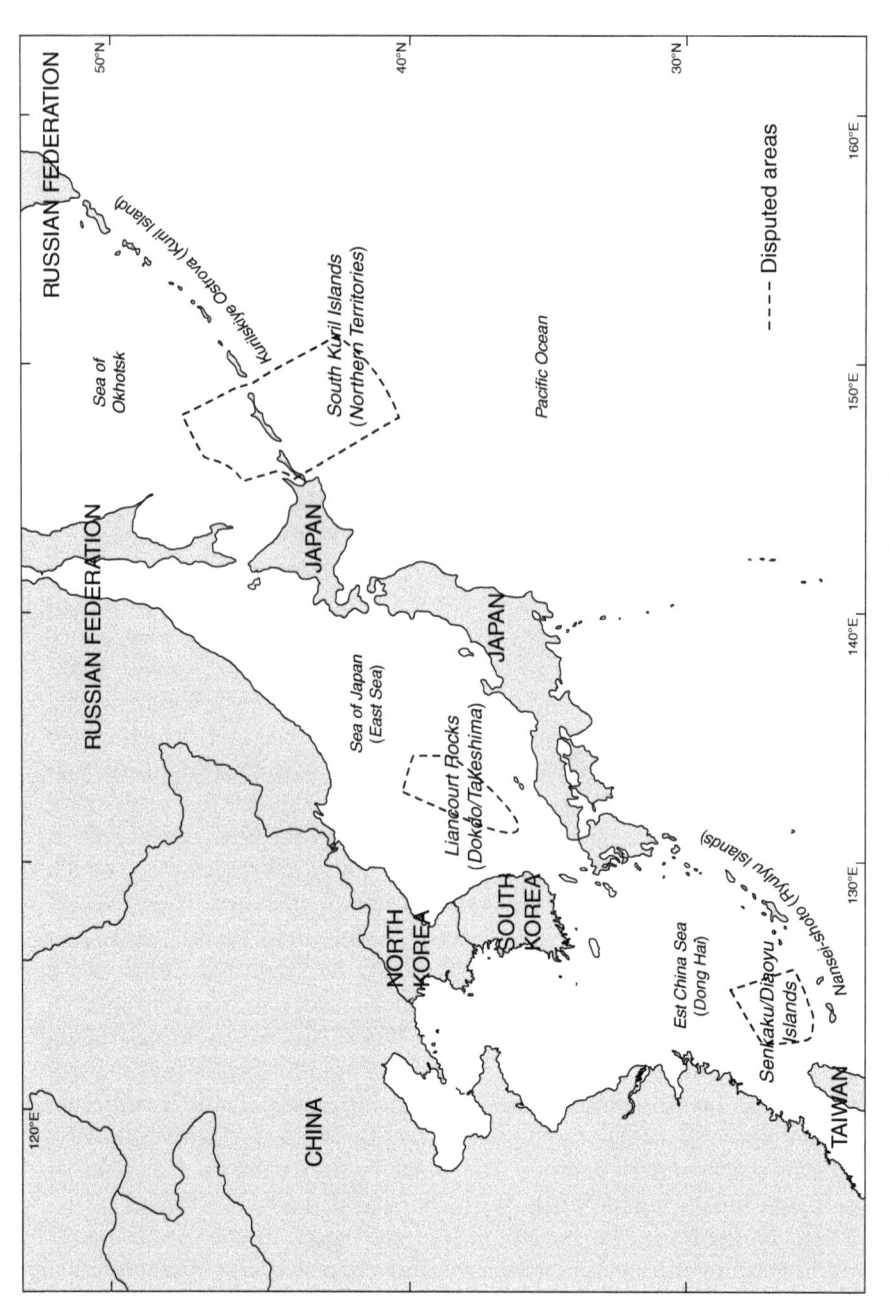

Map 2.2 *Maritime territorial disputes in North Asia*

a 1640 treaty between Spain and two local sultans says about Sabah, which is currently part of eastern Malaysia. Some years ago, Samad and Abu Bakar (1992, p. 567) argued that the 'Malaysian government considers the Sabah dispute a creation of the Philippine side', not unlike Japan's views of China's claims on Senkaku/Diaoyu or Korea's views of Japan's claim to Dokdo/Takeshima.

But disagreements about whether a dispute really exists cannot wish away the appearance of pressing day-to-day problems. Armed groups have from time to time crossed over from nearby Philippines islands to Sabah, creating significant difficulties for the bilateral relationship with Malaysia, whose prime minister has called for a hot-line to be instituted between the two countries to make sure that misunderstandings do not lead to an unwanted conflict.

The complexity of the historical record is partly due to the territorial confusion and contestation ensuing from the era of European colonial control. This is clear for example, in the difficulties experienced by India, Pakistan and China discussed earlier in this chapter. In other parts of the world, and most remarkably in Africa, the European great powers in their competition for territorial possession and economic gain often resorted to the crudeness of straight lines to demarcate one possession from another. As these divisions did not correspond to the often informal boundaries between social groups, future disputation and conflict was almost inevitable. When the influence of the European colonial powers waned, and in some cases collapsed after the Second World War, the effect of such arbitrary divisions became apparent, no more so than in the division of British India (which was discussed in Chapter 2). Despite their very evident appearance as neighbouring sovereign states after their 1947 partition, India and Pakistan have been unable to agree upon the sovereignty of Kashmir, caught up as it has been, in the words of Simon Jones (2008, p. 4) in 'different visions of national identity'. What exists there today is not a formally agreed international boundary but a customary line of control which continues to elude a genuine settlement.

If another serious armed conflict between India and Pakistan were to occur, it would quite likely begin as an escalation from the frequent exchanges of fire which occur across this line of control. India has argued that Pakistan's intelligence service supports Islamic insurgents who move across the line of control, and that regular army personnel from Pakistan are also involved. The frequency of these disturbances is affected by the extent to which Pakistan is preoccupied by events in the areas close to another porous border which was agreed in the 1890s by a representative of British India, Mortimer Durand, and the Afghan ruler of the time. While groups of common ethnicity straddle the Durand Line, it is widely regarded as the boundary between Afghanistan and Pakistan. But as a sign of the difficult relationship between these two neighbours, the government in Kabul has refused to recognise it. The flow of people, arms, drugs and money across this indifferently monitored mountainous area have proven an ongoing complication for international efforts to bring security to Afghanistan, not least because some of the Taliban and other groups have exploited this ease of movement from the other side of the line. This mobility is nothing new. Amin Saikal (2006, pp. 137–8) has observed that:

> While successive Pakistani and Afghan governments may have controlled the main crossing points, such as at Torkham, Chaman, Parachinar and Quetta, the secondary crossing points have by and large remained within the sphere of tribal and nomadic influences. No one – not the British, their Pakistani successors, Afghanistan nor the Soviet invaders – has ever established full control over this border.

But an uncontrolled border need not be a contentious one and some of the boundary disputes that do exist in Asia can recede from view. Countries may have claims to territory that they have long wished to recover but simply lack any capability to enforce their claims. China's rising strength may be one explanation for what has been widely viewed as its more assertive approach in the South China Sea. Beijing's strong attitude to matters of its territorial integrity, which it defines as stretching from Tibet in the west to Taiwan in the east, may strike some as revisionist. It would seem that if there is a territorial status quo in Asia, then China has perhaps been the keenest of all countries to change it. But the return of Hong Kong from Britain in 1997 was the restoration of a part of Chinese territory that had been ceded when China was at a particular point of weakness in 1842. In the decades that followed a series of so-called unequal treaties carved up parts of coastal China into European possessions. This was part of China's century of humiliation, which as Scott (2008) has argued, carved lasting imprints into the country's national identity. In particular, this experience left a strong determination for reunification that is noticeable in Beijing's attitude to territorial matters today.

In this connection, the main unfinished business for China is Taiwan, to which the defeated nationalist army of Chiang Kai-Shek fled in 1949 as Mao's revolution gained control of the mainland for the Communist Party. Initially recognised by many countries (including the United States) as the sole government for the whole of China, the Taiwan-based regime even used to hold the permanent seat allocated to China in the United Nations Security Council. But as the United States and many others recognised communist China in the 1970s, that situation was reversed. Taiwan is now only regarded as a sovereign state by a handful of (mainly small Caribbean, African and Pacific) countries. Beijing continues to insist that Taiwan is part of one China and that any move towards Taiwanese independence will be resisted strenuously. Under an Act of Congress, the United States has remained committed to assisting Taiwan with its defence. Partly because of better relations across the Taiwan Strait between the Chinese and Taiwanese governments, strains between the United States and China over Taiwan are not as high as they once were. But this issue could still create a collision course.

So too could the fact that the United States–Japan Mutual Security Treaty (which will be considered in further detail in Chapter 11) is seen to cover island territories that Japan contests with China. In other words, if Japan and China get themselves into a violent stand-off, will the United States feel obliged to come to the help of the former, its number one ally in Asia? This is not an entirely hypothetical question because in the last few years, Japanese and Chinese vessels (including large auxiliary ships but also naval vessels) have been shadowing each other in close proximity to the Senkaku/Daioyu Islands. These two neighbours lack an effective mechanism for exchanging information in the event of an accidental clash, and a deliberate attack might well result in a more serious conflict. And their expectations of what the United States might or might not do is not necessarily a recipe for confidence either. Sheila Smith (2013, p. 3) has argued that just as 'Beijing could miscalculate Washington's commitment to defend Japan and/or seek to test that commitment' it is also entirely possible that 'U.S. assurances could lead Tokyo to overestimate Washington's response and to act in a manner that would increase the chance for confrontation'.

The jostling looks set to continue, posing demands on all sides to be accurate in their perceptions of one another's moves should a serious crisis emerge. A reminder of that delicacy came in 2013 when China declared large parts of the East China Sea to comprise an Air Defence Identification Zone (ADIZ). China expects to be notified before commercial or military aircraft fly through this zone. This is not a novel approach, and Japan and Korea have such zones themselves. But China announced its ADIZ without consultation and at a very tempestuous moment in its relations with Japan. Its zone covers islands that both Japan and South Korea claim as their own. In a calculated risk, the United States flew bomber aircraft through China's ADIZ soon after it was announced. While this did not precipitate a military countermove on China's part, the potential for serious escalation appears to have grown.

While neither Japan nor South Korea have themselves been willing or able to resolve their own bilateral territorial dispute, they both would have welcomed America's concern about China's new zone. But it is unwise for America's regional allies to expect guaranteed support from their big ally should they get into difficulty over territorial issues. This applies in particular to any hope on the part of the Philippines that in its dispute with

China over South China Sea territorial claims, the United States will be willing to take major risks to support its ally. Indeed Washington has indicated that it does not take sides in these disputes, and even if China does not entirely believe this position to be credible, it does allow the United States some space to manoeuvre in the event a major crisis comes.

But as noted in the previous chapter, the Philippines armed forces are noticeably weak in terms of their ability to project force. This sets up a dangerous gap between Manila's claims and its ability to enforce them. That gap was in evidence in 2012 in a stand-off between the Philippines and China over the Scarborough Shoal, a set of features about 220 kilometres east of Luzon, the largest Philippines island. Having sent an ageing naval vessel to monitor several Chinese fishing boats that had appeared in the Shoal's proximity, the Philippines found itself quickly outclassed. Indeed that fishing fleet alone is not to be dismissed as a security actor in its own right: Dupont and Baker (2014, p. 87) assert that there is 'strong circumstantial evidence that Beijing is deliberately using the fleet to test the resolve of other claimants and demonstrate the reach of China's maritime power'. In this particular episode, a growing group of Chinese maritime vessels quickly gained the upper hand in terms of pressure. And while the Philippines withdrew from the area under a deal encouraged by the United States, China has since stayed on.

Most apparent in this crisis was the evident unwillingness of the United States to become involved militarily, a decision that seems wise from the perspective of avoiding a more serious escalation of hostility with China. But with Chinese vessels increasingly present in contested areas, and with the imbalance growing in China's favour, it is growing harder for the United States to demonstrate to its closely observing other allies that its naval predominance can forestall such incidents. The United States protests what it refers to as China's coercion in these maritime disputes, and conducts exercises with its regional allies to demonstrate its support for them. But it remains limited, for obvious reasons, in the extent to which it will allow itself to be dragged into a conflict. Bonnie Glaser (2012, p. 2) depicts Washington's dilemma very precisely in the event of a serious China–Philippines conflict:

> Failure to respond would not only set back U.S. relations with the Philippines but would also potentially undermine U.S. credibility in the region with its allies and partners more broadly. A U.S. decision to dispatch naval ships to the area, however, would risk a U.S.–China naval confrontation.

In today's Asia, Beijing and Washington have a common interest in not getting drawn into the same contest. But several decades ago this is precisely what happened on the Korean peninsula, which was divided at the end of the Second World War as Soviet troops moved half-way down to control what is now North Korea. Ever since the Korean War the peninsula has remained divided in a territorial sense by a very obvious boundary: the famous Demilitarized Zone where the forces from the two Koreas eye one another off. But this is nothing like a settled border. It is the result of a truce between two antagonists who remain formally at war with one another. And a good deal of China's interest in supporting North Korea also has a territorial dimension. North Korea does not need to be especially friendly with China to still do the latter something of a favour. A North Korea which keeps the United States and its allies at bay, and which prevents the presence of US forces on one of China's borders, provides China a sense of security in one of its most vulnerable external points. You Ji (2001, p. 387) observes the situation crisply when he writes: 'The core of the strategic value of North Korea to China is that it provides a crucial buffer keeping US military presence away from the Chinese border.' This also means, however, that North Korea's leaders know there are limits to how much China is willing to punish it, hardly an incentive to responsible behaviour on Pyongyang's part.

Lighting the fire: nationalism

Even in an era of increasing globalisation, the idea that geography is dead would be a shock to most of the countries of Asia. Yet this picture is not a completely consistent one across the entire region. From the analysis already undertaken in this chapter it should be clear that some territorial questions in Asia have greater significance than others. A number of these challenges could have especially momentous implications for the region's future security environment. The unification of the Korean peninsula, for example, would be likely to change the overall political calculus in the region. As Derek Mitchell (2002, p. 123) has argued, such a development might well 'generate more variables and uncertainties than any other contingency in East Asian affairs'. If that unified Korea was also an American ally, this would definitely increase China's sense of vulnerability for reasons noted earlier. If, on the other hand a united Korea decided that in the interests of reasonable relations with its largest neighbour it needed to distance itself from Washington, the United States would have lost a valuable connection and ally in North Asia. America would then have few if any armed forces left on the Asian mainland, increasing the importance of Japan but also reducing Washington's options and increasing China's freedom from US pressure.

For this reason North Korea is not just a closed-off and mysterious country with few links to the outside world except for the nuclear threats it makes from time to time. It is an important part of Asia's territorial arrangements. And it is not just the ongoing support it has had from China that keeps the regime in Pyongyang going. North Korea's leaders have fashioned for themselves a system of control, through the armed forces and the intelligence services, which has been remarkably effective. The three generations of what has become the Kim dynasty have fashioned an ideology of self-reliance (called *juche*) which holds that North Korea (and especially its supreme leader) is capable of breathtaking accomplishments without any outside assistance. The national myth-making that this involves is unusual in its extent, and carries with it some unbelievable claims about North Korea's innocence and the nastiness of its opponents (including the South Koreans and Americans who are routinely vilified). This is an extreme form of a tradition that is not at all uncommon in Asia or in other parts of the world. This is the tradition commonly known as nationalism which can have a profound effect on regional security.

Nationalism is an ideology: a system of ideas which promotes a particular form of political or social organisation. This specific ideology is reflected initially in the quest for sovereign statehood on the part of a particular national group. And it is reflected eventually in the projection of power by the resulting *nation-state*. This hyphenated term is itself a sign of the complex mix of national identity and political organisation which modern countries as international actors represent. But very little of this process, which relies heavily on the development of feelings of national identity, is pre-ordained or natural. As Benedict Anderson (1983) has argued, a nation is an 'imagined community': a group of people who believe that they belong together and should be politically represented as such.

In a region with a history of peoples and groups who have felt downtrodden and hard-done by, either by external colonialists or by acquisitive neighbours, and in some cases both, nationalism has found some very suitable growing conditions in Asia. Care should be taken in judging all of this as unnecessary, ugly or hazardous. Without strong movements towards national self-determination, the independence movements which transformed Asia after the Second World War would not have been possible. Nationalism thus can be a form of resistance to the hegemonic designs of empires which much of the modern world has come to reject. The conscious encouragement of a sense of common national identity was vital to the emergence of the nation-state of Indonesia, which is in reality a collection of islands and diverse peoples. As Christine Drake (1989, p. 6) has rightly observed: 'National integration is a matter of particular concern in Indonesia because of the great diversity of both its geographical environments and its peoples.' This is not to suggest that all that has happened in Indonesia's relatively short history of independence has been favourable. Indonesian nationalism has been a challenge at times for the country's immediate neighbours, including in the Sukarno years when an emerging Malaysia and Singapore were threatened. But very few of Indonesia's neighbours would prefer disintegration as an option. In that connection, the very small amount of fragmentation which has gone on in Indonesia in recent years, resulting in the establishment of East Timor as an independent and sovereign state, was a reflection of rival nationalisms (both Indonesian and Timorese).

Because nationalism depends upon perceptions of common identity, often abstracted from more complex situations, nationalist causes often involve the selective interpretation of historical events. That this can be divisive and even dangerous is very evident in several parts of Asia, and especially in North Asia, where the nationalism of close neighbours can easily become interdependent and escalatory. In their vocal criticisms of Japan, Chinese nationalists often recount the undeniable acts of brutality committed by Japanese forces after the invasion of Northeast China in the early 1930s. Incidents from the late nineteenth century are also recalled in China. That Japanese textbooks often overlook or downplay the aggression against China is a source of particular resentment. Similar sentiments make themselves apparent in Korean memories of Japan's colonial and wartime conduct, and in concern that Japan's political leaders are not sufficiently repentant about the events of this tumultuous period. But at the same time it is not clear that Japan can ever fully satisfy these criticisms, which are sometimes raised in an attempt to keep Japan in a post-war time warp. There are also versions of contemporary Chinese nationalism which suggest that it is inherently impossible for China to be anything but a defensive and peace-loving country. But is not clear that everyone in the region, including in countries with different perceptions of their national histories, agrees with this pacific interpretation.

These rival nationalisms can be useful for governments in their quest for ongoing domestic legitimacy. This seems especially so in China, where the ruling party is now bereft of a properly believable communist ideology, having embraced a guided form of capitalism some decades ago. In its place, nationalism can become an increasingly important source of unity. But China's popular nationalism, which has been dissected by Peter Hays Gries (2003), is a dangerous tiger for the Communist Party leaders to ride. Once encouraged it can quickly escape the control of even the most single-minded government that Asia has to offer. Despite the existence of a single political party which impinges on nearly every aspect of life in China, the activities of large numbers of private citizens who espouse nationalist causes openly on the internet is both an asset and a liability to the leaders in Beijing. It is an asset when they wish to put pressure on Japan, but a liability when the scores of 'netizens' inflame the situation far beyond that which is tactically useful. That there are rival nationalist bloggers in Japan and elsewhere, each reacting to the lines of the others, is a recipe for some considerable concern.

Internet nationalists are often very active during territorial disputes, and it might be wondered if they push their respective governments into firmer and more unrelenting positions than might otherwise have been the case. The widespread use of social media by individual citizens and political groups to organise and communicate can compress the time between action and reaction, potentially complicating the lives of decision-makers who sometimes come under pressure to respond before an initial provocation can be confirmed. Nationalism in this context is not an aid to the cool-headed decision-making about the use of force which the stability of international relationships in Asia requires. It could lead to what is called crisis instability, where as Schelling (1960) explained some decades ago, the temptation to fire first and ask questions later can reach irresistible levels.

Moreover, because nationalism often feeds off images of the potential adversary as the root of all a country's problems it can become very challenging for leaders who may prefer to walk a moderate path. Each time, for example, the leaders of Pakistan and India are about to meet for peace talks, it is commonplace that shots will be fired across the line of control to disturb that possibility, and that conflicting interpretations of this development will immediately be reported from both sides of the line. Even especially imaginative attempts to break the deadlock, including a 1999 visit by the Indian prime minister to meet his Pakistani counterpart in Lahore, have come adrift because of subsequent violence, as the analysis of Nicholas Wheeler (2010) has demonstrated. In such a tightly interlinked bilateral dispute, those opposed to ceasefires and other peace moves often know that nationalist sentiment on the other side will demand a strong response to any initial violent act, raising the possibility that talks will be scuttled. In other words, it is quite plausible for spoilers of peace processes on one side of a dispute to exploit the combustibility of nationalism on the other side.

It might be argued that the most inflammatory differences between South Asia's two main countries do not stem from rival nationalisms but from religious differences. It is certainly true that the enmity in this fractious relationship is partly based on the violence which occurred during partition, as millions of Muslims and Hindus crossed over in different directions into Pakistan and India respectively. But the Muslim community in India is still one of the world's largest, complicating notions of simple religious differences between the two largest South Asian states. And while Pakistan and Afghanistan are neighbouring Muslim majority states, the difficulties in the relations across the Afghan-Pakistan border are at times even more troubling than those which occur between Pakistan and India.

Islam has at times acted as an integrating factor in regional affairs, just as the spread of Buddhism from India developed connections from South Asia to parts of Southeast Asia. But in the mid-1960s, Southeast Asia's largest Muslim neighbours, Indonesia and Malaysia, were still capable of considerable mistrust and conflict in their relations with each other. In very few places in Asia do there seem to be clear lines of interstate religious difference that would explain occasional security difficulties. Few seem to generate the civilizational cleavages famously suggested by Samuel Huntington (1996). The rivalry of competitive nationalisms seems to offer a better explanation for such cleavages, including on the occasions when multiple national groupings with their various claims for self-determination are present within the same political boundaries. Yet nationalism is not an automatic process which generates and sustains disputes without human agency: it is when nationalism is manipulated for divisive political purposes that things get hazardous.

No real solutions?

Especially when territorial differences are exacerbated by rival nationalist sentiments, it is difficult to see how the compromises can be made which would allow for the peaceful settlement of contested claims. Rival nationalisms seem to encourage a zero-sum game mentality where any gains for one group automatically become equivalent losses for the other. A compromise, where each settles for a share of what is at stake but relinquishes any claim for all of it, would appear to run contrary to this logic. Speaking of Chinese public sentiment as the East China Sea territorial dispute with Japan was gaining a higher profile, Yinan He (2007, p. 21) argued that it had become 'unimaginable that the people would approve Japan to take a share in the resources that they believe should belong to China only'.

But the Asian region is not as zero-sum as these perceptions might suggest, and there are more shared interests between rival claimants than is sometimes thought. For one thing, there is an element of theatre in the competitive nationalism that accompanies tensions over territorial disputes. China and Japan both send coastguard and naval vessels into close proximity of disputed islands, but these gestures do not necessarily imply hostile intent. It is also not unknown for fishing vessels to be detained on the grounds that sovereign claims have been violated. But while tensions almost always rise on these occasions, they have also had a habit of subsiding, even if the events themselves are rarely forgotten. There is a fair amount of posturing and point-scoring involved, which does not indicate a clear intention on either side to break the peace.

Of course it is easy to make this observation in retrospect. Things can appear very dangerous as a situation develops. In 2014, for example, Vietnam and China were involved in what had the makings of a serious crisis when a Chinese oil rig was moved close to a disputed area in the South China Sea which had witnessed a minor naval conflict between the two communist neighbours in 1974. Vessels were rammed and demonstrations occurred in Vietnam against Chinese businesses, which in fact did considerable damage to Taiwanese commercial interests. But unlike forty years beforehand, these events did not result in the violent use of force between China and Vietnam. And even when shots are exchanged in contemporary Asia, hell is not destined to break loose. India and Pakistan have a habit of exchanging fire and mobilising their respective forces behind the line of control, but governments on both sides seem able to choreograph their crises up to a tolerable level of tension before reining them in. The manipulation of these crises becomes a form of influence without war, and rising tensions are not a prologue to catastrophe.

There is no guarantee that such control will always be possible. As both India and Pakistan possess nuclear weapons, the prospects of escalation are especially troubling in any conflict between them. Japan and China may find that in the event of a minor use of force between them in a territorial dispute, there may be vicious incentives to increase rather than to dampen the scale of the conflict, as this author and a colleague have argued (Ayson and Ball, 2014). But Asia's territorial competitors still have strong reasons not to let their disputes go violent in the first place. If such a common interest exists, it might be wondered whether claimants who can agree informally that they have a shared interest in the avoidance of war might be able to agree to share the territory that divides them, or at least to subject themselves to a common authority such as the International Court of Justice that can make that decision for them. In the case of the South China Sea, for example, there have been continuous calls for the claimants to sit down and resolve their differences.

Some hesitant steps in that direction would appear to have been taken. China and the ASEAN claimants have long been talking about an agreement on a code of conduct to establish rules of the game for their activities in these contested areas. That Code would support, for example, freedom of navigation in the South China Sea and prohibit moves to develop settlements on disputed islands as de facto measure to assert sovereignty. Above all the Code would expect self-restraint from all signatories. But even if such a Code could be agreed, it would be a non-binding agreement, and it would in no way mean that the various claims in the South China Sea had been settled. The track record is not altogether encouraging: in the mid-1990s China erected structures on Mischief Reef, a set of outcrops in the Spratly Island group which are also claimed by the Philippines. The two sides did sign a bilateral code of conduct at that time to defuse tensions, but as Ian Storey (1999, p. 98) has recorded, this did not prevent the erection of additional structures.

The fact that a wider South China Sea Code has been hard to come by reflects at least two facts. First, as by far the most powerful claimant (with the biggest claim), China is in no hurry to circumscribe its options. Any sort of early resolution would not necessarily favour China when the passage of time only makes its relative position stronger in terms of the distribution of regional power. Indeed, according to M. Taylor Fravel (2011), China has long employed a delaying strategy while it consolidates its approach and builds its capabilities. Second, even though an early resolution would suit at least some of the Southeast Asian claimants, they have not themselves been able to put up a genuinely united ASEAN front. China has been able to exploit the differences between the claimants themselves and between the claimant and non-claimant members of ASEAN, watering down the pressure it faces.

This is not an unusual situation. Quite separate from their relations with China over the South China Sea, many of the maritime Southeast Asian countries have continued to live with one another despite the unresolved territorial differences between them, some of which have already been mentioned in this chapter. The tendency to sweep these under the carpet in the name of ASEAN amity and cooperation is a long-standing tradition. And it cannot be denied that this approach has been reasonably effective. Living with ambiguity is preferable to dying with clarity, if these are seen as the main options. The notion that diplomacy is about problem solving appears to some regional countries as a western idea that does not belong in Asia. As will become evident in Chapter 10, while there is a degree of solidarity in the multilateral groupings that ASEAN has inspired, the attachment to ideas of consensus and non-interference does not make for a culture of decisive or definitive outcomes. But the region's experience of costly conflicts, caused when one side or another has imposed itself in the search for a decision, may suggest that it is often wiser to be cautious about attempts to reach final and binding settlements.

Conclusion

Asia has not experienced major conflict for many years, and yet many significant territorial differences between regional states remain in play. It is therefore tempting to conclude that these disputes are not really central to Asia's security but are mere sideshows to the real issues. It is certainly true that many of these disputes will rumble on, and that their non-resolution is not necessarily a threat to regional peace in an overall sense. Some of the issues are clearly localised and lack a track record of causing severe ripple effects outwards. It might even be argued, for example, that the India-Pakistan contest over Kashmir, serious though it can be, is really quarantined within a particular part of South Asia despite having the potential to precipitate a truly catastrophic conflict between two of Asia's nuclear-armed states. The boundary disagreement between Thailand and Cambodia is a problem for the ideas of peaceful cooperation which the ASEAN countries support on paper. But it is hardly an issue that sends shivers of concern throughout Asia and its borderlands.

But at least some of these disputes, and especially those over maritime territory in East Asia, are moving in the wrong direction, with worrying wider implications. A few years ago Ralf Emmers (2010, p. 8) argued that the complex triangular interplay of concerns about territory, energy and power could easily bring a 'dangerous and rapid escalation' of the region's main maritime disputes. Signs that this escalation is becoming more likely have been appearing since his book was published. It is somewhat unfair to point to one country as responsible for these developments, but China seems to be a common factor in many of them. This is not necessarily because China has become more assertive about its long-standing claims (which many argue that it has), or that China alone has been unwilling to compromise. It is because China's growing size gives it a major advantage in any pushing and shoving contest. And as China has grown in power, its leaders have had more opportunities to reverse a series of humiliations (imposed in the past by the European colonial powers and also by Japan). This provides an outlet for nationalists in China, but also causes the sort of concern that nationalists elsewhere can seize on. In short the changing distribution of power in Asia, in which China's rise is the central factor, means that old territorial disputes between countries are being revived and revealed.

If China's neighbours feel that only the United States can provide an equilibrium of power in the medium term, Washington constantly faces the risk that it may get unwillingly drawn into contests in which its direct interests are limited. This is a tricky balancing act for the United States. It is made more complex because of the concerns held by many American leaders that China may be changing the rules on how maritime rules of the game in Asia are understood. In turn Washington's commitment to the freedom of navigation is a direct challenge to the arrangements that China would like to see in its own neighbourhood. These disagreements have everything to do with the intersection of principals and power. As James Manicom (2014, pp. 357–8) has argued, 'China is dissatisfied with the possession – manifested by forward deployed forces – and the exercise – manifested by US commitment to the freedom of navigation of US seapower in East Asia.'

This means that otherwise less important disputes in Asia may not be as localised as may first be assumed. In a similar vein, India is simply too big for its own territorial disputes with China to matter very little to the wider region. In the longer-term any Asian equilibrium of power is likely to increasingly involve the interaction between these two Asian giants. And unlike the United States and China, India and China are direct neighbours with their own territorial differences. Again geography is still making a difference in Asia.

Moreover, it should not be forgotten that the contests for sovereignty over a given piece of territory are apt to occur within existing nation-states as well as between them. One of the reasons that secessionist movements have been common in many states in the region is because these states are composed of not one but many nations. As the latter are so often in the eye of the beholder there is no natural limit to this tendency, especially when the appeals of sovereign statehood can be fairly significant. The examples, both historical and contemporary, of Tamil nationalism in Sri Lanka, Timorese, Acehnese and Papuan nationalism in Indonesia, Bengali nationalism in what was once a larger Pakistan before the independence of Bangladesh, Moro nationalism in the Philippines, and Tibetan nationalism within the borders of China, are all indicators of tension. This is just one of the factors pointing to the need to consider questions of internal security and intrastate conflict which will be the focus of the next chapter. Above all, this is one reminder that nationalism, which is sometimes seen as a unifying factor for a country when it exists as a single phenomenon, is often so divisive when it appears plurally.

Chapter 7

Fragmentation: Are Asia's Main Security Problems Domestic Ones?

In thinking about Asia as a region, it is natural to focus on the relations between the region's main political units. This equates in turn to an emphasis on the international relations between the region's sovereign states. Likewise when one is thinking about Asia's security, it is quite normal to emphasise security relationships between these same entities. To at least a significant degree, the security problems and prospects of the international system of states as they apply across Asia and its borderlands constitute an undeniably significant part of Asia's security. This is what a good deal of this book, now at the half way stage, has been doing.

There is nothing wrong with this approach in and of itself. But it does not account for some of the truly significant aspects of Asia's security and insecurity. Asia is an important part of a wider international situation where the fragmentation of sovereign political units is potentially as significant as the clashes between them. On the one hand, the spread of the system of states, which became truly global only after the Second World War, is a story of the gaining of sovereign independence by a whole raft of new nation-states. Many of these still relatively young political entities are located in Asia and its borderlands and there are few more important factors than the achievement of this statehood that account for the region as it is understood today. On the other hand, this same process is also a result of the fragmentation of empires, colonial possessions, and federations. In those periods when several of these sometimes violent transitions have been occurring in the region, it has not been easy to regard Asia's security environment as a positive one.

That's not the only issue. It needs to be recognised that the gaining of independent sovereign statehood does not guarantee an end to internal security problems. To name just a few of the most obvious examples from Asia and its borderlands, post-independence problems have been evident in Pakistan, Afghanistan, Sri Lanka, Myanmar, Cambodia, the Philippines, East Timor and Solomon Islands. In a number of other cases, even where the use of organised violence in the competition for domestic political authority might be rare, leaders of otherwise well established governments may remain more worried about internal cohesion and order than they are about external security challenges. In this second list China, and possibly Indonesia, belong. Even India's leaders have significant ongoing concerns about communal tensions in their vast country. This means that for the three most populous countries in Asia, the risk of internal insecurity may at times seem greater than the risk of an interstate crisis. Bangladesh, Thailand, Malaysia, Tajikistan and Papua New Guinea may also belong in this second grouping.

Asia is replete with countries which have experienced political transformations, including a number of revolutions, within the last one hundred years, and most of these have occurred since the Second World War. It should not be at all surprising if this widespread tendency brings with it significant implications for regional security. But this last point might be challenged by the analytical logic that is being used in this book. One might argue that problems of *domestic* security can have few real ripple effects because by definition they occur within the borders of an individual country in the region. This alone might be seen as preventing the wider regional implications which this book seeks to focus attention on.

There is certainly something to these objections. It is undoubtedly the case that some internal security problems, while locally significant and even very violent and threatening for the people directly affected, may lack a region-wide impact. Sometimes this may reflect the refusal of some analysts to take more seriously security problems that are already crossing national boundaries in the region. Violence in southern Thailand, for example, is also a concern for its neighbour Malaysia, but these developments are not often in the news, although the same cannot be said about some of the cross-border issues between Afghanistan and Pakistan. It can also be said with some certainty that while the residents of Bougainville had some awful experiences of the civil war in which the PNG armed forces were involved in the 1990s, there was little real concern in the wider region that this would have significant ripple effects.

In this chapter, however, there will be opportunities to test the proposition that internal security problems which occur within Asia have little real chance of affecting the wider region. This can be done by considering the often complicated relationships between security and the modern state which have affected so much of Asia at the same time. Also up for consideration are situations where an insurgency in one part of the region may attract the attention of other states in dramatic ways. It is also important to think about the nightmare of a breakdown in order in one of Asia's largest and most significant countries, and what that might mean for the rest of the region.

There is also the important question of whose security matters. If people are considered as world citizens, and not merely as citizens of an individual country defined by national boundaries, security problems which affect the lives of other people elsewhere in the region assume greater importance. It may then be found that that efforts by a particular government to preserve *national* security (as that government defines it) may in fact come at the cost of the security needs and interests of many of that same country's inhabitants. And it may also be found that the psychological effects of these problems, especially if they are on a massive scale, can transcend national boundaries.

Force and politics in Asia

Domestic security problems can often be understood as challenges to the authority of the political unit which has sovereign rights and responsibilities in relation to that country's territory and people. As discussed in Chapter 1, that unit is the sovereign state (or nation-state) which is in turn comprised of those formal institutions which set the terms for the allocation and use of political authority within a given country. In other words, it is through these institutions that governments do their governing. Amongst these institutions of the sovereign state are a country's armed forces, in whom a very unique capacity is vested. This is the ability to use organised violence in the interests of national security, although the decisions on whether to do this are normally not made by the commanders of the armed forces themselves, but by the political leaders (elected or otherwise) to whom they report.

The ability to use force has tremendous political implications, for which reason many countries have constitutions which lay down significant limits on the involvement of the armed forces in domestic affairs. But it needs to be acknowledged that even in Asia's fairly recent history, there are occasions where armed forces have had a significant role in removing a particular government from office through revolution or coup. This is not the norm in today's Asia, where the political role of armed forces has been reduced in so many instances, but it is by no means a completely obsolete tradition.

Because control of the armed forces is central to controlling the power a state has, it should not be at all surprising to see competition for the control of the means of violence in troubled societies. The sociologist of all sociologists, Max Weber (1948, p. 78), argued a year after the First World War had come to an end that 'a state is a human community that (successfully) claims the monopoly of the legitimate use of violence within a give territory'. Yet this appealing formulation is clearly an over-simplification, and it gives the impression that the monopoly on violence is normally absolute and unambiguous.

But, in at least some of the domestic contests for power in Asia, it has been possible for monopolies on violence to be achieved and consolidated. This is one way of seeing Ho Chi Minh's eventually successful quest to unify Vietnam and rid it of foreign armed forces, or of Mao's civil war against the Kuomintang of Chiang Kai-Shek which eventually retreated to Taiwan. Once the communist parties in both of these countries had seized national power there was very little challenge to the new monopoly of armed force that had been created.

But in both cases the armed forces would continue to play an active role in domestic politics. In recognising the reality of this situation, China's leaders have sought to establish strong personal authority over the armed forces. China's president, for example, customarily heads the country's Central Military Commission. An important part of the establishment of Xi Jinping's authority as China's supreme leader was his very early assumption of this role. Indeed, as James Mulvenon (2013, p. 1) has noted, in quickly supplanting his predecessor Hu Jintao as the head of the Commission, Xi Jinping broke a tradition where the incumbent was permitted to stay on in that important political-military role for another two years.

The often very strong role that armed forces played in the domestic politics of a number of Asian countries in the middle of the twentieth century puts a somewhat different spin on Weber's formulation. Here the armed forces were one of the only viable institutions as countries made their difficult and sometimes violent path towards independence in an increasingly post-colonial environment. The state was not a recognisably well established and separate body which claimed and enjoyed a monopoly on violence. Instead, when the non-military institutions of the state were in a very embryonic form, the armed forces almost took the place of the state.

This rather curious variation can be seen in countries such as Indonesia, where, as Harold Crouch (2007) shows, the armed forces played a major role in politics in earlier decades. It was through the manipulation of the armed forces that Suharto, an army officer, was able to seize national power in a coup from his predecessor Sukarno. Given that Indonesia is a complex patchwork of diverse provinces and peoples, and that the institutions of the still emerging Indonesian state were very much a work in progress, the armed forces were given a dual function: in Indonesian *dwi funsgi*. Alongside their normal military responsibilities, the army, seen as one of the only genuine nationwide institutions, was given a social role to help knit together what was still a relatively newly independent country. This is not to suggest that the Indonesian armed forces succeeded in this role or that serious difficulties did not emerge as they sought to do so. But the fact that they had played such an important role in the gaining of national independence was one of their claims to political influence in the post-independence period. This was not the only justification to be found in Asia, however. In South Korea's case, the absence of a genuine peace with North Korea could be used as a rationale for perpetuating a military government which South Korea's main ally, the United States, was perfectly willing to work with.

But now that their critical transitional times have mainly passed, the countries of Asia are more confident about national integrity and the non-military institutions of the state are more established and effective. With these changes, as an important edited volume by Muthiah Alagappa (2001) concludes, the political role of armed forces in Asian government has been receding. In the last thirty years or more several Asian countries once ruled by military regimes have acquired civilian democratic governments. Prominent examples include South Korea and Taiwan in North Asia, the Philippines and Indonesia in Southeast Asia, and Bangladesh and (to at least some extent) Pakistan in South Asia.

However, generalisations about the declining role of the armed forces in Asia's political affairs are hazardous, as two examples in different parts of the region testify. The first and most obvious of these is Thailand, whose political traditions include an almost endless array of contests between civilian and military forms of authority. According to Paul Chambers (2014, p. 113), 'throughout Thailand's history, while civilians and security forces have competed for power, it has mostly been the security forces which have acted as change-agents to increase their power vis-à-vis royal or elected civilian opponents'. Despite recurrent hopes internationally that civilian control could succeed in Thailand, and increased signs of displeasure from a number of other ASEAN countries, the middle of 2014 saw yet another decision by the Thai armed forces to suspend civilian government in the midst of paralysing divisions between the major civilian political parties and their supporters.

The second such example is the small Pacific Islands country of Fiji, whose population is mainly a mix of indigenous Fijians and descendants of Indian workers who were brought to the islands to work the sugar plantations. Fiji has moved in the opposite direction. For many years after its independence from Britain in 1970, there were several peaceful transitions of power. But in more recent years political change in Fiji has often occurred at gunpoint, with a series of military-led coups interrupting attempts at democratic politics, culminating in a 2006 coup, the causes and effects of which have been analysed by Fraenkel, Firth and Lal (2009). This does not mean that Fiji has been scarred by civil war, but by maintaining control in a rather brutal fashion until the 2014 general elections, Fiji's military regime imposed a political order in which intimidation was commonplace.

Yet it can still be said that the experiences of Thailand and Fiji are anomalies in the wider region where armed forces are not the political actors they once were in so many situations. And on the whole there is reason to think that getting the armed forces out of the practice of running countries is good for Asia's security. This is often the case externally because the presence of a military regime in a neighbouring country is generally not conducive to good cross-border relations. India, for example, has regarded Pakistan's army as a source of interstate security concern, even while sometimes recognising it as a stabilizing factor in Pakistan's politics. The trend is good for security internally because military regimes often have a poor record when it comes to treating minority groups and have a tendency to use violence against those who challenge their political authority.

But the transfer to more civilianised forms of government is not always as simple or complete as some would like, and it is certainly not irreversible in some situations. To cite an important example, it remains unclear how far the changes in Myanmar since a new government came to power in 2011 will go, and how deep the constitutional changes will be. Amidst all the international excitement which followed the eventual release from house arrest of Aung Sun Suu Kyi, the famous democracy activist, it is worth remembering the words of Andrew Selth (2013, p. 2) that 'there are still serious problems in Burma. A fundamental transformation of the state, and its coercive apparatus, remains a distant prospect'. There is also some uncertainty about how far political reform in this troubled country will address the full range of civil conflicts involving the national armed forces and numerous groups scattered around the periphery of the country, many of whom have been pushing and fighting for greater autonomy, if not full independence. While ceasefires with many such groups have proved possible, obtaining peace with the Kachin Independence Army, operating in the country's far north, has proven much more elusive.

That challenge is a reminder that some political groups who resort to violence in a domestic setting are challenging the existing monopoly on violence within a particular portion of the country concerned. This was the position of the Tamil Tigers before they were routed by the Sri Lankan armed forces in 2009. The Tigers' violence was designed to put wider pressure on the whole fabric of Sri Lankan society so that they could be free to carve out an independent state in the northern Jaffna peninsula, in which the Sri Lankan armed forces would have no power. That attempt has failed, but while the single Sri Lankan state now has a monopoly on the use of force across the country, a number of the deeper issues that led to this divisive conflict have not been dealt with. One scholar, Nira Wickramasinghe (2013), describes the postwar climate in Sri Lanka as one of 'oppressive stability'. To the extent that this is the case, the seeds for future problems are being planted: short-term military success can too easily give way to longer-term political discord.

Asia's insurgencies

In cases of secession and in the responses to them that governments often make, a violent struggle can have wider national effects which are felt far beyond the particular part of the territory where independence is being sought. This certainly has been the experience of the Philippines, where, as will be shown in Chapter 9, no less than three armed movements have been fighting the government armed forces for independent statehood over the southern province of Mindanao. These intermittent conflicts have directed security attention inwards, reducing the capacity of the Philippines government to provide for the country's external defence at a time when the Asian distribution of power has been changing. These internal struggles have also affected wider regional confidence about the integrity of the country and the writ of the national government. In their southern areas both the Philippines and Thailand face what is commonly referred to as an insurgency. This is a form of internal conflict driven by sub-state political actors who are endeavouring to reduce popular trust in the existing power structure, often with the eventual aim of replacing that power structure with a new one. That word 'often' is important because, as David Kilcullen (2006, p. 112) suggests, some contemporary groups give the impression that rather than seeking to take over a functioning state, they may be focused on 'dismembering or scavenging its carcass, or contesting an "ungoverned space" '.

The best-known insurgency in Asia and its borderlands in recent years has undoubtedly been occurring in Afghanistan. Soon after the United States initiated a military response there to the 9/11 terrorist attacks, the Taliban government in Kabul was forced out of power. This was, in an ironic sense, a response to an earlier and successful insurgency in which that notorious Taliban regime had gained power in the first instance. The Taliban forces, and their Al Qaeda allies, were driven away from the capital city in late 2001 and an interim government under Hamid Karzai was installed in the following year. But while there was some degree of international self-congratulation as Mr Karzai was confirmed as Afghanistan's president in a general election in 2004, by this time regrouping Taliban forces were making their presence increasingly felt in a string of attacks. These were occurring against NATO-led stabilisation forces, the Afghan national forces they were seeking to train, and national and provincial politicians. Within a short space of time, it was not unusual to see dire warnings such as that from Johnson and Mason (2007, p. 84) that 'Afghanistan today is in danger of capsizing in a perfect storm of insurgency, terrorism, narcotics, and warlords'.

Throughout this protracted insurgency, which continues as this book is being written, the Taliban have been neither uniformly strong nor singularly organised throughout the whole of Afghanistan. Their sources of strength have been concentrated in the Pashtun belt which straddles the porous boundaries between Afghanistan and Pakistan, with the latter often acting as a place of refuge when NATO pressure grew, and as a source of a good deal of the violence. But as the international forces have drawn down, the Taliban have remained able to strike targets in Kabul, the most heavily protected part of the country. In doing so they send very strong signals about their intention to contest the authority of the civilian government. It is quite clear that more than a decade of international efforts to bring stability to Afghanistan have met with mixed results. To some degree at least the insurgency is undefeated. William Maley (2013, pp. 264–5) has voiced the concerns of many by suggesting that the US-trained Afghan national army 'is far too costly to be sustainably funded from internal revenue sources, and it is unclear how cohesive it will prove to be once foreign forces (currently providing air cover and medical evacuation capabilities) have left'.

For reasons already mentioned, this does not mean that the Taliban or any grouping will be able to claim a monopoly on armed violence in Afghanistan. The absence of that monopoly is by no means a novel phenomenon. Afghanistan's tribal, ethnic, and religious differences, plus its famously challenging geography and climate, has made it extraordinarily difficult for any external country or coalition of countries to have long-term influence on the security environment. That list includes the British Empire in its nineteenth century 'great game' with Russia. It also includes the Soviet Union, whose 1979 invasion of Afghanistan marked the resumption of the Cold War after the lull of the détente period. Afghanistan's history indicates that there are at least some parts of Asia where it is unwise to expect a government to effect sovereign control over an entire country. Bargains with warlords become a normal part of the allocation of power in Afghanistan, and this automatically reflects and legitimises the absence of a monopoly on armed violence. These deals are unlikely to provide for an ongoing sense of security, as Kimberley Marten (2012, p. 7) argues, but in many cases, 'no one who holds power has an incentive either to replace the bargains with universally applicable state institutions or to consider the long-term pernicious effects of patronage networks on society'.

That the scope of Marten's study extends to warlordism in Pakistan is one sign that the conditions which are bred by and which in turn encourage serious insurgencies are not limited to Afghanistan. And ripple effects connecting domestic instability to a wider neighbourhood are also possible. For a number of decades, as Christine Fair (2011) has shown, Pakistan's security and intelligence agencies have seen a number of armed groups as allies for across-the-border operations in Afghanistan and India. But these same actors have become an increasing threat to the domestic government in Islamabad. The assassination of leaders, including Prime Minister Benazir Bhutto in 2007, is but one indicator of the volatile and often violent environment in which Pakistani political life is conducted. Bombing campaigns, which often take the lives of innocent civilians, indicate the security challenges facing some of Pakistan's cities and suggest that the ungoverned areas, in which insurgent groups operate, are extending their reach. Not all such violence may be directly connected to an organised insurgency in the strictest sense, but in reducing public confidence that the existing state apparatus can provide security to the wider population, the legitimacy of the existing arrangements of power are being sorely tested.

It needs to be remembered that insurgent actors do not normally operate from a position of strength. Rather than engaging the state security forces in large battles, which they would be bound to lose, insurgent forces tend to operate like guerrilla armies, mounting surprise hit and run attacks and then often disappearing into the background. In urban areas the often close connections with civilian populations will sometimes make insurgents hard to identify, and in rural and mountainous areas they will take advantage of inhospitable terrain to find sanctuaries from which they are not easily targeted. The latter was one of Mao's approaches during the early stages of China's civil war, when the Communist Party was relatively weak. As Mao (2005, p. 68) argued when conducting operations against the Japanese forces in the late 1930s, China was

> a vast country with great resources and tremendous population, a country in which the terrain is complicated and the facilities for communication are poor. All these factors favour a protracted war; they all favor the application of mobile warfare and guerrilla operations.

Likewise the insurgents' weapons of choice are often small and light weapons, which as was noted in Chapter 5, are easy to transport and use and cheap to purchase and repair. This is the case, for example for the Naxalite insurgents in India, who, Ahuja and Ganguly (2007, p. 250) suggest, are armed with 'AK-47 rifles, revolvers and guns (many of these forcibly seized from the police)' and whose struggle comes out of a history of rural poverty and complex land problems in some of the poorer states of that vast country. A number of sub-state groups in Asia have also conducted bombing campaigns. But here too the quest for simple means, often made necessary by limited resources, is evident. Improvised bombs, which do not require complex manufacture, are also often preferred. These include the car bombs and suicide vests which have caused so much carnage in Sri Lanka, Afghanistan, and Pakistan. In another ripple effect example, Seth Jones has suggested that on the basis of the tactics of Hamas, Hezbollah and the Iraqi insurgents in the Middle East, and of the Tamil Tigers in Sri Lanka, the Taliban and Al Qaeda became convinced about their utility in Afghanistan: 'Suicide bombs' says Jones (2007, p. 23), 'allow insurgents to achieve maximum impact with minimal resources'. Terrorism (to be discussed in next chapter) allows at least some insurgent groups to intimidate wider audiences, and to indicate the limited authority of state forces.

Insurgencies of one sort or another continue to operate in a range of Asian countries including Afghanistan, Pakistan, India, Bangladesh, Myanmar, Thailand, and the Philippines. For the Indonesian government the separatist movement in Papua constitutes an insurgency, as do some of the activities by Uyghur nationalists for the Chinese government. But some care does need to be taken in throwing this term around. The pejorative connotations of the word 'insurgency', and the claims that are often made that insurgents are terrorists, can reflect a very subjective account that favours the interests of the existing regime. This labelling can also be counterproductive and increase the chances of conflict: a point that Michael Clarke (2010) makes in analysing China's often heavy-handed response to protests and other actions from members of the Uyghur minority in Xinjiang. Few could deny that the armed forces of Myanmar, Indonesia, and Pakistan, for example, have sometimes been directed by their national governments to act in ways that inflame the internal divisions on which insurgencies can thrive.

Sitting back and taking a wider view is important here. At some time or another many of the movements for national self-determination constituted insurgencies. Unless there is a desire to wind back the clock and return to an era of colonial control in Asia, it might be sensible to reserve judgement on the idea that an insurgency is by its very nature unjustified and abhorrent. Or at the very least it might be wise in some cases to make a distinction between the political ends of the insurgent actors and the sometimes violent and indiscriminate means they resort to.

Failed states and human security

Things are therefore not nearly as simple as suggested by the idea that successful statehood connotes control over the monopoly of violence. And two further considerations jump out at this stage of the current chapter's treatment of domestic security issues in Asia. The first is to wonder whether the writ of the established state is limited in some Asian societies because of insurgents and other challengers to the status quo, or whether some Asian polities lack cohesion in the first place. In other words is the internal sovereignty of some states in the region inherently constrained? Perhaps that sovereignty is limited regardless of whether active challenges are being mounted. It may be unfair and inaccurate to surmise that some portions of Asia and its borderlands are destined for continuing and unremitting social and political instability as if there were natural forces at work which make that condition inevitable. States and non-state actors must still make choices about whether they use violence or not, for example. But the rise of global concern about failed states, which was driven partly by worries that such places might be breeding grounds for terrorism, reflects a crucial assumption. This assumption is evident in an essay written in the immediate years after the 9/11 in which Robert I. Rotberg (2003, p. 1) argues that 'Nation-states fail because they are convulsed by internal violence and can no longer deliver positive political goods to their inhabitants.' The crucial assumption is captured in the words 'no longer'. In other words, a failed state is one that at some previous time used to work properly.

But in Asia there have been so many extraordinarily challenging pathways to independent statehood, where some of the newer nation-states are states of many nations, and where the enjoyment of economic progress has been uneven. It would therefore be surprising if all of these new entities were in good working order before they began to 'fail'. Particularly where traditions of strong central authority are weak, expectations need to be adjusted. The absence of such traditions is evident, for example, in a number of countries in Australia's neighbourhood, and this feature is unlikely to change. As I have argued in a separate piece of writing – Ayson (2007) – the depiction of Australia's northern neighbourhood as an 'arc of instability' probably says as much about the outlook of policy-makers and analysts watching the region from Canberra as it does about the real situation in Indonesia and a good deal of the South Pacific. To use Wolfers' formulation, it is as much about subjective interpretations of security threats as it is about objective conditions. But it is very clear nonetheless that the Melanesian countries in that arc, including Papua New Guinea, Solomon Islands and Vanuatu, have histories where power has been dispersed into multiple and changing centres and personalities, making the building of state institutions more challenging than in some other places. This does not mean that state-making is impossible, but it does mean some difficult moments in what Volker Boege et al. (2008, p. 35) depict as the 'marriage' between 'relatively strong customary spheres and state institutions that struggle with problems of effectiveness and legitimacy'.

The second consideration is to question the corresponding argument that security is assured so long as the countries of the region have strong state institutions. Sometimes too much state power at the hands of a central government can be a dangerous thing, especially if substantial amounts of that power are vested in the armed forces. It is hard to argue against the proposition that if security is understood in terms of national cohesion, including the inability of armed groups to mount serious challenges within the country, at least some of Asia's more authoritarian governments boast quite effective records. But, at least internally, these achievements have often come at a high price, as civilians are caught up between the state authorities and the groups they wish to control. Some of Asia's stronger states have had a poor record in observing human rights, including such civil liberties as the freedom of opinion, expression and assembly. As the Indonesian expert Dewi Fortuna Anwar (2003) has argued, some of the Southeast Asian governments who prided themselves on taking a comprehensive approach to security which goes beyond safety against military threats and includes considerations of social cohesion and economic security were still often thinking merely about the security interests of the regime.

This comprehensive approach to security was adopted, for example, by the same Suharto government which in 1975 would invade and annex East Timor, which had been let go suddenly by Portugal as the vanishing European colonial power. The result was catastrophic for the local population. In one study, Ben Kiernan (2003) has suggested that a quarter of East Timor's pre-invasion population lost their lives, a comparable proportion to the impact of the genocidal rule of the Khmer Rouge in Cambodia, which also occurred in the 1970s. After the Suharto era had collapsed in Indonesia in the late 1990s, armed militias opposed to East Timor's independence went on a rampage, with the connivance of significant parts of the Indonesian military. By this time, however, global concern about the plight of local populations in the face of internal violence had increased. This was in large part due to developments beyond Asia, such as in the former Yugoslavia, where the treatment of the Albanian population of Kosovo by Slobodan Milosovic's Serb-dominated military regime drew particular condemnation.

This period saw gathering international interest in arguments about human security, a set of ideas which suggest that individual people and social groups have security interests which no government ought to violate and which governments instead are expected to provide for. As will be seen in Chapter 9, human security logics can sometime be used to legitimise complex and challenging interventions, and leave big questions marks about well intentioned arguments that force can be used for good in these situations. But in the context of the present discussion, ideas of human security can reveal some important insights about internal conflict in Asia. One of these is that steps taken by regimes to provide for national security, including in the way that armed forces are used to respond to internal problems, can often be at the expense of the security interests of other actors within the community. Another is whether the reverse is true: might respect for the human security interests of all of a country's inhabitants actually bring national security into question? The debates which have occurred in a number of countries, including in Asia and its borderlands, about balances struck between liberty and security (including in legislation designed to reduce the chances of terrorism) suggest that this is a live issue.

Understood broadly, human security is not just about the safety of individual citizens and groups from threats of violence (which can often be delivered in the name of the state's interests). In some formulations, human security cannot not be achieved without addressing the deeper problems of poverty, injustice and prejudice. The leading example of such an all-encompassing approach comes from a prominent report by the United Nations Development Programme (1994, p. 23) which argued that:

> Human security can be said to have two main aspects. It means, first, safety from such chronic threats as hunger, disease, and repression. And second, it means protection from sudden and hurtful disruptions in the patterns of daily life – whether in homes, in jobs, or in communities.

As might be imagined, the breadth and ambition of this approach raises some challenging issues for the way that security is defined and considered. A pattern of daily life is of course something that is widely valued, and when it is disrupted, including by sudden and dislocating events (such as a natural disaster) the sense of security enjoyed by many individual and groups is put at risk. To go back to the original formulation presented in the first chapter of this book, if security is the absence of threats to acquired values, then it seems reasonable to conjecture that there is no theoretical barrier to this very inclusive understanding of security. But the problem here is that when security comes to include the absence of any sort of problem, it becomes at the very same time a master concept and an idea which is so broad that it lacks analytical utility. If security is defined as the absence and unlikelihood of harmful things, and the absence of the fear that harmful things of any kind may occur, it becomes almost impossible to find a factor that is not part of the fabric of security. And being everything, security risks becoming nothing.

How understandings of human security can be utilised in understanding Asia's security without succumbing to such banality is an intriguing question. Paul Evans (2004, pp. 266–7) argues that against a view in which human security is the absence of threats to human wellbeing there can be set a 'narrower and more pointed view of the scope of human security, focusing on protection of individuals and communities in situations of violent conflict'. This is closer to traditional understandings of security which focus on the absence of threats involving organised violence. It still means, however, a change to what scholars call the 'referent object' of security: the unit whose security is in question. In most traditional understandings of security, that unit or referent object is the state. For human security analysis, the unit is the individual human being or, more commonly, social groups comprised of individual persons. What remains constant in this picture is the main threat – organised violence.

Such an approach means that security remains focused on a less universal set of bad occurrences. But it can still result in some rather unusual outcomes. The more recent decades of the Communist Party's rule in China have seen great advances in the living standards of hundreds of millions of people. The contrast could not be clearer with the famine that cost tens of millions of lives after Mao's disastrous Great Leap Forward campaign of the late 1950s. Moreover, most of these same citizens today have little reason to live in fear of violent internal conflict (although something more about China's outlook on this matter will be said a little later). Yet, repression is part of the index of human insecurity, and if that repression can take place through judicial means without any obvious physical violence, how then is human security to be measured in an illiberal political environment? This is not to suggest that the human security debate has escaped the attention of China's scholars or its government. But particular assumptions about what a secure life for individuals and groups entails has fed into the way that this concept is viewed. This includes the judgement that economic development is the essential prerequisite for meeting basic human needs, a view that also emphasises the role of the state (and the party) as the driver of these changes. Accordingly Shaun Breslin (2015, p. 257) observes that while a great deal of human security thinking has sought to 'move the focus away from the state and the collective towards the individual...Chinese understandings and discourses often re-blur the distinction and restore the state as the referent point.'

It should come as no surprise that the idea of the individual as the reference point for security has been promoted most strongly in western liberal democracies and received cooler receptions in countries with authoritarian political systems. Yet there is no necessary correlation between political freedom and economic wellbeing. Nor does political reform necessarily safeguard the interests of all minority groups within a country. The reforms which have taken place in Myanmar, for example, have allowed many Burmese people increased scope to enjoy rights of political association. But this process has not been favourable for the Muslim Rohyinga minority, which is concentrated in the border areas near Bangladesh, and whose oppression (in Buddhist majority Myanmar) appears to have become more severe.

Indeed in many Asian countries the experience of human security and insecurity is unevenly distributed. For example, as an increasingly prosperous democracy India has a growing middle class with hundreds of millions of its citizens enjoying the fruits of economic development. In contrast to China, these citizens also enjoy long-established civil liberties. This would appear to satisfy both the freedom from want as well as the freedom from fear (from government oppression) which broader understandings of human security seek to incorporate. But some of India's provinces still feature hazards of privation (from drought and disease) and exposure to communal conflict, often between Hindu and Muslim communities. India's ongoing territorial dispute with Pakistan, covered earlier in this volume, similarly exposes communities in parts of Jammu and Kashmir (and also on the Pakistan side of the Line of Control) to the dangers of arbitrary violence. In regard to some of India's anti-terrorism legislation, Anil Kalhan et al. (2006) point to a further dilemma: attempts by the state to protect citizens against violent threats to their lives, surely one aspect of human security, may also pose a challenge to civil liberties.

The uneven experience of human security within a particular country is nothing unusual. It is especially evident in Afghanistan where the struggle for economic advancement which affects human wellbeing occurs alongside the vulnerability of people in many provinces to organised violence. This experience is far from universal across the country. It also comes in a range of contexts. The violence which has been directed by some Taliban groups against women who wish to play a more active role in civil society and education is a reflection of a gender dimension to a number of human security questions, an area that has received attention in the work of Kandiyoti (2007), amongst others. More generally, particular communal and ethnic groups seem more vulnerable to violence and intimidation than others. This is reflected in the uneven composition of the refugees and asylum seekers who have fled Afghanistan, including many Persian-speaking Hazaras who have been a particular target by some of the Taliban groups. This indicates that some portions of that society feel much more vulnerable to the political changes occurring within it, or are more able than others to leave the country in the first place.

The spread of internal conflict?

The movement of peoples away from conflict zones is one sign that internal insecurity can have genuinely international implications. But it still might be wondered whether Asia's domestic conflicts can have truly transformational ripple effects that are felt across the region. It must be admitted that some domestic conflicts in parts of Asia and its borderlands do remain relatively localised. One such example comes from the South Pacific country of Solomon Islands, which consists of a group of islands which are separated by water from other countries in the region. The Solomon Islands civil conflict, a result of tensions between two main ethnic groups over land and power in the main island, led to a political coup in 2000. The ongoing difficulties there were exacerbated by the government's inability to deliver services to the Solomon Islands population.

For the South Pacific region this was a serious matter, precipitating an Australian-led intervention in 2003 (of which more will be said in Chapter 9). But it cannot be said that any of the countries in East Asia were concerned about what violence in Solomon Islands might mean for their own security. Yet there was a wider connection. When protestors burnt down shops owned by Chinese business-people in the Solomon Islands capital city of Honiara, the government in Beijing expressed concern at the vulnerability of overseas Chinese people. Migration patterns can therefore set in train potential connections that are not easy to see if the region is viewed through traditional security lenses. These perspectives are being challenged. As Fiona Adamson (2006, p. 167) argues, 'international security scholars and policy-makers are finding it increasingly difficult to ignore the relationship between migration and security in a highly interconnected world'.

In more land-locked parts of the region geography is much less of an insulating factor and can act more readily as a transmission belt. To cite one important example, Thailand and Myanmar are separated by a porous and unevenly monitored land border. As a result of many internal conflicts in Myanmar, in which some ethnic groups have come under particular pressure, significant flows of people (as well as resources and guns) have occurred across this boundary. Whether it has wanted to or not, Thailand has for many years inherited responsibility for some of the outcomes of its neighbour's internal instability, with camps of refugees springing up on its side of the border. This is not a new development. As Hazel Lang (2002) suggested over a decade ago, it began to make itself apparent in the mid-1980s. Cross-boundary movements, which will be considered in more detail in Chapter 8, can be lengthy and chronic, disappearing from media view and public opinion. But this does not mean that their security implications are necessarily minor.

Moreover the interaction between domestic security problems in one country and developments in neighbouring states can have more obvious and more violent effects. In this chapter it has already been shown that connections between the government and armed groups in Pakistan have had implications for the security of both Afghanistan and India. Perhaps more worrying still is the possibility that Pakistan's state institutions might struggle in such a way that uncertainties develop about who has control over the country's expanding arsenal of nuclear weapons. A weakened Pakistan, with divisions in the armed forces who have held tight control over the country's arsenal, could increase the chances that for the first time anywhere in the world a militant group might have access to genuine weapons of mass destruction. As Rolf Mowatt-Larssen (2009, p. 10) suggests, there is some chance that 'in extreme circumstances, the Pakistani government would be hard-pressed to reassure the outside world, especially India, that all its nuclear assets are under the full and authoritative control of the military'. Depending on how India might choose to respond, this could most definitely mean that internal security challenges in one Asian country are able to ripple out further afield with potentially catastrophic consequences.

As was noted in Chapter 5, this scenario remains improbable. It is what analysts of international risk would call a low probability-high consequence problem. Because the consequences of such an event could be incredibly dramatic, any tiny chance of it occurring grabs our attention. In Asia most internal conflict issues have higher probabilities but smaller regional consequences, at least as far as their neighbours are concerned. But there is at least one good reason to keep in mind the possibility that the fragmentation of one Asian country could spell very serious difficulties for others. That reason comes into view when we consider the scenario of a collapse of internal order within a major Asian power. The obvious candidate to analyse in this regard is China, not least because of its importance to the wider region but also because over its very long history, China has experienced periods of stark disorder as well as of unity and cohesion. This would seem to fit with the finding by Rein Taagepera (1997) that large polities do not have a habit of remaining intact for very long periods of human history.

Today's China is an historically unique experiment of enormous economic and social change affecting over a billion people in a very compressed time-frame. This experiment is being overseen by a one-party political system which lacks some of the pressure release mechanisms that more democratic, open and multi-party systems provide for. As this book is being written, tens of millions of Chinese citizens are on the move from the interior to the more wealthy coastal regions in search of employment and a better way of life. Should China's leaders be unable to deliver the levels of economic growth that guarantee reasonable levels of employment and prosperity, the social strains will grow. And the pressure on China's natural environment is potentially even greater. The damage to China's ecological system from decades of rapid industrialisation to China's ecological system – air, water and soil – has reduced the country's ability to support its growing population, a situation made worse by rapid urbanisation and the loss of agricultural land (Chen, 2008).

The links between Asia's environmental problems and regional security will be analysed in the next chapter. But it is important to note that China's problems do not stop here. Civil disturbances, which are almost inevitable in a rapidly changing country of enormous size and complexity, worry the Communist Party leadership in Beijing. That leadership also has to contend with perennial concerns about the fragmentation of China. These include, to the far east of China, the diminishing possibility of Taiwan declaring independence, and to the far west the pressures for greater autonomy and independence from Tibetan nationalists and the Muslim Uyghur community in Xinjiang. And China's top leaders also have fresh memories of a more general and serious crisis of legitimacy which was centred on the capital city in 1989. The brutal suppression of the student-led movement for greater political freedom that year in Tiananmen Square is testament to the nervousness of the Communist Party hierarchy to any challenge to the political rules of the game under one-party rule. That this nervousness survives is confirmed by the leadership's considerable concerns about the possible replication in China of the popular activism associated with the Arab Spring in many Middle Eastern countries.

China's leaders probably have less reason than some of their authoritarian counterparts to be worried, and not just because of an impressive record of economic development. Steve Hess (2013) suggests that as much of the daily government in China is decentralised to local leaders, protests are less likely to be aimed at the central leadership than they have been in places such as Egypt. But, however unlikely, the wider Asian region still needs to consider what might happen should China lose its overall cohesiveness. The main regional implications of a seriously weakened China, at least in the short-term, are likely to be economic ones. Most of Asia depends heavily for its own prosperity on China's, and would probably be plunged into recession. This is not a security challenge per se. But as their legitimacy is questioned, it might be tempting for the Communist Party and the People's Liberation Army to exploit an external crisis to boost internal unity around Chinese nationalism. This would be a dangerous thing to do. Suisheng Zhao (2005) argued some years ago that China's leaders tended to encourage a pragmatic form of nationalism, worried about losing control should they allow too much scope to this phenomenon. But desperate times can encourage desperate measures and India, Japan, Russia and Vietnam, some of China's main neighbours, might find this a very challenging period. In other words, the ripple effects might come not in a spread of internal disquiet, but in its implications for China's relations with other regional powers.

The decline of internal conflict?

But the big story about internal conflict in Asia might not be the possibility of devastating ripple effects for the region if the wrong mix of conditions occurs in an especially important country. Perhaps the main issue is the declining incidence of serious domestic insecurity in many parts of what was once a much more violent region. It has become very common for scholars internationally to argue that the incidence of wars *between* states has dropped off markedly. Sometimes these accounts suggest that by contrast there are plenty of messy internal conflicts occurring instead, almost as if there is a set quantum of violence which if not appearing in one form will be found in another. But as Muthiah Alagappa (2011) has argued, there is good reason to think that many types of internal conflict experienced in Asia after the Second World War have also been on the decline. It is therefore necessary to ask whether peace has broken out in Asia domestically as well as internationally.

The possibility that Asia's internal conflicts are becoming less common increases when consideration is given to the reasons they break out in the first place. Thinking strategically about Asia's wars, including internal conflicts, can help here. In such thinking armed violence is not seen as the direct and immutable result of poverty, the collapse of institutions, and even the changes in climatic conditions – all factors which have come to be regarded as a likely cause of future conflict. Some of these problems may increase the chances of war, but they are not in and of themselves necessary for it to take place. Instead it can be argued that armed conflicts occur when political actors decide that organised violence is a way of furthering their ends. Thinking in this manner opens up the possibility that some of the once prominent reasons and rationales for internal conflict in Asia have become less pressing and convincing to the actors who might resort to violence.

Consider, for example, the numerous wars for national independence that were part and parcel of Asia's insecurity in the postwar years. As became clear in Chapter 2, there were plenty of European colonies in Asia (and some non-European ones) in which the often violent struggles for national self-determination erupted as one of the major features of the expansion of the international system of states. But the Indonesian nationalists could only fight one successful war of independence with the Dutch. As Pandey (2001) shows, the partition of British India into India and Pakistan, and the tragic violence with which this was associated, remains something of a conundrum. But that partition was also an unrepeatable event. The same can be said about the subsequent and also violent division of Pakistan to form Bangladesh. The Vietnamese nationalists had at least three bites at the cherry, against Japan, France, and the United States. But unification in 1975 spelled the end for this long era of conflict.

This is not to suggest that there can be no more wars of independence in Asia. Some of the newly independent polities in the region carried with them the seeds of further violent disintegration as subsets of the new country fought or were caught up in secessionist battles for their own independence. That was the case for the violence that engulfed East Timor in 1999 after a referendum delivered a vote significantly in favour of independence from Indonesia. The failed secessionist war in Sri Lanka fought by the Tamil Tigers is another such example, as is the quest for independence from Papua New Guinea in Bougainville, whose future remains somewhat unclear. Neither is it easy to forecast the future of the southern insurgencies in the Philippines and Thailand. At the very least, as Joseph Liow (2006) suggested some years ago, it is wise to be alert to the differences in these situations, with the MILF (Moro Islamic Liberation Front) in the Philippines having a much more obvious campaign for national self-determination than any counterpart groups in southern Thailand.

It is certainly possible to draw up longer lists of quests for independence in Asia. But it would be wrong to conclude that this process is endless. First of all, there are very few colonial possessions left in Asia itself. Hong Kong, which Britain returned to China extremely peacefully in 1997, was one of the last such examples. Some of the former republics of the Soviet Union in Central Asia, including Kazakhstan and Uzbekistan, were in effect internal colonies of a Soviet Empire, and somewhat artificial constructions at that. Olivier Roy (2007, p. xiii) submits that the 'Muslim Republics of the ex-USSR were the creations of decrees between 1924 and 1936 which determined not only their frontiers, but also their names, their re-invented pasts, the definition of the ethnic groups that they were reckoned to embody, and even their language.' But in the years since the sudden collapse of the Soviet Union and the consequent ending of the Cold War they have been independent sovereign states in their own right. And despite occasional concerns about Russia's policy towards its neighbours, they are likely to remain so.

Yet there is one part of Asia and its borderlands in which territorial possessions, including of a colonial variety, are still a significant factor: the South Pacific. Many Pacific Islands countries did indeed gain independence from the European powers and in Papua New Guinea's case from Australia in the 1960s and 1970s, and for the most part these were very peaceful and orderly processes. But France has held onto French Polynesia and New Caledonia. The United States retains American Samoa and has relationships of close association with the Republic of the Marshall Islands, Palau and the Northern Marianas. Niue and Tokelau remain New Zealand territories.

In many of these cases, there is little sign of a significant groundswell of pressure for sovereign independence. But indigenous independence movements in French Pacific territories have been strong in the not-so-distant past. This is especially so in New Caledonia where the contest between indigenous supporters of independence and their opponents spilled over into violence in the 1980s. Towards the end of that decade, however, an accord was signed which allowed for greater regional autonomy. While it may have seemed at the time that this would only delay an eventual independence referendum, it is not clear that this is in fact the case. Over a decade ago, John Connell (2003, p. 141) argued of New Caledonia: 'In an uncertain world a substantial degree of autonomy, where culture and identity are respected and protected, reasonable access to employment and services exists, and security is guaranteed, has weakened the strength of the claim to independence.' Hence the economic arguments for remaining under the umbrella of larger countries has been a powerful protector of the status quo.

If anything, the international climate of opinion on the question of full sovereign independence has become even more conservative. The appetite for welcoming new members of the system of states appears to have reduced by the knowledge that small and newly independent sovereign states often require extensive assistance in their early years. That has certainly been the experience with East Timor, and beyond Asia the entry of South Sudan into the system of sovereign states has been a very difficult enterprise.

This is not to suggest that the political contours of Asia as they currently stand are bound to remain as they are for decades to come. It is just that the major changes to that map have already taken place. In some instances what appeared to be simmering movements for secession seem to have gone off the boil. Indonesia appears to have worked reasonably (although not completely) well with leaders in the Aceh province, located in the far west of the archipelago, with increased autonomy being granted in 2005 in place of complete political separation. This is still a work in progress. As Marcus Mietzner (2007) has observed, an academic debate continues to rage on whether autonomy processes tend to temper or actually encourage the quest for full independence. Indonesia's experiences offer evidence which can be used to support both of these positions.

Indeed, there is unlikely to be a template for autonomy that will work equally well in all situations. This is clear from the uneven success of a similar extension of autonomy to West Papua (or Irian Jaya), which shares a long and porous land border with Papua New Guinea and which features a decades-long independence movement, with an armed wing known as the Organisasi Papua Merdeka. One of the world's leading think-tanks, the International Crisis Group (2012, p. 3) has warned that spirals of violence are possible, given 'the increasing radicalisation of one part of a broad-based pro-independence movement and Jakarta's determination to deal harshly with those it describes as separatists'.

It is clear that the ire of a number of human rights groups has been drawn by Indonesia's sometimes heavy-handed approach in West Papua. But for the sake of good relations with the Indonesian government, and a preference for the territorial status quo, the governments of neighbouring countries have been restrained in their criticisms. Over fifty years ago Australia was deeply upset by Indonesia's takeover of the province from a departing Netherlands. But today Australia wants to avoid the Papua issue upsetting the relationship with Indonesia, which has become an increasingly important partner in the region. This is part of the same conservative international political climate which is less receptive to new claims of national self-determination. This would also seem to reduce the chances that internal conflict in one part of the region will have a ripple effect further afield, unless of course unmet demands for independence boil over into the violent contests that characterised Asia's past.

Similarly, the incidences of insurgency which were discussed earlier in this chapter appear to have less impact on the wider region in today's Asia than in earlier parts of the postwar period. Afghanistan is something of an outlier as the case of a country whose whole political future rests largely on how its complicated insurgencies are managed, although it might be suitable to add Pakistan to this small list. This reflects quite significant change for the region. In the 1950s and 1960s, for example, almost every Southeast Asian government, many of them newly independent, or about to become so, faced genuine fears of a debilitating insurgency, or the after-effects of a war for national liberation, or some other sort of internal conflict. This was also a period of severe internal trauma for China, which was also sponsoring a number of the insurgencies in Southeast Asia.

The ubiquity of internal conflict in earlier eras has analytical significance for the purposes of this volume. It can be argued that in the 1950s and 1960s internal conflict within a large number of Asian countries had a wider regional effect in at least two ways. First, while there is a profound dislike today for the troubled notion of the 'domino effect', where communist revolution in one country was expected to spread to neighbouring places, this sort of thinking helped give domestic conflict an interstate feel in Asia. This was not least because a number of western countries were dragged into the fight, or, as Victor Cha (2010) argues with regard to United States policy, because of a fear that they might be dragged in by the unwise responses of anti-communist allies. Moreover if the communist domino effect was an exaggeration, another idea was certainly infectious in Asia: this was the idea of independence from external colonial control. A good deal of Asia's earlier internal conflict is explained by the spread of this motivation and the temporary political instability it brought with it. That instability was not instantly removed as soon as independence was achieved because newly sovereign states so often faced challenges in bringing a new semblance of unity to countries no longer under external control. Second, even without the spread of new norms of political authority in Asia, the widespread experience of internal conflict meant that it was a dominant part of the security environment through large parts of the region.

These phenomena did not need to be linked to one another for them to be experienced in a large number of regional countries. Moreover, some ripple effects may have been moderated by this common experience. Focused as they were on their own internal struggles, at least some of the region's government's simply had little time to focus on potential adversaries abroad. And when they did work together, as the initial members of the Association of Southeast Asian sought to do in 1967, as Hiro Katsumata (2003, pp. 112–14) suggests, their intentions were to restrict the involvement they might have in one another's domestic processes and to ward off similar domestic political and military interference by larger powers.

Conclusion

One question still hangs over the subject of this chapter: is there a single and overall explanation for Asia's domestic conflicts? The answer is probably No: the diversity of the region is a safeguard against one-size-fits all explanations. But then perhaps it is the appearance of this diversity within so many of Asia's countries that explains why some groups within Asian countries resort to violence against other groups, and against the prevailing government. Perhaps ethnic, religious and, in some places, tribal divisions, mean that Asia is full of countries who are struggling with domestic versions of what Huntington (1996) controversially referred to as *The Clash of Civilizations*.

Looking back on this chapter it is possible to find examples which may fit this bill. The Hazaras in Afghanistan, the Rohyingas in Myanmar, the Papuans in Indonesia, are just a few of the groups in the region which have been caught up in internal violence. Other instances are not difficult to find. There are fairly regular occurrences of low-level violence in the southern highlands of Papua New Guinea, a country in which over eight hundred different languages are spoken (the most in any place in the world). But the link between these differences and the occurrence of conflict is a very hazy one, as the influential post-Cold War study by Fearon and Laitin (2003) suggests. If that connection was strong enough to stand on its own it would be hard to explain why some of these societies are not in a perpetual state of war, and why instead organised violence occurs in some circumstances and occasions in Asian countries rather than others. It would also mean that great fears should be held for the cohesion of ethnically, religiously and culturally diverse large countries such as India and Indonesia where communal violence is in fact far from all-encompassing. This might also be reason to regard the growing Asian populations in a number of western-oriented liberal democracies in Asia's borderlands – in Australia, New Zealand and Canada – as a sign of future insecurity in those places, when there seems to be little evidence that this is the case.

An upswing in internal violence in Asia is not unthinkable. But it is wise not to treat what is happening in some of the less internally stable parts of the region as a signpost showing where the rest of Asia is heading. Afghanistan and Pakistan, for example, are not models for the region's future. Unnecessarily pessimistic assessments for the whole of Asia and its borderlands can be avoided because the domestic political relationships which remain the major drivers of organised internal violence within Asia can also adjust in ways that restrict it. Over time the political motivations for internal conflict in many of Asia's countries have simply diminished. Sometimes these motivations have been satisfied, as in the case of successful efforts at national self-determination. On other occasions they have been managed, occasionally by more enlightened policies on the part of the national government of the day, including through the granting of increased political autonomy for disaffected provinces.

In some cases, other strategies, including by external states and organisations, may have helped. But interfering in another country's difficulties is not for the faint-hearted, even once an invitation has been extended.

There is always the possibility of making an already bad situation even worse. A range of these various attempts to make a difference and heal Asia's security wounds, despite all of the attendant difficulties, are the subject of Chapter 9. But not all of these are responses to the internal conflicts which have been considered in the current chapter. A series of transnational, rather than domestic, challenges and actors also require consideration. So it is to them that this book will next be turning.

Chapter 8

Hazards: Will Non-State Actors and Transnational Challenges Overtake Asia?

The end of the previous chapter was wrestling with one of the central issues in this book: the extent to which security challenges in one part of Asia have effects on the wider region. That discussion focused on a particularly intriguing variation of this theme: can what appear to be specifically *domestic* security problems within particular Asian countries still create ripple effects that others in the region will notice? The answer, not to put it too bluntly, is that it depends on the circumstances. Domestic security challenges can be felt beyond the specific national boundaries which help define them, but this is not the case for all such internal events. Yet one might then wonder about security problems which by their very nature cross national boundaries, and which may even give the impression that the distinction between domestic and international security in Asia is actually irrelevant. These are the sorts of issues that will be considered in this chapter.

Earlier parts of this volume considered a series of *inter*national security challenges: those likely to affect relations *between* states (the main political units of the region). But *trans*national issues are developments which by definition occur across sovereign borders and which can have effects on polities and societies on either side of these political boundaries. Paul J. Smith (2000, p. 78) has defined transnational security issues as 'non-military threats which cross borders and either threaten the political and social integrity of a nation or the health of that nations inhabitants'. As such they appear to bypass the sovereign state actors that have dominated the security analysis to this point of the book.

Indeed, even while the last chapter considered threats to the cohesion of individual nation-states in Asia, these struggling political units were still the predominant actors. Proponents of human security, for example, still expect the nation-state itself to be the main security provider. But the discussion of transnational security may introduce a situation where nation-states are little more than spectators. They may be unable to control, and possibly even to significantly influence, the unauthorised flow of people, goods, weapons and problems across national boundaries.

Indeed if those boundaries themselves are in question, then so too might the state whose geographical authority they define. Making things even more interesting is the possibility that when political units are driving these transnational flows they are often non-state actors who are not generally acting with the permission of the states in which they operate.

As has already been noted, Asia is a region full of countries which are still relatively new to the international system as sovereign states, and who guard that sovereignty very closely. That makes it even more important to consider the impact of transnational challenges on Asia's security and to work out whether non-state security actors might actually affect regional security in ways that no individual country in the region would welcome. This is, in fact, an opportunity to test the proposition that processes of globalisation will define the region's future security agenda. If Asia's security is to witness systemic change, then perhaps it will happen here.

Accordingly, the first task in this chapter is to look at the non-state actors who often operate across national boundaries and challenge the authority of Asian states: transnational terrorists, pirates and criminal groups. These are all groups involved in purposeful behaviour, although it might be wondered whether their motivations are necessarily political. The second focus in this chapter will be on transnational *factors* rather than *actors*. Some problems flow across boundaries not by dint of conscious choice by humans, but are at best the indirect transnational consequences of human activity. Here it is necessary to examine the extent to which climate change and the spread of disease might be considered as an important part of the contemporary Asian security environment. By their nature these developments would seem to satisfy the ripple effect test because of their proven ability to be experienced across the region as a whole. Does this make them the supra-challenges for Asia's security, or is it unwise to give these problems such an analytical stature when no specific or identifiable political actor is directing them?

Transnational terrorism

Of the issues being considered in this chapter, terrorism has especially strong political connotations. That might seem odd at one level since acts of terrorism attract attention primarily because of their violence and the direct effects of that violence on innocent civilian victims. The explosion of car bombs in crowded market squares in Pakistan, for example, or some of the world's earliest suicide bombings that occurred in Sri Lanka, let alone the enormous explosion which ripped through a nightclub in the Indonesian resort of Bali in 2002, all seem to indicate that violence itself is the calling card of terrorism. The ultimate version of this behaviour was planned within the borderlands of Asia by al Qaeda's leadership, which at the time of the 9/11 terrorist attacks on the United States was based in Afghanistan. If nothing else, the crashing of airliners into the World Trade Centre and the Pentagon was potent evidence of the capacity of a non-state actor to cause immense civilian casualties and physical devastation within the territory of the world's strongest country.

But the violence of terrorism, illustrated in each one of these examples, is not the beginning and end of the subject. While there may be some exceptions, acts of terror are not committed for their own sake. Terrorism, like war, involves the purposeful use of violence (the means) for the sake of political objectives (the ends). As Max Abrahms (2008) has argued, individuals who join groups which use terror may do so for non-political reasons, including the comradeship that belonging to these groups may offer. And these groups may use acts of terror partly to keep these bonds intact. But when terrorism is witnessed in any part of Asia it is important to keep in mind the political agenda which is involved. It is this political connection that is the most important feature of any of the various definitions of terrorism that may be encountered. The English word 'terror' comes from the French *'terreur'* meaning great fear. Terrorism then is the use of violence, normally against civilian targets, to induce fear so as to produce a political effect. This means that those who organise acts of terror deliberately seek out a wider target than those who are unfortunate enough to be harmed or killed by the particular event. As Robert Pape (2005) has argued in his study of suicide bombers, these groups are seeking through their conscious exploitation of the fear that terrorism causes, to put pressure on political actors whose policies they wish to change. The immediate act of terror may look physical. But its intended purposes are psychological and political.

Strictly speaking, not all acts of terrorism in Asia qualify as examples of *transnational* security challenges. Some acts of terror in support of the domestic insurgencies mentioned in the previous chapter clearly are designed to influence the political process within an individual country. A number are conducted by nationalist groups whose political agendas do not cross national boundaries. This certainly seemed to be the case in the 1970s with groups such as the Irish Liberation Army (seeking to wrest Northern Ireland from British control) and the Palestine Liberation Organisation (seeking statehood for an independent Palestine from territory controlled by Israel) comprising prominent examples from that period beyond the region focused on in this book. But from Asia at this same time comes an example which confirms that transnational activities are by no means just a recent phenomenon. This is the Japanese Red Army (JRA), which as Patricia Steinhoff (1989, p. 276) explains, was a revolutionary movement which had its origins in a split from the Communist League, the product of an energetic but very divided period of Japanese student politics. One of the JRA factions was responsible for the 1970 hijacking of a Japanese Airlines plane which was diverted to North Korea. The second faction, which had moved to Lebanon, where it was trained by the Popular Front for the Liberation of Palestine, conducted an attack on an airport in Israel in 1972, killing more than two dozen tourists.

Yet it is hard to escape the reality than when people consider the issue of transnational terrorism they are not drawn to these earlier years but to the events of September 2001. The attacks on the United States were a stark reminder that an age of genuinely transnational terror had begun, with groups able to call on cells in various countries to launch attacks designed potentially at the achievement of truly global political ambitions. As a group with affiliates (and copycats) in many countries, al Qaeda was sometimes thought to want nothing less than the establishment of a massive caliphate which would govern much of the world under Islamic law. The apocalyptic religious overtones of the al Qaeda propaganda, and the sheer magnitude of the 2001 attacks on the United States made some wonder whether there was a new age of terrorism in which the limitations of circumscribed political ambitions no longer applied. Even before the 9/11 attacks had occurred, Martha Crenshaw (2000, p. 411) was referring (with some scepticism) to notions of 'a "new" terrorism that is motivated by religious belief and is more fanatical, deadly, and pervasive than the older and more instrumental forms of terrorism the world had grown accustomed to'.

For a time, especially as confusion over al Qaeda's political goals reigned, terrorism appeared to be much less instrumental (as a violent means to a political end). In one example of this reasoning, Morgan (2004, p. 30) argued that: 'Today's terrorists are ultimately more apocalyptic in their perspective and methods. For many violent and radical organizations, terror has evolved from being a means to an end, to becoming the end in itself.' It also became commonplace to argue that if the people carrying out the attack were so obviously willing to give their lives for the cause (motivated by the prospect of a glorious life in heaven) then this was an adversary for whom any amount of violence was valid. This logic in turn fed concerns that the acquisition and use of nuclear weapons by a group prone to terrorism was a matter of 'when' rather than 'if' since, as discussed in Chapter 5, this development was rather less inevitable than is sometimes assumed.

This style of thinking about the new terrorism was strikingly similar to claims about an era of new wars which have replaced the old armed conflicts between states which were once the main source of concern. At the end of the Cold War, Martin Van Creveld (1991) wrote of the relentless rise of wars prosecuted by guerrilla armies and paramilitaries, and the decline of Clausewitzian conflicts fought by the advanced military forces of nation-states for the purposes of policy. The main appeal of these arguments are their simplicity, but war has not changed as much as this logic suggests. Similarly the record of terrorism in Asia suggests a variegated rather than a monochrome picture. National and transnational terrorism exist side-by-side, and groups are just as likely to be motivated by more specific domestic objectives as they are by grander global ones. And if this is indeed the situation, there may be good reasons not to be quite as alarmed about the threat of terrorism across Asia as many were just a few years ago.

This more restrained assessment certainly seems to apply to Southeast Asia, which the George W. Bush Administration labelled as the 'second front' of its war on terror (with the first front being in the Middle East and Central Asia, al Qaeda's main areas of activity). Here there was particular concern about the links and activities of groups such as Jemaah Islamiyah, (JI) which was based mainly in Indonesia, and which was known for having contacts with al Qaeda. These links were emphasised by a number of writers, including Zachary Abuza (2003), who referred to extreme Islamist groups in Southeast Asia as constituting a 'crucible of terror'. In the feverish atmosphere of the time, it was easy to think that JI might be answering to bin Laden's directives, and was the Southeast Asian link in a strong and effective global insurgency.

There was a risk here that, just as analysts decades beforehand had misread nationalist movements in the region as part of a global communist plot directed from Moscow, so they were now tending to overlook the local histories of extremist movements in Asia. As time went on, the notion of a single global insurgency, with regional groups linked in strongly to al Qaeda, began to look rather threadbare. In an important corrective, Sidney Jones (2005, p.172) insisted that JI 'was never an al Qaeda franchise'. JI had a transnational footprint across much of maritime Southeast Asia with its sub-groups working to influence events in Indonesia, Malaysia, Singapore, and also in the Philippines, where they had assisted the Moro Islamic Liberation Front and the Abu Sayyaf Group in bombings. But this did not translate into clear and realisable plans for a wider Southeast Asian caliphate. For JI, Jones argues, the main aim and consistent aim has been the transformation of Indonesia itself into an Islamic state.

It also became clear that al Qaeda was something other than a single global organisation being run from a single location and by a single leader as Osama bin Laden moved from Afghanistan to the Afghanistan-Pakistan border area and then deeper into Pakistan where he was eventually captured and killed in a daring raid by American special forces in 2011. It is more a collection of variants using the al Qaeda brand name in several Middle Eastern and Northern African counties, each focused on violently reversing the political prospects of incumbent regimes, but each also, in the opinion of Daniel Byman (2014, p. 443), adding to al Qaeda's challenges by imposing a greater variety of 'preferences and priorities' as well as 'branding problems, and costly control mechanisms'.

Yet even in cases when the main political causes are local, the results can easily have implications across borders (a clear sign of a ripple effect). Readers may again think that the most significant examples of this exist in the Middle East, including the spillover of the horrendously violent resistance to the Syrian regime into neighbouring Iraq through the activities of the group which has come to be known as Islamic State. But Asia has long featured several examples of this tendency. This is the case for the activities of Lashkar-e-Taiba (LeT), a group based in Pakistan. LeT's most startling terrorist attacks have taken place in India, including an attack on India's parliament in New Delhi in 2001 and the 2008 attack on Mumbai, India's largest city and main commercial centre. This reflects a specific and geographically-circumscribed aim rather than any wider ambition. Stephen Tankel (2009, p. 5) argues that while 'LeT's vision includes establishing a pan-Islamic Caliphate...since 9/11 its primary objective has remained the liberation of Kashmir and the destruction of India.' And yet because activities designed with this primary objective in mind still cross national boundaries (from Pakistan into India), even a local focus can have transnational implications.

Something similar applies in the case of activities conducted by jihadist groups based in Pakistan across the porous border into Afghanistan. These help give Pakistan a reputation for being something of an epicentre for terrorism in Asia. Yet it would be wrong to see this simply as a result of autonomous decisions by non-state actors to take advantage of what many see as non-governed spaces between these two troubled countries. Johnson and Mason (2008, p. 55) assert that:

> The absence of Western state structures of governance in large swathes of the tribal areas should not be conflated...with the absence of governance. Complex and sophisticated conflict-resolution mechanisms, legal codes, and alternative forms of governance have developed in the region over ... millennia.

Moreover, at least some of the insurgent groups operating across Pakistan's boundaries have not been doing so under their own steam. A few years ago Bruce Riedel (2008, p. 31) made the point that 'Pakistan almost uniquely is a major victim of terrorism and a major sponsor of terrorism'. As the reader might expect, this claim is routinely denied by Pakistan's official community, but it is one hint of the strong possibility that states need not always be seen as the consumers of the instability that non-state actors can arrange. They have more agency than some understandings of transnational security suggest.

It would also be wrong, however, to see this situation as typical across Asia and its borderlands. Much of the region is relatively free of the transnational terrorist attacks which have been seen in parts of maritime Southeast Asia (especially in Indonesia) and, in particular, in South Asia. As the 9/11 attacks demonstrated, geography is no automatic protection from terrorist plots devised thousands of kilometres away. But most terrorist attacks are unlike the dramatic events of September 2001. In an early report on the LeT's siege in Mumbai, which killed over 150 people in a range of civilian locations (including a railway station, a hospital and several hotels), Angel Rabasa et al. (2009, p. 22) pointed out that 'one of the most important lessons of this attack is the continuing importance of an earlier operational form: the firearms assault'. Similarly, car bombs, suicide bombs, improvised explosive devices left on roads, and assassinations, all often utilising simple materials, continue to feature prominently.

These violent events do not occur across the region in an even spread but are concentrated in particular places and countries, and normally with specific political effects in mind. In that sense, India has more reason to worry about its direct exposure to further transnational terrorist attacks than Australia or Japan, even though neither of these countries would be wise to rule out the possibility. Indeed Australia also appeared on JI's target list and in 1995 members of Japan's home grown cult Aum Shinrikyo used the nerve agent sarin in an attack on Tokyo's subway. The latter, while not achieving nearly the levels of casualties that its apocalyptically-minded leadership had intended, still managed to create significant ripple effects, at least in the psychological domain that terrorists often exploit. As Box and McCormack (2004, p. 104) note, 'the attack was less successful than initially feared, but it profoundly shocked the nation and the world'.

Indeed while many of the terrorist attacks in today's Asia seem restricted to specific parts of the region, they can still evoke a wider impact. The 2008 Mumbai attack was not just a big story for India and, as suspicions mounted, for Pakistan. Media coverage made this instance of terrorism a global event, as it did for the 2002 Bali bombing. The latter killed scores of tourists, including many from Australia, where Emma Hutchison (2010, p. 77) has suggested in an examination of local media representations, that: 'Fear invoked from the bombing was represented as the product of a potentially wider threat and representations of the bombing evoked a corresponding sense of societal terror.' If it is the case that groups using terror are seeking a wider psychological effect rather more than the initial physical effects of the violence, one might be tempted to argue that transnational terrorism had succeeded. One might also argue that anyone, anywhere in the region can be subjected to the psychological impact of a significant terrorist attack anywhere in Asia and its borderlands, a clear sign of terrorism's ripple effects. If the reader doubts this proposition, they might wish to cast their mind back to remember where they were when they first heard about the 9/11 attacks on the United States.

But does that mean that transnational terrorism has indeed achieved its objectives? The answer to that question need not be in the affirmative. Groups which use terror certainly thrive on the publicity that their attacks create, and then depend upon the wider fear that this broader knowledge generates. But the fear itself, the main psychological impact of terrorism, is still a means to an end. A group using terror will only achieve its objectives if that fear generates the political changes that the group intends. But there are not many cases, including in Asia, where the widespread fear caused by terrorism has created such public pressure that governments have had little choice but to change their policies. In fact, terrorism can often have a counterproductive political effect, generating intense public anger once the initial shock of the attacks has subsided, and strengthening the hands of governments against whose policies the terrorism is aimed. Perhaps the best example of this comes from the 9/11 attacks themselves which fortified American and western resolve.

What comes out of a strengthened opposition to terrorism can of course be a mixed bag. It is easy to argue that al Qaeda's spectacular and highly symbolic attacks were aimed at dragging the United States and its many partners into a long and bitter war in Afghanistan. But this is a questionable example of after-the-fact reasoning. Somewhat more convincing is the broader point that terrorist attacks may allow a non-state group to catalyse an interstate conflict if the group responsible feels that it would gain politically from the ensuing chaos and enmity. Because they are almost guaranteed to galvanise Indian political opinion against Pakistan, terrorist attacks in India by groups based on the other side of the Line of Control have the potential to do exactly that. Another serious terrorist attack in India might just be enough to cause the government of that country to take armed action against Pakistan. For groups who regard an India-Pakistan peace as a betrayal of their interests, including in terms of who controls Kashmir, that prospect might be appealing, raising the prospect of a deliberate attempt to initiate a wider conflict.

Yet even if this attempt was successful and a conflict between India-Pakistan ensured, this would only confirm that groups who use terror are able to cause wider destruction, harm and fearfulness. Politically they would have achieved only negative goals: the absence, for a time at least, of closer cooperation between South Asia's two largest countries. It would not mean that they were able to translate these effects into the sort of political transformation in the region that suits their purported interests. Creating an awful mess in South Asia is not the same as achieving a change in political authority there. As Lawrence Freedman (2007, p. 334) has observed, 'the strategic achievements of terrorist activity have been meager'. Selective terrorist attacks can help raise the prospect that a waning political system will be seen as increasingly illegitimate. But this probably means that acts of terror had more political significance during the quest for political independence from colonial rule than they do in today's largely post-colonial Asia.

Piracy in Asia

If terrorism is an inherently political form of violent behaviour, the same can seldom be said for piracy, another security challenge in the region that stems largely from the activities of non-state groups. Definitions of this practice vary, but much of today's maritime piracy includes the disabling and hijacking of vessels, their crews and passengers and the commandeering of the property that these vessels carry when operating in international waters. Its most prominent examples today occur beyond Asia's borderlands, and are undertaken by Somali pirates off the east African coast of the Indian Ocean. But these attacks on merchant shipping beyond the region still affect the safe passage of seaborne cargo that the globally integrated economies in Asia and its borderlands rely on. Indeed a good number of them, including Australia, India, Japan, Malaysia, Singapore, South Korea and New Zealand, have consequently been involved in counter-piracy operations off Somalia. While a good deal of this effort has been coordinated under a US-led combined task force, it is also notable that China took the opportunity to send naval vessels to the area. This decision has been depicted by Gaye Christofferson (2009, p. 9) as 'a major turning point – China's first operational deployment outside of Asia'.

For some time these pirates had acted with relative impunity because the absence of effective national government in Somalia meant they faced no significant risk of domestic prosecution. Those risks, as Menkhaus (2009, p. 24) has argued, were smaller than they might otherwise have been because local officials were also benefitting from some of the proceeds of piracy. Moreover, the weakness of Somalia's economy, coupled with the decline in the local fishing industry after the arrival of foreign fleets in local waters, generated incentives for piracy as an income-generating activity. Pirates may have indirect political interests in the maintenance of weak rules and enforcement, but their primary motivations appear to be economic ones.

Less prominent globally but of greater direct concern for this volume are the activities of pirates in maritime Southeast Asia. A large part of this picture is the complexity of the archipelagic spaces through which some of the world's biggest seaborne trading routes travel and where some coastline areas (including in parts of Indonesia) are irregularly governed. Among the targets for maritime pirates in Southeast Asia are tens of thousands of commercial vessels which each year ply the Strait of Malacca between the long Indonesian island of Sumatra to the southwest, and peninsula Malaysia and Singapore in the northeast. As much as 40 per cent of global trade, and the majority of North Asia's oil imports, pass through this waterway, which is nearly 1000 kilometres long and which connects the Indian and Pacific Oceans. At its narrowest point, the navigable part of the Strait is little more than two kilometres wide.

The economic incentives for piracy in Southeast Asia grew in the years after the 1996–7 financial crisis in which normal sources of income generation declined. The turn of the millennium was also a time of significant political fluidity in the archipelago which did not help the governance of remote coastal areas. The combination of these factors is captured nicely by Rosenberg (2009, p. 46) who has argued that:

> Sudden and severe impoverishment, especially among marginal coastal seafaring communities, makes piracy a viable way to meet basic needs. For example, the big increase in the number of piracy attacks in Indonesia's waters and ports in the past ten years may be attributed to its sharp economic downturn and domestic instability in the wake of the 1997 currency crisis.

As maritime piracy peaked in these years, the main implications were themselves economic. Shipping companies transiting routes which are vulnerable to piracy can face additional insurance and crew wage costs which are then passed on to the consumers of the products they carry. In one study of piracy on the maritime trade routes from Europe to Asia, Bensassi and Martínez-Zarzoso (2012) found that it is only with more violent incidents, and in particular the hijacking rather than just the boarding of merchant vessels, that trade levels are likely to be reduced. But should shipping companies decide that particular routes had become too vulnerable, the search for alternatives would extend transit times and raise costs further. The economic consequences for a country like Singapore, heavily dependent on its role as a logistics hub for a great deal of maritime commerce, could potentially be severe. At stake might even be Singapore's identity as a stable and prosperous trade enriched city-state.

Along with a number of other countries, including the United States, Singapore also became concerned that shipping containers might be used by terrorist groups to transport explosives which could then be detonated in ports, causing significant damage and casualties, and spooking investors. This sort of worry envisioned a nexus between piracy and terrorism that seems more theoretical than real, a connection which has been rightfully questioned by a number of informed observers, including Young and Valencia (2003). Nonetheless, piracy certainly has some *indirect* political and strategic consequences. As Southeast Asian maritime piracy grew in these earliest years of the twenty-first century, some of the major powers, themselves dependent on the trade through the Indonesian archipelago, indicated a closer interest in patrolling these waters and conducting counter-piracy operations. But deeply sensitive about their sovereign territorial interests in the waters immediately off their coasts, Indonesia, Malaysia and Singapore were reluctant to see the United States, Japan or India take matters into their own hands.

The trilateral cooperation between the three Southeast Asian maritime states that this difference of opinion encouraged, plus improving economic conditions in Indonesia, were two of the conditions which allowed a reduction in piracy in the area. These countries would therefore have welcomed the assessment in a significant report by Bateman, Ho and Chan (2009, p. 20) that the 'situation with piracy and sea robbery in the region appears to be under control'. And they would have been especially gratified to see the judgement that 'There are no grounds for the operational involvement of non-regional countries in providing security at sea against piracy and sea robbery in Southeast Asia.' Before long, the Horn of Africa would become the international piracy hotspot, drawing attention away from Southeast Asia's challenges.

Even so, the experience of the upsurge of this problem in Southeast Asia was a clear indication that non-state groups could complicate sensitive international relationships in the region. They could also lead to new patterns of cooperation, and remind all countries in the region that their prosperity depended on a secure maritime environment that even small actors (without large navies or air forces) could disrupt. In other words, in such regionally significant locations such as the Malacca Strait, even small incidents could easily have ripple effect.

The drugs and guns trades

Some of those same principles are on display in the wider realm of transnational criminal activities in Asia. By international law, piracy is itself a crime, although as Collins and Hassan (2009) suggest, the gaps between existing statutes (including the relevant provisions of the United Nations Convention on the Law of the Sea) do not always allow for a consistent or effective approach to this problem in the region. Even so, there are other transnational criminal activities which need to be brought into consideration in terms of their potential implications for regional security. Here, the smuggling of drugs and light weapons stand out as two of the most significant issues. Like pirates, drug syndicates and illicit weapons dealers and smugglers are more likely to be motivated by economic than political goals, and both exploit weaknesses and lapses in regulation and governance in particular parts of the Asian region.

For example, the porous borders between Afghanistan and Pakistan and between Afghanistan and Iran provide crucial pathways for a significant amount of the international heroin trade in which Afghanistan is the world's leading country of origin. This heroin has its origins in opium poppy farming, a popular means of making a living for many landowners in Afghanistan, where economic problems have made it one of the only reliable sources of income, but which in turn has made the Afghan economy over-reliant on an internationally illegal product. Parts of this trade are controlled by Taliban groups who have a presence on either side of the Afghanistan-Pakistan border, the proceeds from which help fund the trade in weapons and insurgencies in both countries. But much of the trade is controlled by Afghan warlords, some of whom are in cahoots with corrupt government officials. The net effect is a strengthening of the challenge to the national government in Kabul and a weakening in public confidence in the official sector. These domestic implications are not to be taken lightly. As Goodhand (2008, p. 412) has argued:

> A complex pyramid of protection and patronage has emerged, providing state protection to criminal trafficking. This assemblage of actors, networks and institutions, like the opium economy itself, is extremely footloose and flexible, and patterns of capture and corruption can shift across ministries and other institutions in order to evade regulatory mechanisms.

But it is beyond Afghanistan that the real potential for the ripple effects of the drug trade emerge. En route to lucrative markets in Europe the trafficking of heroin from Afghanistan has not only left Iran, and also Tajikistan, with growing addiction problems among their respective populations. In addition, scores of Iranian security personnel have been killed as violence has erupted with the drug traffickers. Experiencing a political vacuum after the collapse of the Soviet Union and flooded with drugs from across the border, Tajikistan has been identified by Paoli et al. (2007) as Afghanistan's fellow 'narco-state'. A great deal of the violence that occurs around the border between the two countries, which runs to over 1000 kilometres, is associated with the narcotics trade.

Similar patterns are seen in Myanmar, which is the world's second largest country of origin for heroin. A major amount of that trade has its origins in the Wa State area on the border with Thailand. This has been a major funding source for the United Wa State Army, a non-state armed force composed of tens of thousands of personnel. At least for a time this non-state military organisation had a gentlemen's agreement with particular members of the Burmese army, who were prepared to sanction the trading of drugs by groups which had agreed to oppose insurgencies against the central government. This is but one sign that transnational security issues result from the conscious actions of decision-making individuals and organisations and are not simply problems that arrive in a random or autonomous fashion. And it does not stop there. The Shan State, in which the Wa State is a sub-region, is also the location for very extensive methamphetamine production facilities. The flow of millions of these pills into Thailand appears to have had similar effects to the Afghanistan cross-border trade: a growing addiction problem in Thailand and local security challenges for the Thai armed forces which complicate existing problems on the Thai-Myanmar border. The so-called Golden Triangle of the Southeast Asian drugs trade which links Myanmar, Thailand and Laos, and which has been examined by Chin (2009), clearly has transnational security implications in terms of the sharing of instability and the intensification of violence.

Indeed the lucrative nature of the drugs trade has had some unexpected implications for regional relations in the wider region. Extraordinarily short on hard currency, the North Korea government is widely thought to profit from the trading of drugs through official and officially sanctioned channels. While its own drug production has been associated with a rising domestic methamphetamine addiction problem in recent times, one of the best-known incidents involving North Korea occurred in 2003 and had its impact felt on the other side of the region. This was when the *Pongsu*, a North Korean vessel, was seized by Australian authorities, who found on board over one hundred kilograms of heroin. Interestingly, the vessel was registered in the tiny Pacific Island country of Tuvalu, a flag of convenience country for foreign vessels. The complexities grow further in the knowledge that, as Sheena Chestnut (2007) recounts, the North Korean government has often worked with criminal gangs in the region who are involved in the drugs trade.

These groups can find some parts of the region easier to penetrate than others. Often short on enforcement abilities, and with many islands scattered over vast distances, South Pacific countries are also sometimes used by criminal syndicates to move and hide drugs en route to market. Indeed when justifying the intervention it was leading in Solomon Islands in the same year, the Australian government argued that the breakdown of law and order there might well offer an opportunity for transnational criminal syndicates to operate with relative impunity. Transnational criminal activity in Asia and its borderlands are not *caused* by state weakness, nor is state weakness in the region caused by transnational crime: each has a tendency to exacerbate the other.

The possibility for a vicious security cycle between state weakness and transnational security problems also seems evident in the case of other challenges. As was clear in Chapter 5, small arms and light weapons, which include the ubiquitous Kalashnikov AK47 semi-automatic rifle, have for many years been associated with a majority of the deaths in internal conflicts around the world, including in some of the most contentious hot-spots in Asia. Within the region, a proportion of these weapons have transnational origins. In the Philippines, for example, insurgent non-state groups including the MILF and the Moro National Liberation Front (MNLF) have acquired some of their light weaponry from overseas sources, helped by that country's challenging coastal geography, which offers many potential points of transit. Even easier, it would seem, is land-based transit in the region. The Thai-Myanmar border area, for example, has been a significant transit point for weapons headed to insurgent groups in Bangladesh.

But the notion that weapons simply cascade into regional countries from vast supply networks is an oversimplification of a much more complex reality. A good many of the weapons possessed by non-state groups are recycled, often internally. The Tamil Tigers' stocks of arms were often seized from Sri Lankan government forces. Underpaid soldiers and military units in Indonesia and the Philippines have been known to sell their weapons to insurgent groups, even those they are purportedly fighting. Old weapons left over from the Second World War have been dug up and used in Papua New Guinea and Solomon Islands. And, as David Capie (2002) has observed, the ending of long decades of civil conflict in mainland Southeast Asia and especially in Vietnam, released a significant quantity of light arms into the region, including weapons originally used and supplied by the United States.

Asia's small arms and light weapons challenges cannot be understood without reference to the role that nation-states, as the main original purchasers of weapons, play in the process. There are very important national dimensions, as well as transnational ones, to this particular problem. The state isn't quite as feeble in all of this as it may sometimes seem, and it may often be responsible for the problems rather than an unwilling and passive victim of them. Unintended effects are also highly probable. When the United States began supplying the Afghan mujahideen with armaments after the 1979 Soviet invasion, little could it know that some of those same groups would be targeting its own soldiers in the first decade of the twenty-first century. 'Blowback' is the term that Chalmers Johnson (2004) has provocatively used to depict the long-term consequences of American attempts to fashion influence in Afghanistan and elsewhere. At least in the case of the supply of weapons, it seems a fitting expression and a warning to others. Indeed it is hard to know what will come in the future from the small arms and light weapons which China has sold, for example, to Pakistan and Bangladesh.

The flow of people

It is too easy then to explain cross-border flows as products of 'globalisation', as if the mere utterance of this over-used word demonstrates a series of inevitable and inexorable linkages beyond the control of the region's most significant political actors. Nation-states, as Ian Clark (1998) has argued in a wonderful book, are not weak victims, let alone are they passive bystanders in transnational transactions. They are instead the main arbiters between global processes and domestic politics. States do get a say in globalisation. And even when the flows of goods, money and persons remain unregulated, they are not necessarily in violation of the agreements that states have between them. This point is highlighted when one considers people flows in Asia and its borderlands.

The millions of people who travel through Asia, crossing borders by land, sea or air, are participants in an increasingly dense network of transnational connections. Facilitated by international laws which are often taken for granted, and overseen by customs departments and, in some situations, border police, the vast majority of these interactions are both legal and regulated. Unless travellers are carrying hazardous contraband with them as they travel, or engaging in espionage, or encouraging an insurgency in their country of destination, the security implications of their transit can be few and far between. In fact these security implications can well be positive, if one holds to the liberal view that increased international contact (or what are often called 'people to people links') boosts common understanding and reduces the chances of conflict. At the height of the Cold War, for example, J. David Singer (1958, p. 95) referred his readers to the 'UNESCO approach' to conflict reduction, which holds that when peoples from different countries have opportunities to meet one another (including through travel) 'this will lead to increased mutual tolerance, understanding, and respect and a consequent reduction in tensions between them'. This philosophy is part of the hope for regional cooperation which has often been vested in ASEAN and its array of offshoots, whose impact on Asia's security will be considered in Chapter 10.

But whether or not they generate peace, flows of people can also be stimulated by violence and the threat of conflict, as groups flee oppression and privation, often in unregulated migration. Asia's recent history includes the tragedy of very large displacements of people, including between one and two million refugees who fled from Vietnam, Cambodia and Laos in the mid-1970s as a consequence of the internal conflicts in mainland Southeast Asia. The Soviet occupation of Afghanistan, which began at the end of that same decade, led to the flight of millions of refugees into neighbouring countries. While many have since returned to Afghanistan, as have many Sri Lankan refugees to their home country, over a million Afghan refugees are registered in Pakistan, a country whose internal challenges have led to a significant displacement of people internally.

Indeed not all displaced persons cross international boundaries, as is the case with the Muslim minority Rohingya people who over several decades have been fleeing Myanmar's impoverished Rakhine state. As Parnini (2013) recounts, earlier waves of displaced Rohyinga people have meant security concerns – both objective and subjective – for Bangladesh. And in recent times, many have sought to migrate by sea, braving the Bay of Bengal. But many have also moved to camps inside Myanmar's sovereign borders. Likewise, while tens of thousands of Uzbeks fled across the Kyrgystan border into Uzbekistan in 2010, even more were displaced by conflict elsewhere in Kyrgystan itself.

Victims of oppression who leave their countries of origin under their own steam, and who have genuine reason to fear that they would be in danger if forced to return, should enjoy rights of asylum under international law. But not all of the countries of the region have signed up to international agreements recognising the rights of such refugees. The reasons for this vary, although Sara Davies (2006) suggests a common thread in the concerns among Asian countries that the main existing international instruments are Eurocentric and did not include their concerns when they were drafted. Moreover, domestic opinion in some countries makes it difficult to accommodate larger numbers of people than have already arrived. The larger commonality, however, is that the security dimensions of the crisis of displaced people in the region, who are numerically concentrated in South and Central Asia, are more pressing in the places from which they have come than the places at which they arrive.

Internal conflict, associated with a number of the domestic insecurities we considered in Chapter 7, is often the big driver here. The flow of people says as much about civil disorder as a cause as it does about the transnational security consequences that may be produced after migration has occurred.

Yet the implications for the countries of destination, and the places of transit in between, should not be overlooked. Australia, for example, has become a favoured destination for tens of thousands of migrants – from Afghanistan, Sri Lanka, and Iran, many of whom move initially to Indonesia before boarding notoriously leaky boats for the risky trip across to the north of Australia. For the Australian government the issue is more than the fact that these persons are seeking to land on Australian territory without permission, visas or passports, the frequency of which has been enough to highlight the challenge of policing the country's vast northern maritime approaches. Concern also relates to the mode of transit as professional people smugglers, including those operating in Indonesia, illicitly organize dangerous transits. This introduces a very obvious transnational criminal aspect into the picture. In fact Ralf Emmers (2003, p. 434) suggested some time ago that at least in the case of Southeast Asia, transnational crime 'could be dealt with more effectively at a regional level if it was approached primarily as a criminal problem rather than as a security issue'.

However one wishes to approach these issues, their ability to affect relations between states in the region is not in question. With the numbers of boats laden with displaced persons fluctuating, Australia's expectations that Indonesia would and should do more to stem the tide has on occasions strained relations between these two neighbours. And efforts to prevent these asylum seekers from landing on the Australian mainland, and instead moving them to neighbouring Papua New Guinea and the smaller Pacific Island country of Nauru has the potential to place additional strains on the already complicated law and order issues in these countries. This is in no way a suggestion that desperate migrants, whether fleeing persecution, poverty or both, themselves have in mind the creation of these ripple effects. But this can be the unintended consequence of their sometimes involuntary movements to different parts of the region in the quest for personal safety and prosperity.

Quite how the human security concerns of the millions of displaced persons in Asia can be reconciled with the national security concerns of countries in the region remains to be seen. In Australia's case, the realisation that these migrants do not pose the security threat that the political debate there suggests may be part of the answer, but as was noted in Chapter 1, security is as much about psychology and perception as it is about concrete challenges, and these perceptions are not easy to shift.

Scholars should keep asking thoughtful questions about these practices. Alison Mountz (2011, p. 118) argues that Australia and other states routinely 'exploit legal ambiguity, economic dependency, and partial forms of citizenship and political status on islands to advance security agendas'. These are of course national rather than human security agendas. But it should not be surprising to see these political choices being made if flows of people are deemed within public opinion to be a national security issue of some concern.

Are transnational plagues regional security issues?

The importance of security perceptions is tested by another transnational challenge: the potential for disease spreading from one part of the region to another, bringing with it a range of damaging consequences. The flow of human beings around Asia is of course the main vector for the spread of disease from one country to another, a fact made clear by the movement of the Severe Acute Respiratory Syndrome virus in late 2002 and early 2003. Originating in southern China, the disease quickly moved across to Hong Kong and cases also appeared, thanks to the speed of air travel, in Vietnam, Mongolia, Singapore, Taiwan, the Philippines, Thailand as well as cities in both Canada and the United States. Of the more than 8000 known cases, nearly 10 per cent of those infected with SARS died from the virus.

The spread of the virus would have been far more severe were it not for strict public health reactions in a number of countries and a degree of cooperation between them, including a level of co-ordination among ASEAN countries that Caballero-Anthony (2005, p. 486) has called 'unprecedented'. Local reactions included the isolation of infected persons, something that on a much larger scale would probably be beyond many of the countries of the region. Issues of public order, and the trade-off between freedom of movement and association on the one hand, and the control of the spread of the virus on the other, suggest particular challenges for some of the more open societies in Asia and its borderlands. The declaration of states of emergencies and restrictions on internal movement and international travel are also conceivable parts of the reaction to a similar or worse health crisis, if and when it occurs, although this is likely to involve a further trade-off between the short-term economic gains from tourism and other travel-related industries and the public health implications of the pandemic. As part of these national responses, the armed forces and police would likely be called on to provide order in an especially severe pandemic.

A new bird flu strain, which might arise in southern China where fowls and people live in close proximity and big concentrations, could spread quickly if transmission were easy and if there were a too brief incubation period. Some scholars, for example Andrej Trampuz et al. (2004, p. 523), compare a possible new bird flu pandemic to the Spanish influenza pandemic which killed as many as fifty million people around the world at the end of the First World War, and which itself may have originated in China. Should even a considerably smaller disease event occur, which still found its way into far flung parts of Asia and beyond, the ingredients would appear to be in place for a very obvious ripple effect. It is quite possible that people in the entire region would be affected by disease aided and abetted by normal transnational connections including air travel. Asia could be the epicentre of a global tragedy which might occur without a single weapon being used.

But it might be wondered if it is wise to regard such an event as a security problem per se. In line with Ralf Emmers' questioning of the wisdom of treating criminal behaviours as security issues, there is an ongoing debate about whether pandemics and other problems should be incorporated within security agendas, both in an academic and policy sense. The specific term introduced to explain this questionable phenomenon is 'securitisation', coined by members of what has come to be called the Copenhagen School of security analysis. '"Security", write Buzan, Waever and de Wilde (1998, p. 23), 'is the move that takes politics beyond the established rules of the game and frames the issue either as a special kind of politics or above politics'. Becoming accepted as a security issue allows, among other things, the allocation of special resources, including the armed forces. And it can denote a sense of emergency and imminence that other challenges are unable to achieve.

Whether the reader wishes to subscribe to securitisation theory is not the point here. The real need is to ask whether the spread of transnational disease would get different, and perhaps more effective, responses if it were to be treated simply as a transnational problem for public health. That is certainly a point that is raised by Christian Enemark (2007) in an important study of disease as a security issue in East Asia. One way of navigating through this maze is to suggest that while not everyone may be convinced that the spread of transnational disease is in and of itself a security issue, in the way that the spread of weapons system might be, it can still be argued that on some occasions a plague may have security implications. For example, if social order was affected in some regional countries as people sought to leave their places of residence or if there was a mass panic for vaccines thought to be in short supply, then it might be held that the disease in question had generated security issues.

But diseases quite different to bird flu might well have these impacts in a more chronic and initially less perceptible way. In some regional countries, diseases with slow incubation rates and long-term effects can also have important social consequences for order. The spread of HIV/AIDs in Papua New Guinea for example, and the pneumonia which often accompanies it, has placed huge pressure on struggling health systems in that country. As Ronald May (2012, p. 55) has suggested, the poor performance of PNG's health system exposes its citizens to yet other diseases. But in these cases as well it would probably be wise to continue to make a distinction between the undeniable human tragedy that widespread deaths from a pandemic might cause on the one hand, and the implications for political order on the other. Some problems can be serious and can threaten lives and livelihoods, but it is not clear that they should always be treated as security challenges.

Climate change and Asia's security

The real test for the extent to which Asia's security challenges are increasingly transnational ones most probably comes from the regional implications of global climate change. Asia and its borderlands include both major and minor emitters of the carbon that international scientific opinion overwhelmingly agrees is responsible for the warming of the earth's atmosphere. And as the region's share of global economic activity grows so too does its share of global emissions. And whether the country in question is a major Asian power (and a significant emitter) or a small island country with little or no direct impact on global emissions, there is a good chance that the effects of climate change are already being felt.

Among the most significant of these will come from the rise in sea levels, which is partly a consequence of the melting of polar ice. This particular change is having uneven effects in the region, partly because some countries have low-lying coastlines and deltas, while others, like Mongolia or Uzbekistan, are completely landlocked. Some of the most severe consequences are predicted to be in store for some of the smallest Pacific Island countries which comprise low-lying atolls, a form of geological feature which is concentrated in the Micronesian area. There is a widespread impression, as reported for example by Nicholls and Cazenave (2010, p. 1519), that the small Micronesian country of Tuvalu, already hard put to support human life, may simply become permanently inundated by the tides and disappear from the map. And if one regards the ability of a human community to continue to exist within its place of origin as a question of security, this would seem to count as an especially challenging example of that view.

Some of these fears are disputed. Mortreux and Barnett (2009) argue, for instance, that some of the direst scenarios for Tuvalu overlook the possibility of local adaptation strategies. Yet it is clear that rising sea levels will have very considerable implications elsewhere in the region, partly because of the significant concentration of populations in some Asian countries in deltas and other low-lying coastal areas. Bangladesh, with its densely populated delta areas, is particularly susceptible to this development, and the loss of arable land, soil and food production is likely to be among the effects. The movement of environmental refugees away from such areas will generate some of the complexities that come from the displacement of persons which have already been described in this chapter. Indeed, Rafael Reuveny (2007, p. 658) notes that Bangladesh has been experiencing similar issues for some considerable time. The negative effects on agricultural productivity from an already difficult climate featuring floods and droughts was a major factor behind the movement of over ten million of its people to India from the early 1950s. If an increase in climate variability led to similar flows of people, it would create major challenges for India-Bangladesh relations and for social stability in parts of east India.

This is a reminder that the effects of climate change extend well beyond sea level rise. Perhaps the most obvious change to effect the whole region, but to different extents, will come from an expected increase in extreme weather events. Already dry areas with low rainfall and limited water sources (including in parts of China and India) are expected to become drier. The reverse is also the case: some already wet areas will experience increased extreme rainfall and associated flooding, which can be both economically devastating and also physically dangerous. But even when rainfall totals remains relatively constant, they may be delivered through a smaller number of extreme weather events, including major storms. The delay in the monsoon, or its disappearance in certain years, would have a catastrophic effect on food production in large parts of South and Southeast Asia, not to mention the difficulties for the people who live there to get water for drinking, cooking and washing. The failure of normal precipitation would also reduce the snowmelt from South Asia's mountains and glaciers, reducing the flow of rivers upon which India's agriculture and industry depends. In one prominently published account, Immerzeel, van Beek and Bierkins (2010, p. 1382) point to the particular vulnerability of the Brahmaputra and Indus basins which they say 'are most susceptible to reductions of flow, threatening the food security of an estimated 60 million people'.

Quite what these developments mean for Asia's security in overall terms remains a matter of some conjecture. Competition over water sources, made scarcer by less reliable rainfall patterns upstream, is often seen as a potential cause for armed conflict. Many of the world's most strained river systems, for example, which cross the boundaries of upstream and downstream states alike, are in Asia. This includes the Mekong River system, which begins in the Tibetan plateau, and runs through China's Yunnan province before entering each and every one of the mainland Southeast Asian countries: Myanmar, Laos, Thailand, Cambodia and Vietnam. It also includes the Ganges River which begins in the Himalayas and ends up in Bangladesh. To the extent that climate change reduces either the regular river flows, or simply makes these flows unreliable and inconsistent, a further point of pressure on an already strained basic resource is added. China's huge dam projects, which are already having a deleterious downstream impact on the Mekong area, are not making things much easier. Alan Dupont (2008, p. 34) has suggested that if China were to go ahead and

> channel the waters of the Brahmaputra to over-used and increasingly dessicated Yellow River... tensions with India and Bangladesh are likely to rise, as existing political and territorial disputes are aggravated by concerns over water security.

The potential for ripple effects here appears to be inbuilt. And given the already challenging political relations between some of these countries who share the same river system, it is clear that water competition can be part of rising tensions. What is less certain is that the regional water conflicts that analysts such as Chellaney (2011) are suggesting will come to pass. It may be safer to think that in this regard climate change is unlikely to become a direct, and most improbably a sole, cause of conflict. For an interstate dispute over water in Asia to contribute to the decision to use force, an existing degree of political animosity would probably be needed. And analysts would then face a difficult after-the-fact analytical job in trying to tease out the particular influence that climate-induced water stresses had played.

If climate change is going to have a serious security effect in Asia and its borderlands it may be more likely to do so by exacerbating some of the domestic strains on political and social order *within* regional countries. The environmental effects that accompany climate change are more likely to put an additional strain on domestic governance, reducing local food supplies, making some areas unfit for human habitation, and thus increasing the internal flows of displaced people. Herman and Treverton (2009, p. 137) put it thus:

> For most states climate change tends to be under the radar, intersection with and exacerbating existing difficulties such as economic weakness, infrastructural shortcomings, communal strife, weak governance and tenuous political legitimacy, often with spillover effects beyond borders.

This certainly seems to apply in Asia. Some of the countries likely to be most affected by climate change, including Bangladesh, some of the poorest states of India, the poorer urban areas in Vietnam and the Philippines, and a number of the low-lying Pacific Island countries, are those for which ensuring reasonable standards of living to growing populations is already proving a substantial challenge. But while climate change may increase some of the existing social strains, once again it will be difficult to isolate it as a determining factor in its own right. In part it will depend on what local political leaders, and their challengers, decide to do. The interaction between two sets of complex systems – natural and social – will be an uncertain one.

To the extent that climate change increases the frequency of catastrophic weather events in Asia – including devastating floods and typhoons – the normal first responders to security problems, including the armed forces and police, are likely to have a role. When natural disasters have a transnational impact, cooperation between the affected countries and other regional states can sometimes occur as a reflection of common interest. New Zealand and Australia, for example, regularly offer assistance to Pacific Island countries who have been affected by extreme events of this type, and that response relies heavily on the armed forces of both countries for transport and emergency relief. But the ease of this assistance depends again upon political relationships, which remain the most important element of our study here. For example, in 2008 Cyclone Nargis struck Myanmar, killing more than 100,000 people, and displacing at least a quarter of a million more. But stuck in its pre-reform era, the Burmese military regime, which remained suspicious of the motives of others, refused most of the offers of assistance from regional countries. This in turn was regarded as a major shortcoming in the national government's appreciation of the immense human security crisis that the cyclone had produced. But as Andrew Selth (2008) has argued, what to outsiders looked like perfectly non-threatening offers of genuine assistance were read by some in the regime through the lens of long-standing concerns about the possibility of invasion. This takes the picture right back to the interaction between objective and subjective understandings of security.

Conclusion

This chapter has been completed by considering the types of challenges which Gywn Prins (1993) referred to some years ago to as 'threats without enemies'. These are still threats to things that communities value, and in extremis the effects of climate change could pose a threat to a community's very survival, if that community is already, for other reasons, very vulnerable. In that sense, in terms of Wolfers' definition of security that was discussed in Chapter 1, it seems quite logical and reasonable to regard at least some of the effects of climate change as potential security issues. As problems that naturally extend across national boundaries in Asia (environmental challenges are rarely respecters of borders), the transnational security label also seems to fit.

But when one compares the transnational effects of climate change in Asia to the problem of transnational terrorism, the stark differences become clear. Threats of terror, and terrorist acts themselves, are deliberate acts by groups of purposeful human beings. They are threats *with* enemies. And unlike disease or climate change, acts of terror are not indirect causes of potential violence (should political relations subsequently break down). They are, in and of themselves, acts of organised violence. Moreover, unlike some of the transnational challenges which have been considered in the second half of this this chapter, terrorist acts are politically motivated. Indeed terrorism is strategic behaviour in a way that many of the other transnational challenges simply cannot be.

It would be difficult to find a security analyst who is willing to leave terrorism outside their basket of security issues in today's Asia. Terrorism seems by its very nature always to be part of the security fabric. But it is easier to have debates about whether some of these other problems really do fit. This is not to suggest that transnational crime or the rapid spread of a virus cannot cause harmful effects. But it is to ask whether insecurity is in play whenever harm of some sort is seen to be occurring. It is not the job of this book to insist on either a narrow or a broad definition of security; but it is important to raise precisely this sort of question. Some writings, including a widely cited study by Daniel Deudney (1990), have proposed that it is confusing to include environmental problems, for example, in one's definition of security. But others are happy to be much more inclusive. In one especially comprehensive treatment of Asia's transnational security issues, Alan Dupont (2001, pp. 14-15) points to a possible middle ground where all sorts of problems can be entertained as parts of our security calculus so long as they increase the chances of conflict. The reader may find this a useful way forward.

But even then there is still the question of what is to be made of the full range of transnational issues which have been considered in this chapter. If they are to be considered as a group, it is not altogether clear that their transnational nature is the main reason for determining whether they have strong security implications or not. What they seem to have in common is their close linkage not to the prospects for interstate conflict, but to the domestic security prospects of a range of Asian countries. Those parts of the region that seem already fragile politically, and for whom issues of social order are constant challenges, seem to be the ones for whom these 'transnational' problems are especially troubling.

That might suggest that it is the role of stronger and more stable political units to help Asia's weaker states with these difficulties. The temptation to intervene in another polity's affairs is a recurrent one in international politics, although it must be said that its record in Asia is at very best a mixed one. And the idea that an outside country, or group of countries, can help two adversarial states locked in tensions or a bitter armed conflict also finds its supporters. These attempts by outsiders (and sometimes insiders) to make a difference are the subject of the next chapter.

Chapter 9

Interference: Can Intervention Work in Today's Asia?

The portion of this book which evaluates the numerous security issues and challenges confronting Asia and its borderlands, and which examines what these problems tell us about the nature of Asia's regional security, is now completed. As has been demonstrated, some of these challenges have a greater capacity to generate security effects in the wider region, while others are more likely to remain localised in their impact. But this is not the end of the road for understanding Asia's regional security. It is also necessary to consider the potential actions which might be taken in dealing with security challenges in the region, from the steps that individual states may take to the possibility of wider regional responses.

One way of thinking about various responses to regional security problems is to conceive of them as strategies. As this author has argued elsewhere (Ayson, 2008) there is a close relationship between the study of *security*, which examines the nature of a particular environment (in this case the Asian region) and the study of *strategy*, which looks at how actors respond to and seek to shape that system. In part strategy is about what a political actor does to achieve its aims: how it uses resources (or means) to pursue objectives (or ends). But strategy is not just about the reconciliation of a particular actor's ends and means. It comes alive when consideration is given to the interactions between the ends–means calculations of two or more of these participants. Sometimes these interactions will be cooperative, with the states of Asia working together to shape the regional security environment and to deal with its problems. But at other times the interactions will be competitive, as in situations when two or more parties find that they have conflicting interests. It is rare for there to be a pure form of either cooperative or conflicting interaction. As Thomas Schelling (1960) has argued convincingly, most relationships involve a mix of cooperation and competition. Even the closest of security partners in Asia can have differing perspectives, and even the fiercest of adversaries can find at least some common ground.

In the light of at least some of the Asian security challenges that have been considered in this volume, it might be thought that extensive strategic cooperation in the region should be forthcoming. For example, some of the threats without enemies which were investigated at the end of the previous chapter appear to be common problems facing almost every country in Asia and its borderlands. Working together would therefore appear to make good sense and there should be few reasons to feel threatened by that cooperation. But that is not always, or even often, the case. Well meaning attempts to solve security problems in the region can sometimes draw the ire of countries who disagree with the diagnosis or with the prescription, who believe that they are being identified as the real threat, or who feel they might be next to come under pressure. This can even happen when countries in the region have grouped together to conduct something as helpful as a peacekeeping operation. Opposition is even more likely when an intervention occurs without the consent of the country in whose territory the action is taking place. This is especially the case when the use of armed force is at the forefront of an intervention, and it is possible even when the countries contributing those military instruments view themselves as working selflessly for a good cause. These are the fascinating issues which this chapter will now address.

Understanding intervention

In a very broad sense, an *intervention* occurs each time a government decides that its armed forces (or other resources) will become part of a conflict, including in an attempt as a third party to escalate, stop or resolve a war that others have begun. Stern and Druckman (2000, p 37) suggest that interventions 'include any action undertaken to change the course of a conflict process'. Hence, when the United States despatched carrier battle groups in Taiwan's direction to resist China's campaign of military intimidation, it can said that it was intervening in the 1995-6 Taiwan Strait Crisis. From China's perspective the United States was interfering in a domestic dispute which had been ramped up by Taiwan's quest for independence.

To the extent that this was an intervention it did not lead to armed combat between any of the participants. War across the Taiwan Strait has continued to be averted since then. But there are also some very clear examples of interventions which have occurred in interstate military disputes in Asia which have involved significant levels of violent conflict. In 1941 the United States was forced to intervene much more directly in the Second World War when Japan attacked Pearl Harbor. James Cotton (2001, p. 208) has called Asia the 'continent of intervention', listing such examples as India's intervention in Eastern Pakistan in 1971 (part of the violent chain of events leading to the establishment of an independent Bangladesh) as well as the Soviet Union's intervention (really its invasion) of Afghanistan in 1979. And should an armed conflict arise between Japan and China, the United States (Japan's main ally) would then face an interesting decision as to whether to intervene in that North Asian war.

But a more restrictive understanding of intervention can be found in most contemporary analysis, including in discussion of events in Asia. Thus an intervention is generally considered to occur when one or more states makes an attempt to change the security circumstances *within* another sovereign country. This domestic interference can take a number of forms, but it becomes especially noticeable and controversial when the use of armed force is part of the action. In particular, when troops move into the territory of another country, crossing internationally recognised sovereign boundaries as they do so, there is an immediate challenge to the principle of non-intervention, one of the building blocks of the international system of states. This issue is evident when S. Neil MacFarlane (2002, p. 15) equates intervention with 'coercive military interference in the internal affairs of another state'. But even some non-military elements of intervention can be coercive, including the removal of economic assistance to encourage a change in behaviour by the target state.

Opposition to intervention has particular resonance in large parts of Asia and its borderlands where many countries were still subject to colonial control only a few decades ago. This is a major reason for the widespread tendency across the region to regard sovereign authority as sacrosanct. As has been noted already in this book, the Association of Southeast Asian Nations was built around the idea that its members would refrain from interfering in one another's affairs. Additionally the memories of the century of humiliation when western empires and Japan carved out special rights in coastal portions of China in the pursuit of economic and political gains have left Beijing with a very strong determination to resist intervention. The Chinese government's sensitivity (shared by many of the more authoritarian regimes in the region) to external criticism of human rights practices and its crackdowns on secessionist sentiment is also part of this picture.

This is not to suggest that the norm of interference has been observed punctiliously in all cases, or that countries across Asia, like those elsewhere, are unable to observe double standards (where they can intervene and others can't). The Cold War experience offers a mixed record in this regard. The superpower practice of exploiting Asia's complex internal wars, including wars for national independence, proved to be an especially hazardous form of intervention. America's bitter experience in Vietnam, one of the most significant examples of that practice, was in the end a lesson about what not to do. While there was a time when some of Southeast Asia's insurgencies received external support from elsewhere in the region, including from China, this practice eventually also came to an end. This trend was partly a reflection of the growing ability of the third world countries (as many parts of Asia were then known) to resist intervention by larger and stronger powers: something that Hedley Bull (1984) noticed had also become the case in Africa and Latin America. It was part of a growing global opposition to intervention, a form of resistance that many western commentators and governments have been seeking to reverse in more recent years.

From peacekeeping to intervention?

For most of the Cold War period, there was little in the way of an active international machinery to support interventions of almost any kind. The United Nations Security Council could in theory utilise Article VI of the charter to authorise a cooperative military action in the event of a serious threat to international security. But breaches to international peace and security tended to be defined conservatively in accordance with the experience of the Second World War: i.e. major attacks by one sovereign state on another. Moreover, given the Cold War divisions that separated the United States and the Soviet Union as veto-holding permanent members, the Security Council was unable to find a consensus when major interstate breaches of the peace were taking place. There was one exception to this rule and it came in Asia in 1950 when the Council authorised a mission to repel North Korea's aggression against the South, although this was only possible because the Soviet Union was absent when the vote was taken.

But the Korean experience did not spell the start of a tide of similar missions. Instead, what the Security Council found possible as a compromise would become known as traditional or 'blue helmet' UN peacekeeping. Here a force authorised and organised by the UN would be despatched to monitor ceasefires or truces between belligerent parties. In other words it would be sent not to prevent or stop a war, but would be put in place once hostilities had been reduced to a less intense level. These missions were lightly armed and faced severe restraints on their ability to use their limited capabilities. 'The force available to peacekeepers' in these circumstances, as Dandeker and Gow (1997, p. 333) have noted, 'is defensive and passive'. Moreover such missions would only be sent if the conflicting parties gave their consent, a reinforcement of the norms of state sovereignty that were written into the UN Charter.

Notwithstanding the scale and intensity of Asia's armed conflicts in the middle of the twentieth century, which have already been noticed in this volume, it can hardly be said that the United Nations was busy authorising missions to the region during the Cold War. One of the lonely examples was the despatch in 1962 of about twenty UN observers to what had been part of the Dutch East Indies and would become the Indonesian province of West Irian Jaya (West Papua). These personnel were deployed to monitor the tenuous ceasefire which had been agreed between the Dutch and Indonesian forces. The latter had infiltrated the area under the military command of Suharto (who would later depose Sukarno as Indonesia's president). This modest deployment was followed by a 1500-strong United Nations Security Force to support the work of the United Nations Temporary Executive Agency, which oversaw the province until it became Indonesian territory in 1963.

But this was not Asia's first encounter with a UN mission. Several years earlier, the United Nations had sent military observers in 1949 to monitor the ceasefire between India and Pakistan in the very sensitive Jammu and Kashmir area, following the dreadful conflict that had been associated with the independence of both countries. This observer force, which went by the lengthy acronym of UNMOGIP (the United Nations Military Observer Group in India and Pakistan), would remain in place for a long time. However, fighting between India and Pakistan would break out again on more than one occasion over Kashmir. The 1965 version of this recurring conflict meant that UNMOGIP was 'rendered useless', according to Virginia Page Fortna (2003, p. 23). But 22 years after it had been first established UNMOGIP was busy monitoring the ceasefire which followed the further resumption of conflict in 1971. Its role continues today, although for the last four decades India has argued that its focus should be directed to a revised Line of Control, a position that Pakistan disputes.

As this challenging South Asian example illustrates, traditional United Nations peacekeeping efforts were concentrated (with mixed success) on the maintenance of the peace between sovereign states. But very little scope was allowed for making the peace in the first place, or for responding vigorously to violence if and when it returned. Either or both of these functions would have required much stronger forces being sent into the field, with more robust rules of engagement. But also missing, including in Asia, was much determination within UN circles to get involved in the domestic security affairs of a sovereign state member, even if this is where so much of the organised violence was occurring. In other words, what seemed to be really needed, if the peace was to be genuinely kept, was an ability to deploy more capable forces and a willingness, in some circumstances, to breach the selective tradition of non-interference.

It is probably unwise to see the ending of the Cold War as a decisive turning-point in all the ways often claimed for it. But it is unquestionable that the immediate post-Cold War years of the early 1990s saw a sea-change for the standard restraints on UN peace missions. The first of these restraints, the reluctance to deploy more capable armed forces, gave way in one of the most famous missions ever authorised by the United Nations. This did not happen in Asia, but was the UN authorised response to Iraq's invasion of its neighbour Kuwait and to its threats against Saudi Arabia. That force, supported by evoking Chapter VII of the UN Charter, destroyed much of Saddam Hussein's fighting power. While this action did not overthrow the regime, it certainly removed Iraq as a threat to its neighbours for years to come. Yet rather like the Korea experience four decades earlier, the UN's deliberations over Iraq in 1990 and 1991 did not begin an avalanche of similar missions. At the time President George Bush referred to a 'new world order' of collective security, in which the UN would act as its founders intended. But as a response to the more classical, but now rare, aggression of one state towards another, Iraq was an exception to the types of conflict that would occupy the UN in an era of more ambitious operations.

Instead it was a UN mission of a different sort, in the middle of East Asia, which would really set the international tone. This was the establishment in October 1991 of the United Nations Transitional Authority in Cambodia (UNTAC). War-torn Cambodia was still reeling from the legacy of the Khmer Rouge, whose murderous reign the world had largely stood by and watched. It had also been subjected to Vietnam's unilateral military intervention, which had ended in 1989. There were familiar elements to UNTAC, including the supervision of a ceasefire which had been observed by an earlier UN mission. But the new effort had a much wider mandate, including the organisation of elections, the disarmament of the armed forces of political movements inside the country, and the control of major agencies of the Cambodian state. This was globally as well as regionally significant: Michael Matheson (2001, p. 77) notes that while the UN had been involved in some very general administration of a number of Pacific Island territories immediately after the Second World War and had administered Irian Jaya for a matter of a few months in the early 1960s, its Cambodia experience was '[t]he first major UN exercise in governance' anywhere in the world.

While this game-changing mission had a suitably strong international feel in terms of the countries which contributed to its total force of over 15,000 personnel, there were also some important regional dimensions. The Chief of Mission was from Japan, and the military commander was from Australia, whose forces made a significant contribution (at times in conjunction with New Zealand's). Of Cambodia's Southeast Asian neighbours, Brunei, Indonesia, Malaysia, the Philippines and Thailand contributed personnel, as did Bangladesh, India and Pakistan, all three of whom have made strong contributions to a range of UN missions around the globe. And China, an erstwhile supporter of the Khmer Rouge, sent a large detachment of engineers to Cambodia in 1992 – its first sizeable contribution to an international peace mission and a major step forward for China's increasingly confident external involvement. It was also a sign that China's commitment to non-intervention will not preclude its involvement in peacekeeping missions when, as Chin-Hao Huang (2011, p. 267) notes, its national interests benefit.

Given the weight of Cambodia's awful recent past, evaluating the success of a mission such as UNTAC probably needs to be undertaken in relative rather than absolute terms. One obvious aim that was not achieved was the disarmament of the Khmer Rouge, although in cooperation with the UN in 2003 Cambodia would establish a tribunal to try the main Khmer Rouge leaders for crimes against humanity, war crimes and genocide. But that there was a government to seek the UN's assistance in this matter was a sign that some form of functioning political authority was acting in Cambodia. Indeed, while he accepts that the UN mission he led had mixed results, the Japanese diplomat Yasuhi Akashi (2012, p. 154) has reason to insist that 'UNTAC played an indispensable role in bringing to birth a new Cambodian polity'. Not all has gone well since. On some human security indexes, including individual freedom from political repression, Cambodia continues to struggle. But it is not hard to think that without UNTAC, the country's challenging political evolution would have been even less satisfactory. And it is not easy to identify better options (at least in terms of measures that had a chance of being carried out) which would have produced a superior outcome.

Regional interventions: East Timor and Solomon Islands

It would be incorrect to see the UN's involvement in Cambodia's complex and sensitive post-conflict internal environment as an isolated instance: at least internationally, it started to take a more ambitious approach to peace missions and nation-building in war-ravaged countries. This shift was made more urgent by the international community's subsequent failure to act decisively over the Rwandan genocide in 1994 and the limited ability of UN forces under a more traditional mandate to stem the violence in former Yugoslavia (including the tragedy in Srebrenica). Yet the more robust approach that was in the making directly challenged the traditional principles of impartiality, the minimal use of force, and the requirement to gain the consent of the conflicting parties. And few examples of this approach were forthcoming in Asia until violence flared after a referendum conducted in East Timor in 1999 found the local population decisively in favour of independence from a post-Suharto Indonesia. Abetted by encouragement from parts of the Indonesian armed forces, anti-independence militia groups, including some from West Timor, quickly wreaked havoc, causing widespread death and destruction.

This violence was too extensive for the UN observer mission that had been overseeing the referendum. But the chances of a larger and more heavily armed UN contingent being despatched quickly were slim, given political sensitivities in the region over interference and the complex and lengthy logistics involved in preparing UN missions. Instead, working on the basis of a generous Security Council resolution which provided a mandate for strong measures to be used if needed (including the use of force), the International Force for East Timor or INTERFET (International Force for East Timor) was organised under Australia's leadership. This was an extraordinarily risky undertaking, illustrating the dangers that more assertive interventions can bring, even when they are shaped for humanitarian purposes. The intervention caused a serious breach in relations between Indonesia, whose government stood accused domestically of losing East Timor, and Australia, which could easily be depicted as a neo-imperialist neighbour bent on undermining Indonesia's sovereignty and security. It was New Zealand, another western-oriented country in the region, which offered the most significant and combat able assistance to Australia. And it was an American president, Bill Clinton, who had been foremost in pressuring the Indonesian government to seek international assistance in the first place. In the view of Wheeler and Dunn (2001, p. 819), the most influential of these moves came in the form of the threat of financial sanctions from Washington.

These largely non-Asian contributions were not all that promising to the idea of a more inclusive regional response to an East Asian crisis. And given the strong commitment among so many ASEAN member countries to the principle of non-intervention, it is not surprising that some of Indonesia's neighbours were reticent about joining in. But even Malaysia, sometimes a critic of Australian policy in the region, provided a small force, and more substantial contributions were forthcoming from the Philippines, Singapore and Thailand, the last of which also provided the mission's deputy commander. This regional participation increased the significance of the Timor deployment for Asian debates about sovereignty, and showed that at least some of the ASEAN countries were willing to take a less than purist approach to the principles of non-intervention. These themes are discussed in more detail below, but it is probably accurate to say that the East Timor intervention was a rarity for the region. Hopes were held out at the time, in the words of Alan Dupont (2000, p. 167), that 'a substantial ASEAN contribution...in East Timor' could act as 'a benchmark for future regional security cooperation'. But this has not proven a repeatable experience. Certainly for Indonesia it represented a line in the sand. Part of the repair work to relations with Australia included the 2005 Lombok Treaty between the two countries, which was designed principally to ensure that future Australian governments would not agitate for the independence of West Papua.

Had things become violent between Australia and Indonesia in 1999, the INTERFET experience might well have been an example of how an intervention in the internal affairs of another country can lead to an interstate conflict. However, the physical resistance that was feared from parts of Indonesia's armed forces was not forthcoming and in time the pro-independence militias were overcome. Replaced by a UN mission in 2002, this Australian-led intervention offers a mixed record of success. But it is indubitable that a newly independent nation-state has arrived, reflecting the will of the majority of its inhabitants. And this new country has had rather better relations with its big neighbour Indonesia than some might have expected. But while INTERFET and the subsequent UN mission did succeed in establishing a more secure environment, the uneven growth of the institutions of the new East Timorese state, and especially some of the more crucial ones, underlined the fragility of that progress. Violence involving the new defence and police forces was one of the reasons for the return of Australian and New Zealand forces in 2006, and an indication that complex interventions can result in much longer commitments than initially anticipated.

Similar conclusions were being drawn at very much the same time in the South Pacific borderlands of Asia, to which another Australia-led intervention, the Regional Assistance Mission to Solomon Islands (RAMSI) had initially been despatched in 2003. Three years later Australian and New Zealand troops returned to the capital city of Honiara on the main island of Guadacanal, where riots underlined the fragility of the new political settlement that RAMSI supported. There had in fact been an earlier opportunity for a regional response. In 2000, the elected government of the Solomon Islands had been ousted in a coup, as an armed group comprising migrants from Malaita (another major island) was struggling for control of land and the political authority that went with it. As this stand-off with Guadacanal island leaders continued, effective government in the Solomon Islands ground to a standstill. And as concerns grew about the impossibility of normal life for the residents of the Solomon Islands, and about the implications of an effectively ungoverned South Pacific country, the Australian government eventually overcame its own resistance to getting involved in a complex internal conflict. This change in view also coincided with growing international concerns about weak and failing states. Notwithstanding the challenge of comparing South Pacific security circumstances to those being experienced in Somalia and Afghanistan, Elsina Wainwright (2003, p. 486) captured that mood by arguing that '[f]ailed or failing states are often

Petri dishes for transnational criminal activity such as money laundering, arms smuggling, drug trafficking, people trafficking, and terrorism'.

RAMSI initially featured a significant military component to deal with resistance from armed gangs and any flare-up in communal violence on Guadalcanal. But it increasingly relied on a strong presence by police from Australia, New Zealand and Pacific countries. Also significant was the presence of seconded officials who were sent to help the main Solomon Islands state agencies be in a position to deliver financial, justice, social and educational services. When Australian and New Zealand forces left nearly ten years after the initial commitment of RAMSI, it could not be said that Solomon Islands was without future economic and social challenges (including population pressures). There is something to the warning from Kabutaulaka (2005, pp. 302–3) that 'intervention might create a quasi-functioning state that is able to restore order and serve the interests of the intervening forces, but it often does not address the underlying causes of civil unrest, nor can it build long-term peace'. The RAMSI mission, which was a response to a request for help from the beleaguered Solomon Islands' parliament, and was supported by the Pacific Islands Forum (the main body for regional cooperation in the South Pacific), appears to have given that country a chance. But on the whole while interventions can create an opportunity to improve difficult situations, they cannot guarantee that the opportunity will be taken. Whether local governments are willing and able to make the best of such chances remains an open question.

Counter-insurgency: the Afghanistan experience

These three examples from different parts of the region Cambodia, East Timor and the Solomon Islands – certainly do not amount to an intervention avalanche for the region. While in each case the benefits of the action seem to have exceeded the significant costs involved, the appetite for similar exercises in interference has actually dropped in Asia, as it has in other parts of the world. It is important not to be too sweeping about this altered view on the wisdom of intervention. But it does seem that there has been a growing chorus of opinion that interfering in the sovereign affairs of weak states can result in long, difficult, and uncertain missions, of which the original aims risk being forgotten as the local situation takes on a life of its own.

A major reason for this change can be found in another experience in Asia's borderlands where external forces sought to bring security to another country's internal circumstances. This was the NATO-led mission to Afghanistan, which followed the 9/11 terrorist attacks on the United

States and growing international concern that weak states could provide breeding grounds for political actors who could reach out and harm other states with relative impunity. Many and varied are the accounts of this long and difficult attempt at bringing security to Afghanistan. But it can be said with some confidence that what was intended at the start of this mission was something other than a protracted counter-insurgency campaign. And the difficulties experienced in this attempt are reflected in the view from Adam Roberts (2009, p. 51): 'Afghanistan may contribute to greater caution before engaging in interventionist projects aimed at reconstructing divided societies.'

It is certainly apt to refer to the war waged by largely western forces and their allies in Afghanistan as an intervention. When they were initially deployed in late 2001 these forces were tasked with making some significant changes to the domestic military and political situation within that country. They were certainly not invited in by the Taliban regime in power at that time in Kabul, so had to enter Afghanistan territory without the consent of what then comprised the national government. Indeed the early aims of the mission included the toppling of the Taliban government, which was accused of offering sanctuary and encouragement to Osama Bin Laden and his colleagues who had orchestrated the 9/11 attacks. Another aim was the military destruction of al Qaeda and the capture of its infamous leader. With the help of the Northern Tribal Alliance, these aims were for the most part quite quickly achieved. However, large portions of the Taliban and some remnants of al Qaeda survived the onslaught as they retreated deeper into the countryside, where, as Seth Jones (2006, p. 116) notes, they would quickly adjust their strategy.

Nonetheless progress also seemed clear in the creation of a sufficiently secure environment around Kabul and some of Afghanistan's other leading population centres to allow for elections to be held. Hamid Karzai, who had been provisional head of the country since 2002, was elected to the first of two five-year terms of office as Afghanistan's president in 2004. On the surface at least, this would seem to be evidence for the argument that military interventions can be strategically successful in an especially direct manner: the ability to translate military success into political progress. And by 2004, other aspects of the Afghanistan mission were having an impact, including the vast amounts of international aid that were flowing into the country and the efforts at nation-building in several provinces. These involved the building of roads, schools and medical facilities and were undertaken by a series of Provincial Reconstruction Teams.

But as these outwards signs of progress appeared, at least two disturbing trends were also becoming evident. First, the United States, which had been the leading contributor of ideas, armed forces and money for the war in Afghanistan, had become distracted by another conflict situation. In early 2003, Washington had led an invasion of Iraq for dubious reasons, including the judgement that Saddam Hussein's regime had strong connections to the transnational terrorist network which had perpetrated the 9/11 attacks. This diminished America's ability to focus on Afghanistan in an important period of transition for that country. The second problem was that while it was certainly true that al Qaeda had been heavily degraded since the end of 2001, the Taliban groups, who had initially been dispersed, were now actively developing the insurgency in Afghanistan which was discussed in Chapter 7.

The aims of that insurgency, to the extent that it can be spoken of as a single effort, included reducing confidence among Afghanis about the Karzai government. Another objective was the removal of foreign militaries, which could be depicted as propping up Karzai. His election, according to Thomas Johnson (2006, p. 22), had been little more than a 'reification of long held ethnic biases and conflicts'. The insurgency presented a clear and increasing challenge to NATO-led forces in the country, which by now were part of a comprehensive and ambitious stabilisation mission. With Iraq, Afghanistan presented one of the most important examples of this approach. Stabilisation missions operated on the beguilingly simple logic that it was the role of the intervening armed forces to create a secure environment in which other agencies could help bring stronger institutions of order to a post-conflict situation. Aid agencies, legal and political advisers, education and medical specialists, engineers, and non-government organisations who were sometimes contracted to provide additional services, were part of this wider package of assistance.

In situations where armed opposition to new political arrangements is a spent force or easily overcome, the logic of stabilisation can proceed reasonably well. But Afghanistan after 9/11 was not like Cambodia after the Cold War and the ruinous rule of the Khmer Rouge, or Solomon Islands on the eve of RAMSI's entry. The fragile attempts at gaining a monopoly on armed violence by the intervening forces and the slowly emerging national army of Afghanistan was being directly challenged by the Taliban's continuing violence. Nation-building in Afghanistan would have been difficult in the best of circumstances, not least because the institutions of the modern nation-state had little to work with. The challenges of filling that virtual vacuum, which included the complexity of coordinating so many international actors, has encouraged the view that the main pathway to stability came from a comprehensive interagency effort, sometimes called a 'whole of government' approach. But the real problem was not whether the intervention in Afghanistan could do a good job of connecting resources to objectives. It came instead from an armed insurgency whose political claims were benefitting from the corruption evident in the Karzai government.

The response was a shift, at least in terms of the rhetoric of the mission, from a mix of war-fighting and stabilisation to a counter-insurgency campaign. That shift was also going on in Iraq, but had clear historical precedents in Asia. This connection became evident when American strategists began to speak of the need for NATO forces to 'win the hearts and minds' of local Afghan communities, wresting them away psychologically from the sometimes beguiling intimidation of the Taliban. That phrase may remind many readers of Vietnam, where the United States had developed an approach to counter-insurgency that experienced some success, despite the eventual political result of the broader war. But it was a phrase coined by Gerard Templer, who had been Britain's High Commissioner to Malaya in the early 1950s. This approach, which sought to win the hearts of minds of villagers from the influence of communist guerrillas in the 'Malayan Emergency', has been heralded as one of the best examples of counter-insurgency practice. Some of its elements, including the forced resettlement of villagers, would have been politically impossible in modern day Afghanistan. But the notion that an external force should focus first on the safety and well-being of the local population, as opposed to prioritising the destruction of the insurgent forces, wove a powerful spell. Singapore-based scholar Kumar Ramakrishna (2001, pp. 91–2) has argued that the main effect of Templer's approach in battling the Malayan communist insurgency was 'not physical but psychological: the creation of the conviction among government officials and especially the rural Chinese that the Communist threat could and would be neutralised'.

At least two main observations pervaded the playbook on counter-insurgency for which the Afghanistan experience was writing an additional chapter. The first was the premium placed on local knowledge. Understanding the grievances of local populations, which the Taliban forces themselves often understood and exploited well, was part of this. So too was understanding local political, economic and social dimensions. A lack of sensitivity to religious practice and belief, or to other local customs, for example, could completely undermine the overall effort. The second, which followed from the first, was that great patience was needed as this expertise was built up and acted on and as relations of trust were developed between the external forces and the populations they were seeking to protect. These are some of the messages which resonate in the best account of Afghanistan-style counter-insurgency warfare, which has been written by a young British officer, Emile Simpson (2013).

At least one immediate challenge to this approach was apparent: the use of force to seek out and destroy Taliban fighters was still regarded as necessary, and the unintended consequences of this military action, including civilian casualties, created negative publicity and drove a wedge between the NATO-led mission and the Karzai government. But it was patience on the home front that was running out. The logic of counter-insurgency which demanded a long mission ran up hard against the desire among western publics and their political leaders for an end to the mission in Afghanistan, at least in terms of the presence of over 100,000 military personnel, the majority of whom came from the United States. In that sense at least, despite significant differences, the domestic political need for the United States to bring its forces home from Vietnam (which meant an acceptance of the unification of that country under communist control), was played out again in Afghanistan with long-term implications which are not easy to foresee. But it was very clear that despite years of training, the Afghan national military and police forces would struggle to deal with Taliban violence in the way that the external forces had done. The conclusion drawn some years ago in a RAND report by Younossi et al. (2009, p. 60) that the Afghan National Army was 'a long way from being able to assume primary responsibility for Afghanistan's security' is likely to remain valid long after the main foreign forces have left the country.

It is unsurprising that the combined difficulty of Afghanistan, in Asia's borderlands, and Iraq, which lies just beyond them, has sapped international willingness for interventions which carry the risk of long, costly and inconclusive missions. After Vietnam, admittedly a much more harrowing experience for the United States, it took many years for confidence in significant military interventions to be rebuilt, and not just in connection to Southeast Asia. The contemporary sense of exhaustion caused by Afghanistan and Iraq is likely to have a profound long-term effect. Beyond Asia, America's desire to restrict its involvement in the limited intervention in Libya and its even greater caution about the Syrian civil war, even after the chemical weapons attacks of 2013, were clear signs of this. So too was the reliance on airpower and the aversion to using ground forces which characterised American thinking as the international coalition against Islamic State was organised the following year. Saudi rather than American leadership of an air campaign which began in Yemen in 2015 might also be deemed a further sign of Washington's limited appetite for getting too involved in responses to local rebellions.

But within parts of Asia, local governments will continue to find they have insurgencies to deal with. That is Afghanistan's lot after the withdrawal of NATO troops, and it is an ongoing challenge for neighbouring Pakistan as well. For these local governments the waging of counter-insurgency is not a matter of choice in a distant country. It can be a question of national survival. Yet the willingness of others to get heavily involved cannot be counted on.

Justifying interventions: a changing canvas

The waxing and waning of political enthusiasm for intervention is not the only mobile element in this discussion. Also changeable have been the arguments that have been employed to justify interventions in the first place, especially when these have involved military force. For example, despite the hopes of George H.W. Bush, if there was to be a new world order after the Cold War, it would not come in the form of consistent international resistance to armed aggression across state boundaries. It seemed instead to be centred on the idea that vulnerable peoples within established national boundaries might need assistance in the event of significant internal violence or some other sort of catastrophe. Accordingly, many of the missions (UN-sponsored and otherwise) that were carried out in the first decade of the post-Cold war period were exercises in *humanitarian* intervention, defined by Douglas Stuart (2001, p. 32) as 'coercive action by an outside Government or an authorized agent directed toward or within another State, in order to alleviate or avoid a mass humanitarian crisis'.

It is quite clear that the attempts to ward off further human suffering in East Timor in 1999 represented an important Asian example of this kind of approach, which the fitful international response to the events in Rwanda and former Yugoslavia had stimulated. The new way of thinking was reflected powerfully in the call from UN Secretary General Kofi Annan (1999, p. 49) at the time of the East Timor intervention for the international community to recognise a second form of sovereignty. As well as the sovereignty of states, around which the UN system had been built, Annan asserted the sovereignty of individual persons which had 'been enhanced by a renewed and spreading consciousness of individual rights'. The implications were clear: governments could not do as they pleased within their own borders but had obligations to promote the individual rights of their citizens. And when they failed to live up to these obligations on a grievous scale, it might be necessary and just for the international community to step in.

It was a small step from this argument to the doctrine that in circumstances of major humanitarian distress, members of the international system of states had a responsibility to protect (R2P) civilians by intervening in the domestic affairs of other state members, including by force, if needed. This notion of R2P was articulated in an influential report by the International Commission on Intervention and State Sovereignty (2001), which also drew on the expanding tradition of human security discussed earlier in Chapter 7. For under the logic of human security, the security interests of human groups and of individual human beings are given precedence, including, when necessary, at the expense of national security. In this sense the responsibility to protect can in fact look like a responsibility to intervene, although with the cautionary proviso that the use of force should apply only when other options have been exhausted. Even so, it might be wondered what this means for the principle of non-intervention which has been such a standard-bearer for the system of states, and which is held in very high regard in much of Asia. It might also be wondered quite how this particular episode of tension between principles of order (non-intervention) and justice (humanitarianism) has arisen.

It may be supposed that these arguments were simply products of major events in which humanitarian distress had been so evident. In Asia this might be seen as going back to the excesses of the Khmer Rouge and moving forward to the violence in East Timor. Ideas about human security in Asia could then be seen as product of that environment. But it is probably more appropriate to regard ideas as conscious rather than automatic responses to the environment, and as deliberate attempts to shape it. If so, one still needs to ask what has made opinion-makers on questions of security and decision-makers on questions of intervention shift the goalposts. One influential view, presented by Martha Finnemore (2004), is that changing attitudes to intervention can result from social processes as elites from various countries exchange views, pressure groups (including international human rights campaigners) seek influence, and as leaders make persuasive claims for the acceptance of new norms of behaviour. In other words, changing perspectives on security do not simply reflect the problems which attract thoughts of intervention. They are also socially constructed. This offers an interesting connection to the argument from Wolfers that security has a subjective dimension as well as an objective one.

Yet at least two further variations still need to be explained. First, the justifications for intervention would change in the very early years of the new century and in a form which overshadowed what had been a growing emphasis on humanitarian purposes. In the wake of the 9/11 attacks, catastrophic terrorist violence tended to be regarded by governments and leaders less as an affront to human security in general and more as a specific threat to individual national security. The George W. Bush Administration certainly did try to posit humanitarian justifications for the 2003 war against Iraq. But, as MacFarlane, Thielking and Weiss

(2004) argued soon after the invasion, this did more to discredit than strengthen the R2P cause. While the use of force in Afghanistan had greater international support, and arguably did less damage to these evolving norms, it is hard to make the claim that humanitarian purposes dominated the decision to intervene there in late 2001. Similarly the new age of counter-insurgency was strongly connected to concerns about the security of states; Afghanistan's national security and the national security interests of the various countries who were contributing forces. The emphasis on understanding the needs of local communities within that logic offered some connection to considerations of human security. But this was a means to an end and not an end in itself. The battle for hearts and minds was seen as a necessary step in reversing the gains that the Taliban had been making so as to ensure space for a moderate and civilian-led government.

The second conundrum is that the social construction of new humanitarian norms for intervention has gone only so far in Asia and its borderlands. In some cases, these ideas came up hard against the reality of quite different political outlooks, which continue to insist strongly on the principle of non-interference. Ambrosio (2008) argues that China and Russia have used the Shanghai Cooperation Organisation to promote authoritarian norms in Central Asia. This is a social construction of values which oppose (rather than support) humanitarian intervention. Moreover questions of power also diminish the idea that there should be a universal liberal principle in favour of intervention on humanitarian grounds. It is doubtful that even a much larger version of the violent political repression that occurred in China in 1989 with the deaths of protestors in Tiananmen Square would see any sort of physical intervention mounted by external actors within the People's Republic. China's material capacity to resist any such attempt, as well as the influence it could wield against others, courtesy of its economic strength, make such a response particularly unlikely. And because China (and its partner Russia) has a veto in the Security Council, there is almost no chance of a UN-sanctioned intervention, unless Beijing surprised everyone by requesting assistance.

Intervention without boots on the ground?

But even when direct physical intervention is regarded as unacceptable, unwise, or impossible, there still may be ways for external actors to influence the course of events within another country. Force can even sometimes be used, for example, without the deployment of a large stabilisation or counter-insurgency mission. For example, the United States has not led the same sort of intervention in Pakistan as it had in

Afghanistan despite the existence of Taliban forces on both sides of the border between those two countries, and despite the fact that Taliban operating from the Pakistan side have routinely crossed that boundary to conduct hit and run operations. But such an intervention has remained unlikely in the absence of a clear casus belli (which 9/11 provided for the campaign in Afghanistan) justifying the deployment of troops to Pakistan, and in the presence of strong Pakistani public opinion critical of American policy and jealous of Pakistan's sovereignty. Yet the United States (and a number of its NATO allies) have remained concerned about the Pakistan government's ability and willingness to exert control over insurgent groups operating from Pakistani territory.

The answer, instead of the despatch of forces – which Pakistan could easily treat as a most obvious violation of its sovereignty enshrined in international law – is the practice of drone strikes against Taliban leaders. These strikes by small pilotless aerial vehicles, controlled from the United States, have certainly succeeded tactically on a number of occasions. Several leaders, including in 2013 the head of the Pakistani Taliban, have been killed. But the legality of this activity remains somewhat dubious and political opinion within Pakistan has been severely inflamed. It has created for Pakistan's government a delicate tightrope to walk between domestic opinion and its quiet interest in working with the United States to combat the Taliban. These attacks may also be politically counterproductive. Sikander Ahmed Shah (2010, pp. 87–8) has suggested that 'the defenselessness of the Pakistan government against US armed attacks bolsters the morale of the extremists as proof of the government's inability to move against them effectively'.

In turn, the reluctance of the United States federal government to reveal as much about its target selection principles for drone attacks as might be expected from an open society has also been problematic. That reluctance has been strongest in the case of attacks in Pakistan, but Trevor McCrisken (2013, p. 107) points to evidence that greater transparency about the decisions to use drones might actually bolster America's reputation there. As one might expect, there is also a body of opinion that very limited military action conducted from the air alone is unable to produce the effects that troops on the ground can sometimes achieve. The result is that drone attacks, which have been known to kill people other than their intended targets, may be the best of a set of suboptimal options in a time when the desire to commit more fully to a conflict has been reduced.

There are also non-military options for policy-makers that might meet the threshold test as cases of intervention. The option that comes most readily to mind, because it has been used prominently in Asia and its borderlands, is the use of economic sanctions, a practice which has already been mentioned in Chapter 4. At times sanctions have been used deliberately in a quest to influence domestic political arrangements within another sovereign state. A favourite target for such measures have been military regimes in Asia and its borderlands which have had a reputation for restricting human rights and preventing democratic political change. In 2003, for example, the United States imposed stiff economic and travel sanctions against Myanmar's military junta which had been in charge of the country since 1988, and which had placed under detention that internationally celebrated political activist Aung San Suu Kyi. These sanctions came on top of a ban on US investment in Myanmar which had been imposed in 1997.

While these moves allowed the United States to send a very clear message about its unhappiness with the regime's domestic policies without any hint of military coercion, a good deal of controversy has arisen about their net effects. First of all, the ban on US imports of goods from Myanmar clearly affected employment conditions in one of Southeast Asia's most impoverished countries. This led Donald Seekins (2005, p. 444) to argue that the US 'sanctions disproportionately impact the people of Burma, not its military'. Depending on how broad a view is taken of human security, it might be wondered whether the benefits of highlighting political rights in Myanmar were outweighed by the cost to economic livelihoods.

Secondly, differences of opinion within the region about the wisdom of punishing Myanmar, and the principles of non-interference, meant that the US action was not properly part of a united approach. Among the other leading proponents of sanctions against Myanmar was the European Union. But this risked generating the perception in some quarters that western countries were adopting a colonial mentality in dictating a particular view in Southeast Asia. This is not to suggest that Myanmar was completely escaping criticism from its ASEAN colleagues. That pressure certainly grew as the first decade of the twenty-first century went on. But it was still an uneven spectacle. Jürgen Haacke (2008) has noted significant differences between the activism of democratic Indonesia at one end of the spectrum and the conservatism of authoritarian Vietnam at the other. This makes it much harder for any united ASEAN front to be more than window dressing, and reflects continuing disagreements about how much, if any, intervention is justified.

Thirdly, the isolation of Myanmar by some western governments also left plenty of space for Myanmar's great power neighbour China, which has been happy to work with governments of all stripes (authoritarian as well as liberal). This is one example of the choice that those who impose sanctions normally have to make: cutting economic ties is designed to shape domestic events in the target country but can also remove a point of influence. This is precisely one of America's difficulties in its relationship with North Korea, where a history of narrow links offers Washington few levers of direct influence. This is not to suggest that sanctions are automatically and always ineffective. Sufficient political reform had occurred in Myanmar by 2013 to allow the United States to remove sanctions and engage with the changing government there. Yet it is not entirely clear how much credit can go to the sanctions themselves as an effective form of external pressure.

Governments need to be careful about the unintended consequences that sanctions may bring. Unhappy with the latest in a series of military coups in Fiji, and with the 2006 event which brought an oppressive regime to power under the leadership of Frank Bainimarama, Australia and New Zealand imposed some limited sanctions on their South Pacific neighbour. These included travel bans on leading figures in the military government. In part because of the severe resistance that Fiji's army would offer, but also because of the precedent it would set, New Zealand and Australia generally concluded that the use of force against Fiji was not a realistic option. Even the scope of measures short of the use of violent force has been circumscribed. As Matthew Hill (2010, p. 112) argued:

> an escalation to overt armed pressure against Suva has been ruled out publicly and repeatedly, a reflection of Canberra and Wellington's belief that the menace of armed intervention could only harden the stance of the military regime, and entrench its public support.

Yet Australia and New Zealand became concerned that as their own connections with one of the South Pacific's central countries became more distant, Fiji was taking the opportunity to build a closer security relationship with a rising China. And in another sanctions-related question with unintended consequences involving China, Beijing's unhappiness over an EU arms embargo that was imposed after the Tiananmen Square repression in 1989 has not entirely redounded to Europe's benefit. As European countries have come to see China as an important economic partner, the lack of a more rounded relationship with Beijing which these sanctions have encouraged has not always been helpful.

Mediation

While sanctions do not involve the use or threat of force for political purposes, they can still generate resistance because of the coercion they involve. They still cause harm, and despite all the attempts at making economic sanctions more specific and targeted at elites and their interests, wider damage can still be caused. One alternative to the various forms of coercive intervention is mediation. This is where a third party, which lacks a direct interest in the dispute at hand, seeks to work diplomatically with the involved parties to help them resolve at least some of their differences. Of such efforts internationally the best known are probably the enormously challenging external mediation attempts which have been undertaken in several bids to bring Israeli and Palestinian negotiators together and to narrow their differences. But there are also several important examples of mediation which have focused on unresolved conflicts in Asia and its borderlands. Some of these initiatives have even had successful outcomes, something not to be overlooked given the complexity of many of the security problems which have been examined in these pages, and the sometimes counterproductive impact of other approaches.

The external mediation of the Acehnese conflict is one such case. An independent sultanate before it became part of the Dutch East Indies, Aceh was then incorporated into newly independent Indonesia, where it enjoyed decreasing levels of autonomy from the central government in Jakarta. Internal conflict broke out as an Acehnese independence movement fought Indonesian forces from the early 1950s. From there followed cycles of promises of autonomy and disappointment and a recurrence of conflict involving what became the Free Aceh Movement. The collapse of the Suharto regime in the late 1990s did not instantly resolve the conflict, but left greater room for international efforts. These came first from a Geneva based non-governmental mediation organisation and then from the former Finnish president, Martti Ahtisaari. The devastation caused by the Indian Ocean cyclone of December 2004 certainly generated a greater sense of common cause among the Acehnese and Indonesian negotiators. But it is also clear that the repeated efforts over several years by European-based mediators helped identify the possibilities for an accord which went some way to greater autonomy without going as far as the creation of an independent Acehnese state.

This was a fairly rare example in Asia of the successful employment of what Michael Vatikiotis (2009, p. 31) has called 'the tools of mediation – dialogue and compromise and binding agreement'. But it also required particular conditions to be in place. These included a situation where the various political actors had decided for themselves, or could at least be persuaded, that it was better to agree to a settlement which was less than optimal for any of them than to continue a conflict which had already been mutually damaging. In other words, there was a need for a recognition among the adversaries that further fighting would not pay. That seems also to have been the case in the attempts by external mediators to encourage a settlement between the Papua New Guinea government and representatives of the province of Bougainville, where in 1989 secessionist leaders had proclaimed a republic. In the following year the Endeavour Accord, mediated by New Zealand, was struck, which promised the return of public services to Bougainville which had been withdrawn by the central government in far away Port Morseby. Unlike its close neighbour Australia, New Zealand was not PNG's former colonial power, nor was it a major supplier of arms to the PNG government. As Jim Rolfe (2001, p. 46) has argued, this gave New Zealand a natural advantage as a mediator which its bigger and stronger neighbour could not match.

But New Zealand's good offices were not enough in themselves at this first attempt, and for the first half of the 1990s a violent civil conflict would rage on the island. It was only when a relative stalemate had been reached, and both sides had become aware of the futility of further violence that a more durable accord was possible. In 1997, following renewed efforts, a truce was agreed and signed at the Burnham military camp in New Zealand. This was followed by the deployment of a Truce Monitoring Group, which took advantage of New Zealand's understanding of Pacific cultures, and which also was assisted by the peace efforts of women's groups on Bougainville. That the Monitoring Group was unarmed said something about the intent of its sponsors, but also indicated that the armed conflict had by that stage almost completely ceased. What followed, however, was a rather uneasy process towards Bougainville's increased autonomy, where a smooth outcome was far from guaranteed.

This condition of ripeness for the cessation of conflict, which according to I. William Zartmann (2001, p. 8) relies 'on the parties' perception of a Mutually Hurting Stalemate', is not as ubiquitous as might be hoped. Signs of a stalemate can also be undone as the politics of a situation evolve. That certainly seems the case for efforts to bring an end to the conflict in the southern Philippines, where no less than three separate (and often feuding) insurgent groups have been seeking independence. The first of these, the MNLF, signed an agreement for limited autonomy after years of fighting with the Philippines government in 1976. This deal, brokered by none other than Libya's Muammar Gaddafi, split the MNLF and was also rejected by a second group, the MILF. In 1996 the MNLF signed a subsequent peace agreement, mediated by the Organisation of Islamic Cooperation, with Indonesia taking a leading role, and the following year the MILF signed a further agreement with the Philippines government. But neither these efforts nor Libya's and Malaysia's roles in encouraging a unity agreement between the MILF and MNLF in 2001, brought an overall end to the conflict. In 2012, the

Philippines government and the MILF signed a framework agreement for autonomy, but a third group, Abu Sayyaf, which has employed terrorist attacks and kidnappings in the name of an overlapping secessionist cause, remains outside that process. Significant international and regional initiatives have also been at work, including an International Contact Group featuring Japan, Britain and two Muslim majority countries from outside Southeast Asia, Turkey and Saudi Arabia. But the situation has remained difficult to resolve.

Conclusion

This brings the discussion back to the original point of departure for this chapter. Rather like traditional peacekeeping, with its restrictive principles of consent, impartiality and the minimal use of force, mediation in Asia has been limited in what it can do unless the parties themselves have been ready to end their conflict, or at least to take a break from it. In contrast to more robust military interventions, these approaches cannot push actors around physically, nor can they threaten to do so. They rely much more extensively on the goodwill of the former combatants, and if that goodwill disappears so too does the effort to keep them peaceful. By comparison, forceful interventions have brought with them the advantages but also the clear disadvantages that the obvious capacity for coercion can bring. The use of force is itself a political act with unclear consequences, and missions that start out with hopes for either a speedy exit or effective stabilisation can end up encouraging the very insurgency that they are expected to extinguish.

It might be wondered in this record whether intervention in Asia and its borderlands has been a largely western endeavour. That would certainly seem to be the case in some of the sharper military interventions, many of which have been led by the United States, or in the South Pacific, by Australia. Given their strong adherence to the principles of non-intervention, it is perhaps unsurprising that many of the states of Asia have not taken a lead in these activities. But Asian countries have by no means taken a completely unified position on these matters. South Korea, Malaysia and Singapore, for example, all deployed forces to Afghanistan, although the two Southeast Asian countries tended to send medical personnel rather than infantry. And in the debates over non-intervention, leaders in some but by no means all ASEAN countries have at times called for a shift from that body's traditional principles of non-interference. Indonesia, moreover, has sought to present itself as an impartial mediator beyond the case of Mindanao, seeking to reduce the temperatures between Cambodia and Thailand in their long-standing boundary dispute.

One of the related calls in this debate has been for a more active role to be taken by regional groupings, partly on the basis that a unified regional approach would give greater legitimacy to an effort to resolve a conflict, or to punish a transgressor of agreed norms and standards. As will be seen in the next chapter, it is now much more difficult to argue that Asia lacks cooperative organisations from which such a combined approach might eventuate, at least in terms of the array of groupings that can now be seen in the region. But how much Asia's security can realistically rely upon these gatherings may be a different matter entirely.

Chapter 10

Solidarity: Can Asian States Work Together On Security?

Some of the interventions considered in the previous chapter have certainly been undertaken by groups of states with a shared interest in the reduction of conflict or in the punishment of a threatening party. But this does not mean that groups of states within Asia and its borderlands necessarily took the leading roles in these actions. A number of the efforts to mediate conflicts in Asia, to cite but one form of such activity, appear to have been mounted from beyond the region. In general, it is unusual to find the majority of states in Asia working together in coordinated responses to the region's security problems. This results in something of a paradox. In recent years ASEAN has become the fulcrum for a series of regional forums designed to promote cooperation among the states of Asia. The remit for, and membership of, these new groupings extends beyond Southeast Asia into the wider region. Does this mean that while these organisations are in place, no meaningful Asian security cooperation is actually occurring?

Sceptical views about regional cooperation in Asia, which are not difficult to find, would answer that question in the affirmative. In one authoritative essay, Michael Leifer (1999, p. 26) asserted that ASEAN 'has never been effectively responsible for regional peace-making as opposed to helping to keep the peace through exercising a benign influence on the overall climate of regional relations'. This assessment remains accurate. But to the extent that effective peace-making requires intervention, this would involve a form of interstate behaviour that the founding members of the Association of Southeast Asian Nations had agreed to avoid. It is perhaps unsurprising, therefore, that ASEAN as a body was not at the forefront of the regional response to the violence in East Timor in 1999, even though some of its members did contribute individually to the Australian-led mission.

By comparison, three years before the 2003 Solomon Islands intervention, the Pacific Islands Forum – ASEAN's counterpart in the South Pacific – had agreed to a code of practice supporting intervention in exceptional circumstances. The 2000 Biketawa Declaration, signed in the small island state of Kiribati, would be cited explicitly in the Forum's endorsement of the RAMSI (Regional Assistance Mission to the Solomon Islands) intervention, another Australian-led effort. Differing historical experience and strategic circumstance matter here. In general, unlike so many of the ASEAN countries, the island states of the South Pacific have not endured violent wars of independence from colonial control, or insurgencies which could attract the interest of neighbours and strong powers. The small Pacific states are also rarely able to pose any sort of threat to one another.

All of this misses another important point. Interventions, which challenge one of the most fundamental norms of the international system of states, are by no means the only response to security problems. Neither are internal conflicts nor territorial disputes the only security challenges for which a regional response might be needed. One of the possibilities considered by the current chapter is whether there are other security problems which encourage a more unified regional approach in Asia. Given that the members of these regional groupings are normally sovereign states, perhaps they are better at responding collectively to challenges from non-state actors whose activities may put at risk the state system they all have an interest in upholding. By this reasoning, some of the transnational and non-state challenges explored in Chapter 8 may provide the best chances for collaborative regional action. It might then be wondered whether the cooperation that is more readily available in response to these issues offers a foundation for states in the region to work together on more sensitive questions. And perhaps some of that rarer collaboration has already been occurring.

Yet before considering some of Asia's potential avenues for security cooperation, it is important to have a sense of what regional cooperation really requires. Do Asia's nation-states need to make sacrifices in order to work with each other, or can such cooperation be consistent with their various individual goals? This requires some honesty about whether some of the more ambitious ideas of regional cooperation actually make practical sense.

Understanding cooperation

Having discussed a significant range of conflicts and points of tension in this book, cooperation is not always the first thing that springs to mind in considering strategic relationships in Asia. In fact, partly because a major war between Asian states still appears conceivable, it is customary to argue that the traditions of cooperation in this region are shallow and fragile by comparison with cooperation in some other parts of the world. These points of comparison include post-war Europe, which has been determined to escape from its ugly history of great power conflict. By contrast, in the late twentieth and early twenty-first centuries North Asia has commonly been regarded as among the most likely locations for serious geopolitical competition between the major powers. Indeed such comparisons, which do not run in Asia's favour, have become commonplace in the academic debate about Asia's security, including a widely cited article in which Aaron Friedberg (1993/4, p.7) argued over two decades ago that 'Europe's past could be Asia's future'.

Asia therefore seems a good candidate for the ideas of perpetual strife between political actors that were mentioned in the introductory chapter to this book. But, as was demonstrated in Chapter 2, in recent decades war has not been the prominent challenge to Asia's security that it once was. And even in times of war, Asia's adversaries have been both willing and able to restrain the level of hostilities between themselves. Pakistan and India may well have fought a series of wars since their establishment as independent states in the 1940s, but their armed conflicts have been limited in terms of duration, geography and the intensity of military action. It might be thought that this logic of restraint only applies to more recent interactions in South Asia. But as a splendid book by Srinath Raghavan (2010, p.3) indicates, India's first prime minister, Jawaharlal Nehru, was confronted with 'events that unfolded in the twilight zone between peace and war'. These were as much diplomatic crises as they were significant wars, and such phenomena have been apparent elsewhere in Asia. They include the limited armed conflict which took place between India and China in the early 1960s. The war between China and Vietnam in 1979 was also a limited affair.

The limits observed in these conflicts were due more to tacit understandings about the need for restraint rather than formal processes of negotiation around a conference table. But cooperation can take on many guises, formal and informal. A mix of these has been evident in the relationship between Australia and Indonesia over the last decade and a half, following their serious crisis over East Timor in 1999, which brought them close to blows. Sometimes the restraint that makes for implicit cooperation is the product of unilateral restraint. For example, in its conduct towards the war which broke out on the Korean peninsula in 1950, the United States avoided some options which would have made for a more catastrophic contest. This included eschewing the use of nuclear weapons against Chinese forces when they joined the conflict. And ever since then, North Korea and South Korea, which are still technically in a state of war with one another, have so far managed to keep themselves from stumbling into a major new conflict. Even a nuclear-armed North Korea has avoided the use of these catastrophic weapons in anger. Indeed since the first and only use of nuclear weapons in an act of war in Japan in August 1945, those countries in possession of nuclear weapons – many of whom are part of Asia or deeply involved in Asia's security – have stopped short of resorting to these weapons of mass destruction.

This last example of cooperation through restraint is an important one. In their relationships of mutual nuclear deterrence (including those which have existed between China and the United States and between India and Pakistan, and the much older relationship between the US and Russia), nuclear-armed states in the region have essentially observed tacit agreements with each other not to resort to catastrophic war. They have, in other words, restrained the competition that clearly exists between them in the interests of their mutual survival. It is uncertain whether some of these deterrence relationships are as clear and robust as they should be (including the rather unclear deterrence linkage between India and China). But it would be churlish to suggest that the absence of a catastrophic conflict between these nuclear-armed countries is simply a matter of luck. At least some good management, as well as good fortune, has been involved.

If these larger powers can restrain the deadliest elements of their military competition, it is likely that even the most difficult relationships in Asia are *not* situations of pure conflict. As Thomas Schelling (1960) argued more than five decades ago, these are not zero-sum-games where the only way for one side to gain anything is for the other side to lose by an equivalent amount. This judgement may be a bit of a surprise given the frequency of references to zero-sum security thinking in Asia. This style of thinking seems to apply to some of the territorial competition in the South China Sea and to the tense stand-off between Japan and China in the East China Sea. It seems to be the case that China cannot grow without Japan feeling nervous and that Japan cannot relax some of its post-war constraints on military activity without China raising the alarm in response. But even here, the two sides would prefer something other than an armed conflict to decide the future of their territorial and geopolitical differences. The tensions between them may rise and they may both deploy naval vessels and aircraft to send strong messages about their interests. But these signals of potential armed action are not being converted into actual physical conflict and the damage and suffering this would bring.

By the same token, it is neither wise nor accurate to think that relationships in Asia, or anywhere else, are characterised by pure and constant cooperation. A pair of countries who have a relationship of mutual deterrence, for example, may be restraining their competition on a daily basis. But any nuclear deterrence relationship between the United States and China still relies on the threat of catastrophic harm they could cause one another should war break out. And when these same two great powers agree that economic cooperation (including the relatively free flow of goods between them) is mutually beneficial, this does not hide the fact that they are quite serious strategic competitors. This is one reason why Chapter 4 found that economic interdependence is no guarantee of peace. The influential scholar Wang Jisi (2005, p. 47) is absolutely right to argue that 'trying to view the Chinese–US relationship in traditional zero-sum terms is a mistake, and will not guide policy well'. But in many non-zero-sum games, conflicting purposes can co-exist with cooperative ones.

In fact the potential for discord is also evident in the closest and most friendly of relationships in the region. Japan, for example, feels vulnerable whenever it thinks that the United States, its long-time ally, and China, its long-time adversary, are getting too close. To the extent that the United States wants a good relationship with both countries, it knows it cannot have a perfect relationship with either of them. This easily creates difficulties between the United States and Japan. Australia and New Zealand are closer to each other than almost any pair of countries in the wider Asian region and any violence between them is unthinkable. But while they have a strong record of cooperating on South Pacific security issues, including in Bougainville and East Timor, they are also capable of at least a little competition in their immediate region. New Zealanders like to think they understand Pacific Island peoples better and they have a more sensitive approach to addressing internal conflict issues in that part of the region. Australians know that their bigger budgets, representing about half of all the development assistance spent in the Pacific, make a difference.

There are other examples of the potential for diverging views among close allies and friends in the region which put a dampener on full cooperation. North Korea views China as its major benefactor and China regards North Korean territory as a buffer against American military power in North Asia. But China's patience with North Korea's provocative nuclear diplomacy often seems on the point of breaking. In 2006, Zhu Feng (2006, p. 47) wrote that China was 'deeply frustrated by North Korea's intransigent behaviour and thinking'. Several nuclear tests later, and with a young North Korea leader who appears to have turned his back on the possibility of reform, that frustration in Beijing has been running even deeper.

This leaves Asia in the middle of the two extremes of a region of perpetual, complete and unrelenting conflict on the one hand, and an utopian dream of pure collaboration on the other. Relationships in Asia, between friends as well as adversaries, are more often than not a mix of competition and cooperation. The implications of this seemingly obvious point can be quite profound. First, it means that cooperation in Asia is indeed possible, including on even the most challenging security issues and among the most divided parties. But second, it means that even when cooperation is occurring, the countries who are working together on an issue are still likely to have different points of view as to how complete their cooperation might be, and on what goals their part of the cooperation is designed to achieve. This means that each party in even the most cooperative endeavours will be wanting their partner(s) to cooperate in particular ways. And as Schelling (1960) argued, these situations between pure competition and pure collaboration are ripe for bargaining.

In taking inspiration from the superpower strategic competition in the Cold War, scholars like Schelling focused much of their work on bargaining relationships between just two states. In two-sided or *bilateral* disputes, the options seem relatively clear. The two parties can choose to work together, both can choose to abandon the idea of cooperation, or one can be ready to cooperate and the other can defect. There are only two sets of objectives to consider here. But things get much more complex analytically when more than two parties are part of a bargaining process. Indeed, the more great powers that are competing for influence in Asia, the more scholars like Friedberg (1993/4, p. 10) worry about Asia's stability because of the complexity that multi-player situations create.

Cooperation among multiple players can be difficult for other reasons. As the proceedings of any large committee often demonstrate, decision-making in bigger groups can be a lot more tedious than the agreements which can be struck from a single interaction between a pair of actors. It should not be surprising if *plurilateral* cooperation between several regional actors has a habit of producing watered-down outcomes if those actors have the ability to hold up progress. This should be especially likely in a more universal notion of *multilateral* cooperation in which all possible partners are involved. That is one reason why in addition to the General Assembly, where all the sovereign states in the world have a seat, the United Nations was established with a much smaller executive committee in the form of the Security Council. And in the latter, permanent seats with the right of veto were only given to five states. If cooperation is to work, trade-offs between equality and effectiveness are probably going to be necessary.

Most of Asia's main groupings which are designed to facilitate security cooperation, and which will be considered in this chapter, lie somewhere between the organisational simplicity of bilateral collaboration and the logistical complexity of properly multilateral grouping. But wherever on that spectrum they sit, they are composed of states with their own agendas and differing ideas on how close and complete their cooperation should go, and on which issues are too sensitive to be raised. These forums often reflect the paradox that something designed as a cooperative mechanism where the smaller states engage the larger states as sovereign equals can easily become a venue for great power competition. Even the decisions on which states should belong to Asia's increasing list of diplomatic groupings can become an indicator about how the region is defined, a recipe for further contestation. And yet, despite all of this jostling and intrigue, there is still a chance that habits of regional cooperation will become established and that individual motivations of states will give way to collective ways of thinking. Indeed if the states in Asia did not have at least an inkling of a common view on how the region should evolve, they might have been unwilling to meet together in the first place.

The ASEAN family of groupings

There is a very good reason to start with ASEAN in an analysis of Asia's cooperative mechanisms. The Association of Southeast Asian Nations not only holds the historical record for being, in the words of Capie and Evans, (2007, p. 161) 'the only governmental multilateral institution to thrive in the Asia-Pacific during the Cold War period', it is also the foundation for a good deal of the subsequent efforts to build traditions of cooperation in the region. ASEAN has certainly attracted mixed reviews since its initial establishment as a small grouping of states in 1967, but it is unclear whether its shortcomings are a commentary on the grouping itself or on the challenge of effective regional security cooperation in Asia in general. It is probably as much a case of the latter as it is of the former.

In ascertaining what ASEAN actually is, it may be easiest to explain what it is not. ASEAN is not an increasingly binding federation of states who are progressively ceding their sovereignty to a central entity. It is not in that sense an Asian version of the European Union. Indeed ASEAN's supporters are generally very resistant to comparisons being made with the European experience, in part because of the chequered history of many European states as colonial powers in Southeast Asia. Unlike the European Community, an extensive bureaucracy has not been built up around ASEAN. Instead there is a small secretariat which struggles to manage the thousands of meetings which now happen each year on all sorts of topics; economic, social, cultural, as well as diplomatic. Nor is ASEAN a political-military alliance, a different type of beast requiring a different type of cooperation between its members, which will feature in the next chapter of this book. The members of ASEAN are under no obligation to offer military support in the event one of their colleagues is attacked and there is really no such thing as a Southeast Asian military planning staff.

As Amitav Acharya (1992, pp. 12–13) noted some years ago, questions of sovereignty and political flexibility play into this reality, as do the military limitations of many of ASEAN's members. Over twenty years since this observation was made these constraints remain in place. ASEAN is therefore not an Asian NATO, and never will be, and comparisons of that sort are once again often regarded as odious in parts of Southeast Asia. Attempts to downplay ASEAN's importance because it is neither a currency and political union nor a military alliance tend to fall on deaf ears in the region.

Two things stand out in ASEAN's approach to regional security issues. The first is its survival despite the ongoing tensions between some of its members, who have yet managed to avoid serious armed conflict with one another since the 1960s. That in itself is quite an achievement when looking back on Asia's violent epoch in the middle of the twentieth century. The second is a series of broad, and not especially binding, agreements which establish standards of behaviour, or norms of practice, which are designed to perpetuate peace in Southeast Asia. The most significant of these is the Treaty of Amity and Cooperation (TAC), signed in Bali in 1976 by the five initial members of ASEAN: Indonesia, Malaysia, the Philippines, Singapore and Thailand.

As might be expected from its title, the TAC attaches importance to the avoidance of armed force as a means of addressing any disputes between its parties. Especially when any such problems are 'likely to disturb regional peace and harmony', the signatories of the TAC 'shall refrain from the threat or use of force and shall at all times settles such disputes among themselves through friendly negotiations'. But in addition to this expected commitment, which is a Southeast Asian twist on similar undertakings in the United Nations Charter, the TAC says something very important about the security concerns of the members of ASEAN. This understanding goes far beyond the notion that security consists of the avoidance of armed conflict between states. In Article 11 of the Treaty, signatories are also committed to:

> strengthen their respective national resilience in their political, economic, sociocultural as well as security fields in conformity with their respective ideals and aspirations, free from external interference as well as internal subversive activities in order to preserve their respective national identities.

This passage offers a clear reflection of the internal preoccupations which at the time of its writing still dominated the priorities of most of

ASEAN's founding members. Their quest for national cohesion put a premium on economic development and political and social stability. This reflected Southeast Asian interpretations of the idea of comprehensive security which, as Capie and Evans (2007, p. 65) have noted, conceives of security in a 'holistic way – to include both military and non-military threats to a state's overall wellbeing'. For the countries of ASEAN, this more inclusive approach to security reflects a preoccupation with internal security challenges. Such an emphasis can be detected in the reference in the TAC to freedom from internal subversion. In the mid-1970s memories of civil conflict remained fresh, and in some cases (including the Philippines and Indonesia) these were still more than memories. Yet one further aspect of this article also deserves attention: the desire for freedom from external interference. This might be seen partly as opposition to some of the practices of intervention that were analysed in Chapter 9. But the members of ASEAN were also determined that their part of Asia be kept free from strategic competition between the major powers, and from the spread of conflicts from the less settled parts of their own region.

The aim of insulating Southeast Asia from the superpower contest had already been reflected in an earlier agreement which the original ASEAN members has signed in 1971: the rather convoluted sounding Zone of Peace, Freedom and Neutrality (ZOPFAN). Strictly speaking, neutrality would mean the absence of overseas bases and even of military alliances with outside powers. That was the definition which Indonesia favoured, but Thailand and the Philippines were formal treaty allies of the United States. Moreover, Singapore and Malaysia were partners with the United Kingdom, Australia and New Zealand in the Five Power Defence Arrangements, an exchange of letters between the five British Commonwealth countries which in 1971 replaced Britain's formal alliances in Southeast Asia. To get around these differences of view, but also to retain consensus, ZOPFAN reflected what would become time-honoured ASEAN practice: the dilution of strong positions to an interpretation that the most cautious member finds comfortable. This approach has led to some of the most stinging criticisms of Southeast Asian security cooperation – that it reflects the lowest common denominator and as a consequence has no real bite. In one such attack on this approach, Jones and Smith (2007, p. 155) have observed that: 'Consensus encourages a shared appreciation of the problem without necessarily producing a shared approach to it.' But, as Shaun Narine (1998, p. 202) has suggested, the very same tendency towards compromise has also been one of the main reasons for ASEAN's survival.

That need for compromise became even more evident as ASEAN expanded. This enlargement occurred initially and straightforwardly with Brunei upon its independence from Britain in 1984. But the real test would come with the admission of Vietnam in 1995, which by removing its forces from Cambodia in 1989 had paved the way for its eventual membership. Vietnam was now a partner when in earlier years it had been excluded as a common threat, a factor which had united a much smaller ASEAN in earlier years. Laos and Myanmar would then accede to ASEAN in 1997, and Cambodia became a full member in 1999. ASEAN thus became a house for an even more diverse set of Southeast Asian polities, from the more liberal democratic (but internally troubled) Philippines to the reforming Indonesia, and from what might called the guided democracy of Singapore to what was until recently an unreformed military regime in Myanmar.

Norms of political organisation, including attitudes towards multi-party competition for power, relations between the state and the army, and the observance of human rights, all differ significantly within this more inclusive swathe of Southeast Asian countries. As might be expected, in a grouping that put a premium on consensus and on agreements that were non-binding, this diversity made it even more difficult for any common ASEAN position on foreign and security policy issues to have any teeth. And yet, unlike in North Asia, where any sort of comprehensive regional body proved impossible to establish due to major power competition and mistrust, ASEAN became a magnet for other countries in the region and a building block for further dialogue. The first of these other portions of the ASEAN family of groupings was the ASEAN Regional Forum or ARF, established in 1994, and devoted specifically to the discussion of security issues in the region. Like the very inclusive grouping of economies called APEC (Asia-Pacific Economic Cooperation), which had been established in 1989, the ARF had members from far and wide. This included not just what would become the ASEAN ten, but also many governments in North Asia and South Asia, the United States, Russia, Australia and New Zealand and more besides. And like APEC, one of the main unspoken aims of the ARF was to remind the major powers of their interests in a stable, prosperous and secure region. As Yuen Foong Khong (2004, p. 199) has eloquently argued, the ASEAN countries had in mind both a rising China and an already strong, but potentially distracted, United States.

With a membership nearly three times the size of ASEAN's, but with the requirement for consensus on any decisions still operating, it is not surprising that it has been difficult for the ARF to become a basis for active security cooperation. Hopes that the ARF might move from security dialogue to preventive diplomacy, where crises are defused before they mature, have borne few fruit. But the sheer scale and diversity of the ARF's membership has not been the only obstacle. There has also been something of a struggle for influence within the grouping. The ASEAN countries like to view themselves, to use the euphemism repeated endlessly at regional meetings, as occupying the 'driver's seat' of regional integration. But rather than the Southeast Asian countries managing the major powers, signs of the reverse have been appearing. Within the first five years of its existence, Robyn Lim (1998, p. 115) was already asserting that 'the ARF can do little to promote security because ASEAN insists on its primacy in it' and that 'ASEAN cannot set the security agenda for the major powers, nor reconcile differences among them'. This criticism is harsh, but it carries more than a seed of truth. Indeed the bigger players have often seen vehicles like the ARF as venues for their own strategic competition. In some cases they have taken opportunities to exploit divisions between the ten ASEAN members.

Finding common positions in the ARF and similar mechanisms on contentious security issues, including maritime territorial disputes, has been made even more difficult by the disunity among the ASEAN countries themselves. For example, as became clear in Chapter 6, several Southeast Asian countries have competing claims in the South China Sea, and histories of reciprocal mistrust among themselves. Cooperation has been easier when it a common non-state problem has faced them jointly, including on transnational health issues, counter-terrorism and disaster relief. Even here, however, there have been limitations in terms of the adoption of a unified position. Just a few months after the 2002 Bali bombings, an annual ARF meeting on Counter-Terrorism and Transnational Crime was initiated, producing a series of declarations urging much closer cooperation. But, as Jonathan Chow (2005) noted just as these efforts were evolving, divisions between the Southeast Asian countries themselves, including tensions over whether some countries were taking their responsibilities seriously enough, were always going to be a barrier to full collaboration. In a more recent study, Haacke (2009) has suggested that the members of ARF have only really found it possible to work together on relief efforts in response to natural disasters. Yet while this is the ultimate threat without an enemy, even here active cooperation under the auspices of the ARF was circumscribed.

Similarly, while the proliferation of further groupings in this plurilateral family might be viewed as building on the precedent established by ASEAN and the ARF, it may alternatively be seen as a sign that the ARF has not met its expectations. In 2005 the East Asia Summit (EAS) came into being as an annual meeting of regional leaders. Its main advantage over the ARF was its more exclusive membership, consisting of the ten ASEAN countries and a core of eight of their main dialogue partners. This partnership built on the ASEAN+3 grouping which included China, Japan and South Korea. This emergent grouping reflected a desire for greater East Asian unity after the 1997 financial crisis, but was still limited, as Stubbs (2002, p. 452) has noted, by the divisions between its larger North Asian members. The EAS was first extended to include India, Australia and New Zealand, and then more significantly Russia and the United States, with an American President attending a summit for the first time in 2011. This depiction may sound like hyperbole, but as China's power increased, a number of the ASEAN countries regarded America's balancing role as essential. Moreover, as the condition for entry to the EAS was a signature on ASEAN's Treaty of Amity and Cooperation, those major powers were now committed on paper to the leading values of international behaviour which the countries of Southeast Asia themselves were obliged (at least in theory) to share.

But it is important not to regard this newer grouping as a supercharged forum which somehow decides on the region's main problems. The EAS is based on an annual meeting of only a few hours between the leaders who attend, and has very little in the way of an ongoing organisational presence in the region. At most, it is an annual signalling device on who is active in the region and which countries appear to be driving the overall agenda. When President Obama missed the 2013 East Asia Summit to attend to the domestic political challenges of a Congress that could not agree on budgetary measures, it was taken as a sign that America's interest in the region was waning. This was in clear contrast to the signals from the previous year's meeting, which was dominated by Mr Obama's push for greater American regional leadership in the face of China's rising influence. But the symbolic significance of the EAS does not translate into an active and ongoing agenda, or a capacity for regional leaders to act in the long gaps between the annual summits.

This is not something that can be said of a further and more recently established grouping which has the delightful title of the ASEAN Defence Ministers Meeting Plus (ADMM+). As a mainly diplomatic forum the ARF has had little direct connection to military operational matters, and little chance of encouraging practical defence cooperation in the region. As a leaders' forum the EAS covers a wide swathe of political, economic and security questions without much ability to focus on any of them. The annual Shangri-La Dialogue, hosted in Singapore by the International Institute for Strategic Studies, has offered a venue for bilateral meetings between regional defence ministries. But until very recently, a plurilateral instrument through which Asia's defence forces might work collaboratively in practice has been absent. The ADMM+ is an attempt to go some way to filling that gap.

Four years after the ASEAN defence ministers began meeting together annually in 2006, the ADMM+ itself was inaugurated. With the same membership as the EAS, this grouping offers another common venue for

China, the region's rising military power, and the United States, still by far the strongest military actor in Asia. Between these meetings, which have been taking place every second year, the core of the work programme of the ADMM+ has been occurring through a series of specialist groups. Their subjects, which include peace-keeping, military medicine, and maritime security, are certainly all worthy of consideration. But the ADMM+ is a long way from having any real time ability to respond to crises in the region, let alone to anticipate them and head them off. For example, the holding of disaster relief exercises should not be mistaken for a real capacity to act. This was confirmed by the absence of an ADMM+ response to the devastating Philippines typhoon in late 2013. That gap would also seem to be an obvious warning against holding out hopes that cooperation on non-traditional security questions might translate into regional action on more difficult interstate questions. Tan See Seng (2012, p. 233) has written of 'the desire among ASEAN's defense practitioners to "talk their walk" – exercise a visionary restraint that is more or less commensurate to actual practice – rather than attempting, and likely failing, to walk their talk'. If he is right, and there is good reason to suggest that he is, it is only sensible to have modest expectations for these groupings.

Coherence: a fruitless quest?

A common complaint heard several years ago was that Asia was under-institutionalised in terms of the mechanisms that were available to support regional security cooperation. Given that many of the forums that have grown up are not heavily bureaucratised and rely on consensus rather than binding commitments, at least part of that criticism may still hold true. But it is now impossible to claim that what is sometimes referred to as the 'alphabet soup bowl' of Asian regional organisations reflects a shortage of gatherings which bring leaders and officials together on a regular basis. Asia is now awash with meetings, including at the leaders level.

The list that has already been constructed in this chapter could go on. For example, in addition to the numerous forums involving government officials that touch on regional security questions, there are many ongoing dialogues among research organisations in Asia which often involve officials participating in their private capacities. The idea with these 'second track' meetings, including the region-wide Council for Security Cooperation in the Asia-Pacific (whose membership mirrors the parties to the ARF), is that they should be able to allow for the discussion of issues that are too sensitive for 'first-track' (or officials level) dialogue.

As Ball, Milner and Taylor observe (2006, p. 180), among some of the more optimistic observers in the region, Track 2 'security dialogue is seen to have offered an alternative diplomatic route when progress at the first track level has stalled or become deadlocked'.

But the evidence suggests a different story. These non-official dialogue processes may have been needed rather more in the mid-1990s, when habits of communication were still in their infancy. In today's Asia there is a danger that these plurilateral second track meetings will simply get overshadowed by the sheer variety and frequency of first track gatherings, where the reluctance to discuss issues is not always what it once was. Perhaps more significant at times are bilateral second track meetings, including between American and Chinese or Chinese and Japanese research organisations, which may provide back channels for governments to signal their preferences and to hear the concerns of their counterparts.

All of this would appear to provide a good reason for the countries of the region to introduce a degree of coherence to this expanding array of organisations, some of which overlap on similar topics. For example, discussions about security of cooperation on some of the less sensitive areas already mentioned above, including disaster relief, occur in multiple groupings. Governments, especially those from smaller countries in the region which lack large bureaucracies, can find it difficult keeping up with the sheer volume of meetings. Forums which appear to have lost their effectiveness are seldom retired: they keep marching around in ever-decreasing circles.

But exactly how to measure that effectiveness is open to debate. There is a strong argument, put forward tirelessly by the backers of ASEAN, that the process of dialogue is an important achievement in and of itself in a region where patterns of communication were once tenuous. The thousands of yearly meetings, including those which specifically deal with regional security issues (even if they do so obliquely and without decisive conclusions) can then be defended as inherently important simply because they take place. But these arguments are becoming harder to defend, especially when tensions involving the major powers grow, and especially when some of the ASEAN countries themselves get into periods of strife with each other. Yet the momentum for organisational reform is almost non-existent. One reason is that while western oriented countries involved in these grouping often want to see results, their occasional criticisms are often seen as unwanted echoes of a colonial past.

In fact the desire for coherence in what is sometimes called Asia's regional security *architecture* is not as widespread as might be thought. Nearly every grouping has at least one champion who will be upset if it disappears. Coupled with the consensus principle, this makes the rationalisation of Asia's regional groupings an almost impossible cause. The ARF, for example, may not achieve much, but it allows a wider range of players, including the EU and Canada, to feel some connection to the region's security discussions, and the countries of Southeast Asia never quite know when wider support may come in handy. The risks associated with moving to a more selective grouping view of Asia's main interlocutors became obvious in 2008 when Australia's regionally ambitious prime minister, Kevin Rudd, proposed the establishment of an Asia-Pacific Community as a one-stop-shop for regional discussions on all sorts of matters. The backtracking began very soon thereafter. First, some of Australia's Southeast Asian neighbours were furious that only the larger powers (including Indonesia) were to get a seat at this new table alongside the big brokers, including China, the United States, Japan and India. Second, the idea that the new Community would enjoy a privileged status beyond the existing cacophony of groupings was anathema to many of the Southeast Asian countries, which insisted that ASEAN remain at the centre of regional security dialogue. Eventually the Australian government found a way of squaring the circle, by quietly dropping the Asia-Pacific Community proposal, insisting (rather unconvincingly) that its main aims had been met by the widening of the East Asia Summit to include the United States and Russia. And in his post-Prime Ministerial phase, rather than repeat his call for a new body, Mr Rudd (2013, p. 14) has himself suggested that the US and Russia should use the EAS, as well as the ADMM+ to 'develop a series of confidence- and security-building measures among the region's 18 militaries'.

Beyond Southeast Asian regionalism

A properly coherent and unified architecture in Asia is a distant dream, not least because ASEAN is actually far from central to all of the important groupings that operate. One such example was mentioned in the previous chapter. This is the Shanghai Cooperation Organization (SCO), which brings together China and Russia along with several of the Central Asian Republics: Kazakhstan, Kyrgyzstan, Tajikistan and Uzbekistan. Established in China's main commercial city in 2001, the SCO partly reflects the common desire of its two larger members – Russia and China – to moderate their own competition for Central Asian influence and to resist the prospects of America's dominance. Indeed it is no accident that SCO activities were being beefed up just as the United States was establishing bases in Central Asia to help it transfer forces into nearby Afghanistan following the 9/11 attacks. That imperative is strong enough for China, but has been even stronger in the Putin era for the other major member of the Organisation. In the words of Isabelle Facon (2013, p. 465): 'Advertising the strength of the SCO is, for Russia, a way to rationalize its rejection of a strong Western presence in Central Asia.'

That competitive element foreshadows some of the issues still to be considered in this volume in terms of the roles of alliances in regional security cooperation. But the SCO is not equivalent to the US system of treaty-based alliances in Asia, which will be studied in the next chapter. The SCO's members certainly do have a joint declaration and a common charter, and a council of heads of government. Military exercises have been conducted, sometimes to send a signal to Washington. But a good deal of the SCO's activities focuses on cooperation between its members in dealing with the transnational challenges which directly concerns its members. These issues include drug trafficking, transnational crime, and terrorism, many of them far from the exciting world of interstate rivalries.

This cooperation has encouraged a degree of solidarity within the SCO on some of the national policies adopted by its members that often attract international criticism, including China's handling of the independence movement in Xinjiang and Russia's robust approach to Chechen separatism. As Stephen Aris (2009, p. 466) has written rather diplomatically: 'the leaderships of the members are in agreement when it comes to questions of terrorism and the sovereign right of governments to pursue a security agenda as they deem fit'. Cooperation between sovereign states without relinquishing a yard of sovereignty seems to be this organisation's catch-phrase and some of it has been facilitated through a mechanism known in English as RATS: the SCO's Anti-Terrorist Regional Structure. If the ARF is sometimes known as the dog that does not bite, perhaps RATS is the mouse that occasionally roars. Moreover, the SCO is not just about security cooperation. Its members have also been seeking to liberalise trade within the grouping, an indication of a much broader remit than would befit the threatening military bloc that the SCO resembles in the eyes of some western analysts.

There are also cooperative security processes in the region that focus on one particular issue and cannot simply be folded into broader gatherings. A leading example of this tendency is the Six Party Talks, a series of meetings hosted by China which are designed, in theory at least, to deal with North Korea's controversial nuclear weapons program. The participants in these talks are a very select bunch. In addition to North Korea and its main ally China, the other members are Russia, the United States, Japan and South Korea. But even among this limited number, the differences of perspectives among the participants, as John Park (2005) indicated in a much earlier stage of proceedings, have remained significant. The Six Party Talks have been held intermittently, and it is all too easy to write them off as nothing more than a face-saving device for Beijing against the backdrop of North Korean game playing. A standard pattern of behaviour seems to occur: North Korea is periodically persuaded back to the negotiating table and signs on to freezing its nuclear programme in exchange for economic assistance, only to upset this possibility with subsequent nuclear and or missile tests or other threatening actions. Then begins another cycle of temporarily improved behaviour in exchange for money and energy.

This has bred extensive cynicism about the Six Party Talks as a process, but the problem is that there seem to be very few better alternatives. Moreover, the fact that the United States and China, which in the early 1950s were fighting each other in the Korean War, can sit down and achieve at least momentary progress on a common strategic problem is a not inconsiderable achievement. This fact once fuelled optimistic views that the Six Party Talks could be the embryo for something much more ambitious – a proper North Asian dialogue process dealing with other contentious issues. In more optimistic times when the Talks were still fairly young, Jaewoo Choo (2005, p. 40) argued that host country China had consistently seen them as 'a gateway to realize the long-sought goal of building a multilateral cooperative security institution in Northeast Asia'. This was partly the ambition behind Mr Rudd's Asia-Pacific Community idea referred to earlier, not least because it is among the North Asian powers that the most serious rivalries in the region exist.

But these very rivalries (including between China and Japan), and the very mixed record of influencing North Korea's nuclear developments, mean that hopes have now disappeared of building greater things on top of the Six Party Talks. As much by default as by design, the ASEAN-centred gatherings that occur further south in the region continue to be the main venue for wider discussion on security affairs. In this respect the Southeast Asian countries have a significant advantage over their larger northern neighbours: none of them are strong enough to threaten – or significantly to influence – the major powers. This makes them acceptable hosts. And because they could be easily squashed should the major powers come to serious blows, the medium and small powers of Southeast Asia also regard their convening role as a strategic necessity. They are compelled to continue these processes even if they are aware of their limitations. But because they are not great powers in their own right, it remains unclear how much the ASEAN states can themselves influence the larger powers to their north.

An Asian security community?

As should have become clear from the preceding analysis, it is unlikely that Asia's sovereign states will be able to find a single arrangement allowing them to hammer out the security problems which still divide them. This is no small matter because, in the extreme case, these problems could still lead to a major war in the region. But even if such a central and single venue was possible to establish, it might well be a sign that rather too much faith was being invested in questions of organisational structure and process. The quest for an effective Asian regional security mechanism is a fool's errand unless the participating countries are genuinely willing to recognise the problems which confront them, and are willing to make the concessions that a properly cooperative strategy would require.

Before giving up on this idea, it is necessary to consider whether some of these more cooperative attitudes have been slowly forming, reinforced by regular interactions and the multiplicity of meetings that now occur. The countries of ASEAN, to use the most obvious examples, do not have a spotless record in avoiding intraregional disputes, as the occasional flare-ups between Thailand and Cambodia demonstrate. But this seems a far cry from the situation when Vietnam occupied its neighbour Cambodia in the 1980s after a bloody fight with the Khmer Rouge. Indeed Vietnam has been a member of ASEAN in good standing for many years, and holds few fears for its partners. It appears, at least on the surface, to be part of a collection of Southeast Asian countries with a common commitment to the avoidance of major conflicts between themselves.

The possibility of a collection of interdependent sovereign states which identified with each other to the point where they had ruled out the use of force in their relationships was noted in the 1950s by Karl Deutsch (1957). But this work on what Deutsch called security communities did not take its inspiration from developments in Asia, which was riven by armed conflict in the first two decades after the Second World War. Europe's post-war experience was distinctly different. A mutual commitment not to repeat the disasters of the first half of the twentieth century underpinned the increasing integration between old enemies Germany and France. This laid the foundations for today's European Union. For some decades now the idea of any two Western European countries going to war with each other has seemed especially unlikely (although in international politics most things cannot be ruled out completely). Likewise, the United States and Britain, which fought one another in the early nineteenth century, are the leading members of a group of English-speaking countries (extending to Canada, Australia and New Zealand), among whom war is virtually impossible to imagine.

Whether it is a prior commitment to avoid conflict that leads to greater cooperation or whether working together encourages peaceful norms of behaviour is a fascinating chicken or egg question. The wise view is that it is likely to be a combination of the two. A positive attitude towards the benefits of cooperation is probably needed before any significant collaboration occurs. This positive view is exactly what India and Pakistan have often lacked in their assessment of each other, to cite but one example from Asia. But the practice of collaboration may itself have flow-on effects. As they work together, some partners may find that they are taking on a collective view of a problem that differs from their starting positions. This underlies some of the constructivist arguments about security cooperation which suggest that norms of behaviour are socially constructed (a point linked to Finnemore's explanation of the changing purposes of intervention which was noted in Chapter 9). And in at least some situations, states may develop a loyalty to the group which actually outweighs the direct benefits they can gain individually from working with these same partners. They may even feel and act as if they are part of a community with its own collective ways of behaving.

The standards of behaviour set by ASEAN's Treaty of Amity of Cooperation, which have already been considered in this chapter, may point to a trend of this nature. Drawing on Deutsch's earlier North Atlantic-focused work, some scholars have suggested that this is precisely what has been happening as the countries of Southeast Asia have set out to build a security community of their own. Among them, Amitav Acharya (2001, p. 8) offers the following analysis, in which a constructivist logic comes quickly to the fore:

> Institutions provide crucial settings within which states develop their social practices and make them understood, accepted and shared by others in the group. ASEAN is not moulded exclusively by material conditions such as the balance of power or material conditions such as expected gains from economic interdependence. Its frameworks of interaction and socialisation have themselves become a crucial factor affecting the interests and identities of its members.

It is possible to quibble with some of the detail here. For example, the balance of power might be viewed as an informal institution, a set of regular practices among states, rather than as something that is just about the concrete elements of power. This was certainly the view of Hedley Bull (1977), who wrote some of the very best work there is on regional security in Asia. A more significant issue is that while the workings of ASEAN have been a necessary condition for peace in Southeast Asia, they have not been a sufficient condition. Other factors, including perhaps the contributions of other and larger states to the balance of power, are required.

Moreover these ASEAN standards of behaviour becomes less influential the further one gets from Southeast Asia and the more one considers the challenging interstate rivalries in North Asia which affect Asia's security. Japan and China, to cite an especially important example, are very closely connected economically and geographically. But there is little evidence that they regard themselves as part of a security community, whether this is an extension of the Southeast Asian experience or a separate North Asian variant. More often than not, Tokyo and Beijing define their security and status vis-à-vis one another in competitive terms. Their identities are much more zero-sum than shared. As this author and Desmond Ball (2014) have argued, this can be a very dangerous situation. One also needs to ask similar questions about the extent to which China and the United States, and also Japan and South Korea, see themselves as part of a security community. It is not at all clear that they do in any real sense of the term. And given that a number of the Southeast Asian countries are divided on how they see relations between the great powers, the greater the competition there is between the United States and China, the more pressure may be placed on the notion of an ASEAN security community.

There are additional obstacles to effective security cooperation in the wider Asian region. The states of South Asia have had their own regional grouping since 1980 when the South Asian Association for Regional Cooperation (SAARC) was established. With the addition of Afghanistan in 2007, SAARC now has eight members; the others being Bangladesh, Bhutan, India, Maldives, Nepal, Pakistan, and Sri Lanka. With regular meetings and its own charter which contains some of the same principles embodied in ASEAN's main declarations and agreements, SAARC is on paper a security community in waiting. But in practice at least some of South Asia's most important security relationships barely conform to the lofty goals of peaceful and cordial relationships. As has been noted in earlier parts of this volume, India and Pakistan, the leading South Asian states and the most essential members of SAARC, have an undeniable track record of tense and sometimes violent relations. The notion that they comprise a security community whose members do not regard armed violence as an appropriate form of social intercourse is denied by that experience. That Pakistan often defines itself as an actor against its stronger and larger neighbour India also puts paid to the idea of shared identity. Indeed, Lawrence Sáez (2011) has argued that the growing military imbalance between India and Pakistan has encouraged a situation where SAARC's limited discussions on security issues have been diverted into less controversial areas. However, a security community could not be said to exist unless the big issues are squarely dealt with.

The notion of security-building in the South Pacific portion of Asia and its borderlands poses a different set of issues again. The Pacific Islands Forum can be said to reflect a shared sense of Pacific identity and a desire to overcome some of the disadvantages of smallness and remoteness by collaborating on common challenges. This included the valuable work of the Forum Fisheries Agency in presenting a united front to larger and more powerful countries (including from North Asia) whose fleets are working within Pacific waters. And, as was seen in the previous chapter, the Forum also gave its approval to the Solomon Islands intervention led by Australia in 2003. The Pacific Islands Forum also embodies an approach to negotiation and decision-making often called the 'Pacific Way', which has many features in common with the emphasis on informality and consensus in ASEAN.

But it is difficult to argue that the almost complete absence of armed conflict, and even threats of conflict, between the Pacific Island countries, is proof of the existence of a security community. In the first instance, most of the very small countries of this part of the region simply lack the ability to exert themselves externally. Few even have armed forces. There are also divisions among Pacific countries on the attitudes that should be taken towards Fiji (where the Forum is located) on the numerous occasions when that country has been led by a military regime. This also suggests that there are clear limits to the tradition of consensus. Indeed Fiji is part of a separate grouping called the Melanesian Spearhead Group (MSG), whose other members include Papua New Guinea, Solomon Islands, Vanuatu and representatives of the Kanak independence movement from the French territory of New Caledonia. As Ronald May (2010, p. 6) suggests, the MSG reflects 'a desire to assert a Melanesian voice among the members of the Pacific Islands Forum, which some island countries perceived to be dominated by Australia and New Zealand'. There is also little evidence that any of the Pacific Islands themselves are part of what remains an incomplete security community further to their north, and PNG's potential membership of ASEAN would probably dissipate the sense of South Pacific unity even further.

Conclusion

Regional cooperation on security issues does occur in Asia. But the region still seems rather too diverse and, at times, competitive, to expect a single coherent process to which all of the countries of Asia and its borderlands submit themselves and their concerns. To the extent that the makings of a security community have emerged in Southeast Asia, not all of the wider region is engaged in this process. A number of larger states beyond ASEAN may have signed the TAC and other declarations of fidelity to the idea of peaceful international relations, but quite how much these commitments can be relied on in the heat of a crisis is open to question. Hopes that security cooperation on some of the more obvious common challenges facing regional states, including disaster relief, pandemics and transnational terrorism, could provide a ready basis for collaboration on the more divisive issues in the region have so far not been met in reality.

It might be wondered if Asia's security situation is simply not worrying enough to demand the greater level of cooperation, let alone any sense of a unified approach. But given the often serious challenges which have been considered in this book, that line of argument doesn't hold. Perhaps then it is the diversity of these many security problems which is one of the main obstacles to closer cooperation. Is it possible that if there was a single over-arching threat that worried everybody in Asia intensely, deeper and more coordinated levels of security cooperation in the region might naturally result?

But one of the problems with common and especially serious threats is that they often are seen to derive from one or more of the states in the system. That was the way of the Cold War where America's allies were united by a common threat from the Soviet Union and the wider communist bloc. These rival alliances were not without defections and moments of disunity. But within themselves they involved intense amounts of security cooperation, including the planning for combined military operations. This is something that the ADMM+ will find difficult to provide for in a serious sense. But as the next chapter will show, while alliances may be good examples of unity within the group, they can be divisive beyond it.

Chapter 11

Division: Will Alliances and Partnerships Split the Region?

Asia's array of regional groupings, many of them still developing, provide at the very least a venue for the discussion of regional security problems. This offers something of a counterweight to the thesis that Asia is light on cooperative mechanisms. But well before even ASEAN came into existence, a set of formal interstate relationships had been established in the region which involved much more substantial levels of security cooperation. Beginning in the early 1950s, as the Cold War spread to Asia, the United States established a series of treaties in which it formally committed itself to the defence of several alliance partners in Asia and its borderlands. With one or two exceptions, these relationships remain in place today.

This does not mean that there is anything today in existence which comes close to resembling an Asian NATO. Most of these alliances are between the United States and just one other country in Asia. But like the North Atlantic Treaty, the texts of the written agreements at the centre of these alliances provide for security cooperation against a common external threat. These alliances might therefore be viewed as having a positive security effect because of their capacity to deter conflict through common resolve, with benefits extending to the neighbours of America's allies, who would also be concerned. Moreover, the knowledge of American support also reduces the need for allies in Asia to enter into costly, and potentially unsettling arms build-ups. Victor Cha (2010, p. 163) has gone so far as to describe America's alliances with Japan and South Korea as 'pacts of restraint' which are deliberately asymmetric (in America's favour) and designed to remain that way.

Even so, these close alliance connections might not necessarily reassure other states in the region as much as Washington would like. They can also set up a wider and sharper contest with major actors who are excluded from US alliances and thought to be the major challenge to them. In this way alliances can bring the direct members together but also divide the region as a whole. The ripple effects here can be considerable. As might be imagined, the country that feels it has most to lose from the US system of alliances in Asia is none other than China. This sets up the real possibility that the closer the security cooperation is between the United States and its Asian allies, the less secure China's government feels about its regional circumstances. Given China's growing power, this seems bound to cause some real problems, including the potential for the security of one major actor in the region only increasing (or at least being perceived to increase) at a cost to the security of another.

This may be one main reason why in the current era, few if any of Asia's states would want to develop new formal alliance treaty relationships with the United States. In their place, other forms of inter-state military cooperation have been emerging. This includes an increasing number of security partnerships, which generally do not involve the same obligations that are carried in formal alliances. But it still needs to be wondered whether some of these closer partnerships are alliances in disguise, and whether they also carry the risk of dividing Asia. Indeed there may be little to stop Washington regarding these relationships as part of an effort to keep China's regional influence from growing. And there may be little to stop China from fearing that this is exactly what is intended.

The manner in which the United States cooperates with its leading Asian ally plays directly into this possibility because Japan also happens to see China as a security threat. As has already become clear elsewhere in this book, a crisis leading to major Sino-Japanese war could very easily have devastating consequences for the region as a whole, especially if it also involved the United States (as the main ally of Japan and the main competitor of China). It is therefore important to consider whether there is a way for the great powers themselves to overcome some of their divisions and establish some rules for the game they play. This is the hope, also to be assessed in this chapter, of those who argue for an Asian concert of powers.

US alliances in Asia

The set of American-led alliances in the region are more than sixty years old. While they continue to have a significant impact on Asia's regional security today, they were established under quite different conditions. In the years immediately after the Second World War, one of Washington's aims was the reintegration of a defeated Japan into what was supposed to be a more peaceful era of Asia's international relations. In September 1951, nearly fifty countries were present in San Francisco for the signing of the peace treaty with Japan. In that same year Washington and Tokyo also signed a security treaty which, among other things, provided for the stationing of United States forces at bases in Japan. This was the first of a number of commitments to the region which involved explicit American backing. Designed to promote Asia's security and prosperity (on terms friendly to western interests), these relationships together have been dubbed the 'San Francisco system' by Kent Calder (2004).

Alliances are about obligations to help, and there might be a normal expectation that this applies in both directions. Stephen Walt (1997, p. 157) has argued cogently that 'the defining feature of any alliance is a commitment for mutual military support against some external actor(s) in some specified set of circumstances'. But in the case of the US–Japan alliance most of the traffic was expected to be heading in just one direction. A central purpose of this relationship was to assure Japan of America's protection in the event of an external attack. This was important not least because Japan emerged from the war under American tutelage with a commitment not to build up its own armed forces to the extent that they allowed for the independent projection of military power. But in providing forward bases near the Asian mainland for US forces, Japan also became known as an unsinkable aircraft carrier for the Pentagon.

Bringing Japan back into the regional fold was not reassuring for some of the other countries in the region. This included Australia, one of America's closest colleagues in the Pacific theatre of the Second World War, which was also increasingly relying on America's might as Britain's regional power declined. Concerned about the possibility that Japan might remilitarise and return to aggression, Australia (together with New Zealand) was also fighting alongside the United States in the Korean War. Washington's appreciation for these efforts, and its awareness of Australia's concerns, were reflected in the signing of another San Francisco treaty in 1951, known from then on as ANZUS (Australia, New Zealand, United States). Here the security obligations were more genuinely mutual. Each one of the three parties were obliged to act to meet a common danger should their forces be attacked in what was called the 'Pacific area' (and which in reality was a significant amount of maritime East Asia and the South Pacific). An armed attack on the sovereign territory of any one of the three parties would also be a trigger for a combined response. While military action was not guaranteed, it was most certainly expected.

The Korean War was also a factor in the development of another leg of the San Francisco system of US-led alliances. Two months after the Korean armistice was signed, noting the end of fighting but not a proper peace on the peninsula, the United States and South Korea signed their own 1953 security treaty. Like the US–Japan Treaty before it, this agreement allowed for the forward stationing in Asia of American forces, in this case designed to deter further aggression from North Korea. While this involved American land forces being present on the peninsula (and still does), for several decades it also involved the presence of American nuclear weapons. And in an undertaking that was clarified by the United States in a separate note in 1954, both South Korea and the United States committed themselves to coming to each other's assistance in the event of an external attack on either party.

By this time the United States had also entered into a mutual security treaty with the Philippines, which was to provide important basing facilities for American ships and aircraft at Subic Bay and Clark Air Force Base respectively. These facilities saw extensive use by American forces during the Vietnam War, but a major earthquake and domestic political opposition to the US presence combined in the early 1990s to substantially reduce the American connection. The treaty itself remained on the law books, but the alliance was in difficulty. As Cruz de Castro (2003) explains, the reduction in Washington's security guarantee and the winding down of military exercises came just before a major increase in the Philippines' concern about its territorial disputes with China, a trend which has stimulated renewed interest in Manila for the old relationship.

By 1955 the United States also had a mutual defence treaty with what was then called the Republic of China, reflecting Washington's recognition of Chiang Kai-Shek's regime in Taiwan as the government of all of China. This timing reflected the Chinese mainland's growing campaign of military pressure against Taiwan. But what was essentially a US–Taiwan treaty was not to last. In 1979, as the United States recognised mainland China, it replaced the treaty with a congressional recognition act.

Finally, the United States also held Thailand to be one of its alliance partners in Asia. But unlike the other relationships which formed part of the San Francisco system, Thailand and the United States did not have their own separate post-war treaty. Instead they were part of what was known as the Southeast Asia Treaty Organisation, the result of a treaty signed in Manila in 1954. Its eclectic membership – Australia, France, New Zealand, Pakistan, the Philippines, Thailand, the United Kingdom and the United States – represents a group of countries which were jointly concerned about the spread of communism in Southeast Asia. This may give the impression of the closest that Asia has come to a NATO-like structure, but as Henry Brands Jr. (1987, p. 269) suggests, this is not what SEATO was. Moreover, unlike the NATO experience, where the Soviet Union was the clear and obvious origin of the threat to western interests, in Asia there was also concern about China, which like the Soviet Union was a supporter of North Vietnam. By 1977, two years after the unification of Vietnam, SEATO ceased to exist. But the US commitment to Thailand remained on paper, even if Washington was far less willing to commit itself to a ground war in Asia after the Vietnam experience.

Yet even as it was licking these wounds, the United States still held together a significant array of treaty alliance relationships which had a growing impact on Asia's security as Britain's power waned. As British forces withdrew from Southeast Asia in the early 1970s the United Kingdom moved out of the alliance business in Asia. The Anglo-Malaysian Defence Agreement was succeeded in 1971 by the Five Power Defence Arrangements (FPDA), whereby the United Kingdom, Australia and New Zealand exchanged letters with Malaysia and Singapore promising to consult in the event of an external attack on either of the two former British colonies. The FPDA is not without its merits, offering a way for Singapore and Malaysia to work together in a limited way (despite their suspicions of one another) and as Bristow (2005) argues, using its consultative nature to adjust flexibly to changing regional security conditions. But, somewhat ironically, one of its advantages has been to offer a second-order link to the United States, given the membership of the UK and in particular Australia.

In other words, there is no question that in the second half of the twentieth century, the United States had become the undisputed alliance champion in Asia. The somewhat messy amalgam of alliance relationships is often referred to as a 'hub-and-spokes' arrangement, with the United States as the common centre and its Asian allies as the spokes, radiating outwards. And neither Washington's rapprochement with China in the 1970s nor the collapse of Soviet power at the end of the Cold War spelled the end of this alliance system. Occasional challenges were managed without too much wider effect. In the mid-1980s, for example, New Zealand elected a new government committed to a nuclear weapons-free policy. This was the result of years of concern about western nuclear weapons-testing in the South Pacific and the more recent revival of the nuclear competition between the United States and the Soviet Union. While many New Zealanders wanted to retain their country's active participation in the alliance with the United States, the domestically popular prohibition on visits to New Zealand ports by nuclear-armed or nuclear-powered vessels led the Reagan administration to suspend one leg of its security commitments under ANZUS. But the 'kiwi disease' (as it was often referred to at the time in the United States), did not spread, including to Japan where anti-nuclear sentiment was also very strong. Moreover the Australia-US leg of ANZUS would grow stronger as the years went on. This is a further sign that, with a few exceptions, America's alliances in Asia have been remarkably resilient.

Alliance benefits and doubts

Evaluating the security implications of this set of alliance relationships is not altogether straightforward. For its part the United States can hardly imagine that its own direct security relies upon the obligations that its Asian partners may have signed up to. But in Korea and Japan, the United States has two long-standing allies which allow it to remain militarily connected in a direct physical sense to the security of North Asia. The ability to maintain forces forward in that part of the world and to reinforce them quickly if necessary supports Washington's long-term position that it would oppose any continental power (first the Soviet Union and then increasingly China) that might want to dominate the regional distribution of power. In other words, there is a link between these alliances and American vital interests. The United States can also count reasonably well on the political, if not always the military, support of at least some its regional allies should a serious crisis come. However, some allies (including Australia) are more likely to be materially helpful in this regard than others (including Thailand).

For America's regional allies, having a treaty-based security relationship with the world's leading military power tends to be an asset in a challenging strategic neighbourhood. In South Korea's case, the alliance with the United States and the presence of American forces were a lifeline, providing support as its own armed forces became more sophisticated and reversed the imbalance that once favoured the much larger North Korean armed forces. For Japan, constitutionally and politically limited in the ways it can use its own military forces (which are consequently called the Japan Self-Defence Force) and living next to a rising China, the United States alliance has remained the starting point of its security policy. An understanding that United States nuclear deterrence has been extended to cover both Japan and South Korea in the event of a nuclear threat (from either China, Russia or North Korea) has been a crucial part of the relationships that these two North Asian allies have enjoyed with Washington. This may seem odd for an era when nuclear weapons are supposed to be moving to the margins of strategic policy-making internationally, but as Andrew O'Neil (2011) indicates, worries about the threat environment in East Asia have caused Japan and South Korea to seek greater assurances from Washington about the credibility of its extended deterrence commitment. These promises also function further south in the region. Having toyed with the development of its own nuclear weapons in the 1960s, Australia opted to rely instead on the protection offered by American extended nuclear deterrence.

Other security benefits have also arisen for close allies of the United States in Asia. These include ready access to US military equipment, opportunities to exercise with the world's most advanced forces, and varying levels of information exchange and intelligence cooperation. Yet these alliances only extend to a small percentage of the region's sovereign states, and not just because of choices made in Washington. Ambivalence about America's alliance network has not been uncommon in significant parts of the region. A very cautious approach to alliances is reflected, for example in ASEAN's Zone of Peace, Freedom and Neutrality which was discussed in the previous chapter. Of course, even when that Zone was established the United States already had two Southeast Asian allies: Thailand and the Philippines, both of whom were founding members of ASEAN. But for an important country like Indonesia, which has adopted a stricter view of neutrality, alliances with a Cold War superpower would have spelled the end of the policy of non-alignment. In part because Britain had also built for itself a set of alliances in the region, (including with Indonesia's neighbours) this form of statecraft could easily be depicted as a relic of the colonial era in Southeast Asia, and as a vehicle for western control that needed to be abandoned. Hemmer and Katzenstein (2002, p. 588) have gone as far as to suggest that in their approach to alliance-building in Asia, American leaders were influenced by what were essentially colonialist viewpoints: 'Southeast Asia, in the view of U.S. policymakers, was constructed as a region composed of alien and, in many ways, inferior actors.'

Washington's outlook has evolved in subsequent decades, but in today's Asia, many countries continue to prefer keeping a much lower strategic profile than an alliance with Washington would allow. This does not prevent significant levels of security cooperation occurring between the United States and partners which, strictly speaking, are not formal allies. Singapore, for example, has a defence framework agreement with the United States which allows for military cooperation, including the porting of US naval vessels. And Singapore is a more important partner for the United States than its formal allies in Southeast Asia. But the Singaporean facilities that the United States navy uses are not referred to as American bases, and if there is an unwritten alliance between the two countries it is very much a de facto connection rather than a formal set of security obligations. Denny Roy (2005, p. 313) was probably right a decade ago to observe that 'Singaporean leaders are arguably America's most supportive quasi-allies in Southeast Asia'. But these leaders watch the regional distribution of power very carefully and would not want their country to be seen as a formal ally of the United States because of the future traps that such a connection might produce.

Further west, India's reputation for non-alignment has also made it a very unlikely formal ally of the United States. Indeed during the Cold War years, India's foreign policy settings and closed economy drew it much closer to Moscow than to Washington. This has not prevented India and the United States from developing a mutual interest in security cooperation. Their agreement on civilian nuclear cooperation, which was approved by the US Congress in 2008 despite India's continuing non-membership of the Nuclear Non-Proliferation Treaty, was widely regarded as a breakthrough. But India is determined to ensure that it is not seen as part of a US-led balancing operation against China. A formal alliance between the United States and India, which would offer precisely that signal, is highly unlikely. Daniel Twining (2007, p. 83) has written: 'India's historical wariness of China, as well as its ambitions to match China's rise to world power, give U.S. officials confidence that it will emerge as a friendly, independent pole in Asia's emerging security order.' But even if this optimism is realised, this is by no means the same as India being regarded as part of the United States-led alliance system in Asia.

This is not to say that America's non-allies in Asia necessarily oppose the alliance preferences of their neighbours, and the effects these other relationships have on wider regional security. Indonesia, Malaysia, and Vietnam, for example, are generally in favour of a strong American commitment to and presence in the region which its formal alliance relationships with other countries help underscore. Public protestations about America's intentions can sometimes be made by politicians for show to convince a domestic audience, but private understandings with Washington can be different. In various ways then, ranging from formal alliance commitment to more informal means of encouragement, many of the countries of Asia have deliberately sought to retain a strong American presence in the region. They have been willing to set up what Evelyn Goh (2013) depicts as a social contract with the United States as the welcome leviathan which offers a benign form of hegemony in Asia.

This view is in contrast to the alternative view of an Asian hierarchy promulgated by David Kang (2007), which sees the same states perfectly prepared to come to terms with a rising China at the apex. Reconciling these viewpoints is difficult and suggests that at the very least the region can have one but not two hegemons. But even China's attitude towards a strong United States presence in Asia has hardly been categorically hostile. China appears on the surface to have been a steadfast opponent of America's alliance relationships in North Asia, but at least until recent years, the real story has been more complicated. The Communist Party leadership has maintained an interest in the perpetuation of America's alliance relationship with Japan to the extent that it has reduced Tokyo's need to build up its own forces.

As Alistair Iain Johnston (2003, pp. 40, 42) pointed out some time ago, 'Chinese attitudes are exceedingly complex' towards the US–Japan alliance whereas the US–South Korea alliance 'serves Beijing's interests in stabilizing the Korean peninsula's division'. The premium China has placed on a stable external environment that allows it to progress economically, which was touched on in Chapter 4, has also translated into a willingness to put up with parts of America's forward military presence in the region. This has been changing with American encouragement for a stronger military role on Japan's part and as China's military self-confidence has grown, but it is still true that there are natural limits to Beijing's hostility towards the San Francisco alliance system.

The United States also faces little direct competition in terms of the erection of a set of counter-alliances in Asia. The collapse of the Soviet Union over twenty years ago put an end to earlier such efforts, which had been declining given the limited value that Moscow was obtaining from its strained relationships with North Korea and Vietnam. The most obvious contender for the role of chief counter-alliance builder is of course China. But Beijing has only one formal treaty-based ally in Asia. This is North Korea, which is as much a liability to Beijing as it is an asset. China has also developed a close military partnership with Pakistan, including, at least in past years, the exchange of materials relating to nuclear weapons programmes. But Pakistan is also a partner of the United States, and as Afridi and Bajoria (2010) have pointed out, the Communist Party leadership in Beijing has become concerned about the growth of extremist groups within Pakistan and their potential connections to groups with independence agendas in western China.

As indicted in the previous chapter, China can rely on the political support of the other members of the Shanghai Cooperation Organisation (SCO). This has come in handy for Beijing's very broad definition of terrorism. The SCO also tends to oppose American claims to leadership in Asia, but it provides at best a thin veneer over an historically difficult relationship between China and Russia. Either with Russia or on its own, China would struggle unable to develop a counter-alliance grouping along the lines of the Warsaw Pact which faced off in Europe against NATO. This comparison may be one reason why Beijing insists that alliances are old-fashioned and dangerous mechanisms which should be left behind in the Cold War. If China had a better chance of developing an equivalent system of alliances it might be less opposed to the concept. But it would be too cynical to regard this as the only reason. China's leaders genuinely seem to hold to the view that regional security ought not to rely as heavily on alliances as the United States and many of its closest friends in the region are inclined to. There is of course some self-interest in this argument, but this also reflects views within China on international relations.

The surprising revitalisation of Asia's alliances

It is possible that America's alliances in Asia are increasingly part of a security problem rather than being a reliable security answer. China's view of them at any rate has been hardening since the end of the first decade of the twenty-first century. And this is largely because the United States and some of its leading allies have been consolidating their defence relationships at a time when China's rise has become more obvious. Of course China has been rising for some time. But for a few years at least, the United States seemed distracted from the region and less focused on the way its alliance relationships contributed to its aims in Asia. Indeed, as the United States became embroiled in Afghanistan from 2001, and especially in Iraq from 2003, it sought to make use of its Asian alliance relationships to support these more distant missions. Australia in particular gave consistent help, but South Korea also played a noticeable role in the provision of forces and Japan provided financial and some logistical assistance. New Zealand's contributions in Afghanistan laid the basis for an eventual warming of its own strained relationship with Washington, leading not to a resumption of formal ANZUS ties, but a closer military relationship nonetheless.

Always conscious of the gap between itself and the world's leading military power, China seemed not too worried at all that the United States was bogged down in Central Asia and the Middle East. Indeed the widespread perception that the United States was too busy elsewhere to focus on East Asia's changing diplomatic and trade arrangements seemed to offer China an opportunity to quietly gain regional influence. But then at least two things happened to bring this happy coincidence of factors to an end. First, China's rather successful diplomatic attempts to reassure some of its neighbours in East Asia, analysed at the time by David Shambaugh (2004/5), were subsequently overshadowed and undone by a much more robust approach to its territorial claims in the South China and East China Seas, which became obvious from around 2010. As became clear in Chapter 6, these are far from trivial contests. It is possible that in pushing its regional maritime territorial claims so strongly, China had become overconfident after years of impressive economic growth. Its leaders may have concluded that US influence in Asia was waning, not least after the global financial crisis which began with the debt problems affecting America's banking system. Some of the decision-makers in Beijing may simply have felt that China's time had come and that long-held ambitions were now being matched by at least some military capacity to project power. But as Thomas Christensen (2011) has wisely observed, the worries of the Communist Party's leadership about domestic politics, including the rise of popular nationalism, also seem to have been involved in this more active external policy era.

Almost as if to compensate for China's increasing assertiveness, the second change came in the form of a shift in focus from the United States. As the Obama Administration wound up America's costly involvement in Iraq, and looked to do so as well in Afghanistan, it was guided partly by a determination to direct more of its attention to Asia, fearing that China had been stealing a march. Many of the smaller and medium powers in Asia shared some of these concerns about China and stood ready to welcome any additional attention from the United States. The result of this mix was the Obama Administration's pivot to Asia in which alliances and other forms of security cooperation would play a substantial role. When Washington found that 'pivot' sounded too much like abandonment to its allies and friends in the Middle East and Europe, it changed the term to the more politically correct 'rebalancing', but the overall message was still the same.

As it began, including with a prominent journal article from US Secretary of State Hillary Clinton (2011), the rebalance took up some less controversial diplomatic themes. First up was the decision to commit the United States to presidential-level involvement in the East Asia Summit. Later on would come the Obama Administration's determination to be a full part of the negotiations for the Trans-Pacific Partnership trade agreement. This was a sign of Washington's concerns that China had too much of the inside running in Asia's economic integration (as examined in Chapter 4). But while these wider aspects of US policy were significant, the military dimension of the rebalancing remained central. Within time the Pentagon announced that an increasing proportion of America's navy (as much as 60 per cent) would be based in Asia. Washington sought to allay South Korean and Japanese fears that the old practice of extended deterrence was insufficiently specific for the new challenges that were arising. For Japan at least, these challenges included the modernisation of China's nuclear weapons, the growth of China's conventional military power, and Beijing's more robust stance in the East China Sea. One of the most visible, although highly symbolic, military aspects of the pivot would come with Australia including an agreement to send US marines on a rotational deployment to Darwin in Australia's far north.

It was certainly not a case of all push by Washington and no pull from the region. Having been alarmed by China's maritime pressure, including in the Scarborough Shoal in 2012, the Philippines looked to revitalise its alliance relationship with the United States. But much of the story was also being told by the responses of regional countries which are not formal allies of the United States. In 2012 Leon Panetta became the first American secretary of defence to be welcomed to Vietnam's important port of Cam Ranh Bay in over three decades. Singapore agreed to the rotational presence of American littoral (shallow water) vessels. US defence cooperation with Indonesia and Malaysia was stepped up.

Moreover at least some long-standing American allies found room for closer security relationships among themselves, challenging the old hub-and-spokes model in which the only strong links are the bilateral ties with the United States. Japan and Australia have been the most potent examples of this trend, and in 2013 the newly elected conservative government in Australia gave unstinting support to Japan's opposition to China's sudden declaration of an Air Defence Identification Zone in the East China Sea which included the disputed Senkaku/Diaoyu islands. The strengthening of the Australia–Japan–United States strategic dialogue almost suggests that a non-treaty alliance relationship now exists between these three countries. This seems at least a partial fulfilment of the idea of a trilateral alliance which was raised over a generation ago by William Tow (1978).

Even when considered in combination, all of these developments have not shifted the distribution of power in Asia markedly in Washington's direction and away from China. At least this is not the case in a military sense. Despite all the negative publicity about Washington's commitment to Asia after 9/11, the United States never left Asia militarily. Its Seventh Fleet, the bedrock of its maritime presence, remained focused on the area that the United States likes to call the Western Pacific. The strengthening of US facilities on Guam, much closer to the Asian mainland than Hawaii, was occurring well before the pivot was announced. And Washington had certainly been paying attention to its leading Asian allies, principally Japan, South Korea and Australia. The reorientation of US basing arrangements on South Korea and Japan, a very slow process in both cases, had been in play for some years.

Moreover, not all of the security relationships that have been strengthening in Asia have the United States as a member, nor are they necessarily being developed specifically in reaction to China's rise. Much of the impetus for a stronger security partnership between Indonesia and Australia, for example, stemmed from a joint concern about tackling terrorism, and a desire to exclude the possibility of Australian intervention in West Papua. And not all of America's allies have been happy to work with each other, reducing the prospects of joining all of the spokes in the wheel into a rim. Japan and South Korea have remained at loggerheads, partly over territory, but also over Japan's record of conduct in the Second World War and competition for status, not dissimilar to that which exists between Japan and China. The visits by Japanese political leaders to the controversial Yasukuni shrine have exacerbated these tensions, and have created a further point of cooperation between South Korea and China. Even the encouragement that Japan and South Korea have both received from their chief ally to boost their own military capabilities in the context of China's rise and North Korea's threats have not brought the two allies together. Instead, as Sheryn Lee (2013, p. 103) explains, Japan and South Korea have developed an increasing tendency to see one another as military rivals.

This is not good news for the long-term prospects of the American rebalance, because of that strategy's heavy reliance on relationships with allies and partners and the limited extent to which the United States can come up with additional resources of its own in Asia. Whether other major powers are ready to step up remains open to question, partly because of their sensitivity to China's likely reaction. India is a case in point here. In 2007, Shinzo Abe (then in his first period of office as Japan's prime minister), proposed a quadrilateral form of security cooperation between the region's leading democracies, extending to India, Australia and the United States as well as Japan. But none of the latter three at the time were especially enthusiastic about the proposal. That hesitation was partly because of concerns that China would read such a development as an attempt at containment. Indeed as David Brewster (2010, p. 3) has explained, China had already expressed its serious disquiet about some maritime military exercises that the four had been involved in off India's coasts. The prospects for such quadrilateral cooperation improved somewhat with the election in India of a new nationalist government in 2014, especially in light of Prime Minister Modi's enthusiasm for closer ties with Japan. But these do not amount to an alliance, nor to a forswearing by India of its interests in maintaining amicable relations with China.

Alliances and security dilemmas

This means that it is important to be wary of sweeping generalisations about the division of Asia and its borderlands into two alliance blocs. There is no question that the rise of China is the primary reason for the strengthening of the US-led system of alliances and the development of new partnerships between Washington and some countries in the region. But there is no equivalent Chinese-led block providing resistance, and many of the countries of East Asia are reluctant to get caught up in strategic competition between the United States and China. Yet they still want a closer relationship with the United States partly because they are genuinely concerned about China's rise and assertiveness.

Keeping these aims in balance can be a tricky tightrope walk. Many Southeast Asian countries, for example, are aware that their incipient security community offers little real protection in the face of serious major power competition. Those states with claims in the South China Sea are concerned about China's more robust assertion of its own maritime claims, and conscious of the limitations of the array of regional security forums which were discussed in the previous chapter. But this does not mean they are willing to side with the United States on any sort

Division: Will Alliances and Partnerships Split the Region? 259

of consistent basis. Singapore and Indonesia, for example, have sought to encourage an Asian strategic equilibrium and balance China's growing influence by reaching out to engage the United States. But not for a minute do they wish to abandon the benefits of their relationship with that same growing China, including the gains which closer integration with China's economy can bring. In one provocative account, Kishore Mahbubani, (2009, p. 9) one of Singapore's leading public intellectuals, has recommended that:

> In reflecting on future strategic opportunities, Western minds should reflect on the Chinese wisdom in translating the western word 'crisis' by combining two Chinese characters, 'danger' and 'opportunity'. Too many Western minds are looking at dangers; few are looking at opportunities.

While an increased American focus on the region has been welcomed in much of Asia, Washington being unable to accumulate a long list of all-weather friends whose support is guaranteed in more difficult times. In the event of a serious crisis with China, the United States would expect Japan's moral support, but Japan is unlikely to involve itself militarily unless its own direct security concerns are involved. By the same token, Japan would want the United States to come to its assistance at the first sign of a serious deterioration in its own relationship with China. South Korea may have supported the United States in Afghanistan, but closer to home it values its strong relationship with China. The US–South Korea alliance is much more attuned to the possibility of aggression from North Korea, in which case the United States would be expected to meet its alliance obligations to South Korea rather than vice versa.

Some of Washington's other allies in the region are more inclined towards zero-sum calculations about China's rise. As it has become increasingly concerned about China's South China Sea ambitions, the Philippines has sought to encourage the United States into a renewed alliance relationship after years of drift. But the armed forces of the Philippines are so weak that there is very little, if any, support they could provide to the United States. Instead, the United States has to watch that it is not dragged into a conflict with China that may not be worth the costs. Thailand, America's second formal ally in Southeast Asia, is focused on its own internal political future, and has a good and promising relationship with China to protect. Washington can expect very little in the way of support from Bangkok in putting pressure on Beijing.

Perhaps the closest thing to an unconditional ally that Washington has in the region is Australia, which might be the only other country in Asia to come to the help of both the United States and Japan in the event they are caught up in escalating tensions with China. Australia might do the same in the event of a serious crisis on the Korean peninsula, and would probably support an American stand, should it come, on the South China Sea. But while Australia has defence capabilities which are technologically superior to the militaries of nearly all of its neighbours, it does not have at its disposal numerically significant maritime forces for any major conflict in North Asia. And like all countries in the region, Australia would need to consider the value of its long-term relationship with China, a consideration which has led Hugh White (2011) to suggest controversially that Australia should be willing to reduce the closeness of its alliance relationship with the United States.

Neither Vietnam nor India are bound by an understanding that in a security crisis in Asia they would assist the United States, which in turn is under no obligation to come to their assistance in the event of an armed conflict with China. This is not to say that these two independent-minded countries would be inactive in such a circumstance. Vietnam could be expected to push back on China in the event of a serious crisis over their mutually exclusive South China Sea claims. Quite what India would do in a similar crisis in its land border disputes with China is more difficult to forecast. But neither of these countries can be seen as guaranteed members of a wider grouping led by the United States in order to balance China's rise. Their difficulties with China remain in a largely bilateral and local context. Much of the rest of the region might well wish to stay on the sidelines, and many Asian countries, which generally lack any significant capacity to project military power, could offer no more than political support to the United States if they really wanted to take sides.

In an *objective* sense then, there is little to suggest that China is being encircled by a vast and growing US-led alliance system in Asia. The Cold War contest in Europe between the western bloc, organised around NATO and led by America, and the eastern Warsaw Pact block under the control of the Soviet Union, is not bound to be replayed in Asia. As Rosemary Foot (2009, p. 141) has argued, the US–China relationship is positioned 'somewhere between a cold and warm peace, or between cooperation and rivalry'. As was clear at the start of this volume, even in its own Cold War past, Asia was never a region divided along clear demarcations. Nor is it likely to be. The notion of a line being drawn down the middle between Chinese and American blocs runs counter to many of the issues that have been discussed in this book.

But at the psychological level, things can seem different. When security is also seen in terms of *subjective* judgements about threats to things that are valued, it is easy to see why many in China remain concerned that their country is being strategically encircled by a United States-led coalition. That perceived encirclement has many elements. Part of it is unilateral American action, including the insistence by the United States in flying its aircraft and directing its naval vessels too close to China's coasts for Beijing's liking. But a good part of it is about America's alliance relationships. This includes Washington's very strong support for Japan, which, as Michael S. Chase (2011) notes, feeds in particularly strongly to views in China that US alliances are being enhanced with containment in mind. Similar messages are heard or misheard in China about the strengthening of triangular alliance links among the United States, Japan and Australia. America's willingness to provide extra assurances to South Korea about extended deterrence, which while ostensibly aimed at North Korea, can also be seen as a response to China's growing heft. Closer US security cooperation with almost all the countries in Southeast Asia who have competitive maritime claims with China is also a problem for Beijing, as is ongoing US interest in closer cooperation with India. This is by no means a complete list, but it already provides a sense as to why some of the more pessimistic analysts in China have plenty of material to work with.

Moreover, non-military developments can also be perceived as part of a strategy of encircling and marginalising China. These include Washington's renewed enthusiasm for fuller participation in the ASEAN family of multilateral security forums which were considered in the previous chapter, especially when the United States encourages discussion of South China Sea issues. They also include America's involvement in the Trans-Pacific Partnership trade negotiations, which exclude China. Washington's criticism of China's human rights record can also come across as part of a co-ordinated program of pressure. Even America's greater interest in the South Pacific might be depicted as part of that contest, noting China's growing involvement in the provision of development assistance in that remote part of Asia's borderlands.

As might be expected, the United States (and its allies and partners) deny that they have the encirclement of China in mind. In part because of the close economic relationships that all of the region's members have with China, the common mantra is that nobody desires the containment strategy which the United States and its Cold War allies adopted to cope with the Soviet threat. Instead, at most the United States and its close friends would depict their ambitions in terms of maintaining a balance of power in Asia, although if the depiction of balancing without containment in Tellis (2013) is any guide, the gap between these approaches can look very fuzzy. Yet adamant voices continue to be heard that there is no intention to trap China, cut off its promising outward links, and wait for its economic, social and political systems to collapse. That would, the standard argument goes, be impossible given China's participation in the global economic system. And there seems to be little public appetite for the much more robust response to China's military and economic rise which has been advocated by Aaron Friedberg (2012), who argues that the United States and China are locked in nothing less than a great power battle for regional supremacy.

But even if the intentions of the two largest powers are less confrontational than some think, there is still a potential problem of competing misperceptions. Many in China may still feel that their country remains vulnerable and that, by contrast, the growth in China's power, including its ability to influence regional security, is simply a return to the natural order of things which should not be regarded as threatening by any means. This is reflected in the efforts by scholars such as Yan Xuetong (2011) to draw lessons for China's current trajectory from the wisdom of ancient dynasties. Yet what to some is a logical philosophical exercise is to some others the evidence of a quest for regional dominance by China. In turn it is possible for analysts in China to regard the regional status quo as a device to subordinate China's rightful ambitions towards a stronger position in the region, because that situation reinforces the imbalance that remains in America's favour.

At least in theory, the strengthening of alliances seems a perfect fit for the creation of an Asian security dilemma principally because alliances are exclusive groups which run the risk of perpetuating an 'us and them' mentality. The obligation to come to an ally's assistance in the event of an external attack is premised on the notion that there is a potential external attacker in the neighbourhood. The US-led alliance system was not built in the 1950s on the assumption that China was that threat. And America's commitment to Japan's security under the most important component of the San Francisco alliance system did reassure China by reducing the prospect of a more strongly rearmed and assertive Japan. But in today's Asia, America's alliances have different implications for a rising China. There is also no question that a number of Asian states have been keen to build stronger partnerships with the United States at precisely the time when the growth in China's power has become more obvious. This doesn't mean that China is somehow to blame. But it does mean that China is involved. Becoming a great power has its advantages, but it also has its complications, as China and the rest of the region is witnessing right now.

A concert of powers in Asia?

This raises the intriguing prospect that Asia's major powers may be stumbling unwittingly into an ever more serious and dangerous strategic interaction. Might the region be on the cusp of a hazardous situation

Division: Will Alliances and Partnerships Split the Region? 263

where the next crisis, or the one after that, could lead to Asia's first serious interstate war for many decades? Such a march to conflict may not be nearly as automatic as security dilemma thinking sometimes suggests. Governments do have choices, and it is not inevitable that they will assume the worst-case scenario about the real intentions of their strategic competitors. As Lawrence Freedman (2013, p. 622) has written in a enlightening book on strategy, while control of the future may be impossible, 'in the end all we can do is act as if we can influence events. To do otherwise is to succumb to fatalism.' The states of Asia have choices to make, and one of them might be for the United States, China and the other major powers to decide to work more closely and cooperatively with one another.

An improved regional security environment might be possible if the United States prioritised its relationship with China over building its alliance relationships in Asia. Several of the adjustments indicated by White (2012) would be necessary. A strong commitment to Sino-US cooperation would require Washington to avoid casting doubt about the legitimacy of China's rise, including the stronger capabilities of China's armed forces and the purposes for which these capabilities are intended. For that to happen China would need to be more willing to regard the United States as a security partner, and would need to rule out the notion of somehow taking over the top spot in regional affairs. Japan would need to resist the temptation to drag the United States into its ongoing tensions with China, come to terms with China's growing power, and see cooperation between its large neighbour and its close ally as something other than a threat to its own objectives. In turn, China would have to accept that a stronger Japan in the twenty-first century need not be a replay of the militarist power of the first half of the twentieth century. And as it assumes a greater role in Asian security affairs beyond its own South Asian neighbourhood, India would need to see all three of these major powers as partners, just as they would in their respective views of India.

Getting to this level of mutual respect and cooperative intention among the region's big four would not be easy. But this combination would likely be required if there is to be an Asian concert of powers. The idea that major powers can work in concert to manage some of their differences has an important historical precedent. But like so many of the ways of thinking about security cooperation, this model comes from Europe. After France had sought hegemony in Europe through a series of major wars in the late eighteenth and early nineteenth centuries, the other major powers came together upon Napoleon Bonaparte's final defeat to see if they might be able to restrain the territorial competition between themselves. The participants in this concert were Britain, Prussia,

Austria-Hungary, Russia and, in a wise act of coopting the defeated power, France as well. Their objectives were realistic. One of these was not perfect justice and freedom. Russia, Prussia and Austria-Hungary were hardly models of liberal democracy and were interested in the concert partly for the support it gave them in sustaining their autocratic political systems. (One might think of China in this context as a modern representation of this domestic political intention.) Nor were the major powers particularly devoted to the concerns of the minor players in Europe. The sovereignty of these smaller states were hardly guaranteed in the European concert whose members were meant to consult one another in the event of territorial changes but not expected to be especially sensitive to their small neighbours.

But the concert agreed on the important principle that it was unwise to allow any one of the major powers to repeat the Napoleonic quest and seek hegemony on the continent. Their most important achievement – the avoidance of major power war – is not to be underrated for an era in which armed conflict was often regarded as a legitimate means of pursuing national objectives. The concert was not permanent. The war between Russia and Britain in Crimea in the 1850s indicates that the concert did not last quite as long as some of its supporters have suggested, although as Schroeder (1986, pp. 5–6) argues, even this did not bring all of Europe's other great powers into a truly colossal conflagration. But before long Europe would experience significant changes, including the rise of a united Germany, and would be increasingly divided into the two blocs who would later fight the First World War. Well before that happened the European concert had irretrievably broken down. But for the best part of half a century the Concert of Europe had given Europe a rest from the worst of major power tensions and from the major power wars that could follow from them.

As Amitav Acharya (1999) noted some years ago, there are several problems with the idea of a Concert of Asia. For the purposes of this book, at least four stand out for consideration. The first is that, unlike the European great powers which met in Vienna in 1815, their Asian counterparts in the early twenty-first century do not have the fresh memory of devastating armed conflict which inclines them to see collaboration as a necessity. Perhaps China, Japan and the United States could be more accurately compared with the European major powers ninety-nine years later, which by 1914, far too optimistic about their abilities to avoid catastrophe, had drifted into a war. Indeed, as the one-hundredth anniversary of the outbreak of that global conflict approached, scholars such as Margaret MacMillan (2013) were making precisely this more worrying comparison.

The second problem is that, unlike the European major powers, which could draw on a similar cultural heritage, the United States and China have such different histories and outlooks as to make such a convergence unlikely. Questions of common culture will be addressed in the concluding chapter, but at least for now there are reasons for being wary about this criticism. Those same European major powers, despite their common approaches, found themselves in a dreadful war a century after Vienna. So it may well be that the lack of a common culture is not as significant as authors like Hedley Bull (1977) have suggested it might be.

The third obstacle is the unevenness of the Asian major power picture. On the one hand the United States is unsurpassed as the strongest great power influence on Asia's security. On the other hand, China's rise is the big issue that dominates all the others. In reality, the United States and China crowd out the other potential members of a great power concert. In the most compelling contemporary argument that there is for an Asian concert, White (2012) includes Japan not so much for its own power potential but because of its capacity to trigger a wider conflict between China and the United States. Indeed this recipe for a concert really comes down to a single choice by a single great power: a decision by the United States to fully accommodate China. India's participation in an Asian concert depends upon whether it can be consistently confident in realising its potential to affect matters in East Asia. For now, at least, India is falling significantly short of just that vision which C. Raja Mohan (2006) and others have suggested. And if it was to play such a role, this still might be in support of a region which remains dominated by two other larger powers. This is evident in the suggestion from Rajagopalan and Sahni (2008, p. 20) that: 'By building robust political and economic links with both China and the US, India could be the catalyst in bringing both countries together in a new cooperative Asia.' At least initially, India's contribution to the concert could therefore be more about allowing China and the United States to work together than about India's own independent role.

Might then an Asian concert be built on just two legs: China and America? This seems to be the picture that advocates for a 'G2' relationship for global governance are suggesting (a boiled down version of the G7 grouping of the world's leading economic powers). It is also reflected in the Chinese idea of a 'new type' of power relationship between the great powers, which fed some very optimistic reports around the time of the 2013 summit meeting between China's new president, Xi Jinping, and America's second-term president, Barack Obama. At the time some of the signals back from the United States seemed promising. In one essay on the subject, David Lampton (2013, p. 52) suggested that this 'this vague but potentially useful concept has been generally endorsed by leaders in Washington'.

But significantly closer cooperation between China and the United States as the region's benign duopoly would exacerbate a fourth problem which would be bound to complicate the idea of an Asian concert. The early nineteenth-century European example was built on the basis of an elitist view of international politics, in which the major powers were willing to recognise the importance of each other but thought little of the interests of their smaller neighbours. Tiny principalities, and there were many at the time, could easily find themselves being sacrificed for the sake of major power peace. Many of the smaller and medium powers of Asia have clear memories of a not dissimilar experience. A number have had the experience of being subjected to colonial control as the European empires competed for imperial possessions, and several later became proxies in the Cold War competition between the United States and the Soviet Union. Independence for these countries is not just a matter of national authority over territory and people. It also translates into avoiding invisibility in regional affairs lest their interests be overlooked by the larger countries of Asia and its borderlands.

It almost goes without saying that there would be a chorus of disapproval from much of Asia if the major powers established an exclusive club of which they were the only members. That chorus would be even stronger if the concert were to be a club of just two. That would be taking the elitism of concert behaviour to its highest level. While the smaller and medium powers would be likely to welcome the prospect that the two major powers might avoid a truly catastrophic war, they would nonetheless resent the decline in their own voice on security matters. An Asian concert would, for example, be a direct challenge to the role of the ASEAN family of regional security groupings, as Lee Jones (2010) points out. Such an unlikely development might also make Australia's alliance with the United States effectively null and void if the price of the United States getting on with China was the disbandment of some parts of the San Francisco system which Beijing saw as amounting to its virtual encirclement. Vietnam and the Philippines would fear that their South China Sea concerns and interests would be completely ignored as a price for China's collaboration, and that at the extreme, the United States would be willing to come to terms with China's dominance in parts of maritime Southeast Asia.

A three- or four-power concert, extending to Japan and India, would not necessarily remove these problems. South Korea would be likely to believe that its interests were about to be sacrificed in favour of Japan's. Pakistan would probably seek to make India's life more difficult and drag its attention back into South Asia. Indonesia might seek to organise a counter-concert of medium powers in Asia to challenge the region's power elite. If the major powers needed the cooperation of their smaller regional colleagues, for example in the avoidance of further nuclear proliferation or in the establishment of clear rules of the game for avoiding maritime disputes, they might find that cooperation very hard to get.

Yet these complexities should not stand in the way of considering the advantages of more concerted behaviour among the major powers. Some of these sacrifices might just be warranted if a major power war were a real possibility in Asia and if only the major powers working together could prevent that from happening. Given the limitations in the region's more inclusive groupings for security cooperation which have already became clear in this volume, the disadvantages of a more deliberately elitist approach to Asia's security might begin to fall away. Indeed, it might well be thought that there is already a clear hierarchy of power in the region, and that it is best to recognise that for what it is and let the major powers get on with it. But it will not be as easy as that, because it remains unclear how many sacrifices the major powers are themselves willing to make in order to work more closely together. Alliances or not, there are still some fairly formidable divisions in the region, including the contest for regional leadership between China and the United States.

Conclusion

It may seem disappointing to complete the penultimate chapter of this book on such a relatively unpromising note. But one of the important consequences of studying Asia's security is the realisation that there are no magic wands lying around which can absolutely guarantee a bright future of security cooperation for the region. Yet this does not mean that all is lost. The region may not be unified and serious differences between the major actors remain. Stronger alliances reflect concern about China's rise and can also increase regional divisions. An Asian concert may have its merits (and also some real problems) but seems unlikely any time soon. And yet the point that was made at the beginning of this book still holds. The region has not experienced a serious armed conflict between the major powers for several decades. And it cannot conclusively be argued that the tensions and misperceptions between the major powers, which the strengthening of alliances and the building of new partnerships may indeed fuel, are bound to change that situation. In short, while there is every reason to watch these major power dynamics closely, and some reasons to wish for a tradition of cooperation between them to be established, outright panic is neither necessary nor helpful.

This leaves this analysis in a curious situation. It is likely that the whole region would suffer in the event that major power relations in Asia were completely to break down. But doesn't Asia's relatively peaceful record of relations between states over the last several decades mean that the main concerns should lie with many of the other security challenges and issues that have been considered in this book from problems of domestic order to questions about transnational challenges? Alternatively, if the avoidance of major armed conflict remains the sine qua non of a positive Asian security environment, can a way be found of thinking beyond alliances and even concerts, and beyond the region's almost endless security dialogue? The next and final chapter will address both of these important questions.

Chapter 12

Conclusion: Towards a New Asian Security?

This final chapter will take stock of Asia's security and then look forward into what may be next for the region. The first of these tasks is essential in light of the many security challenges and strategic responses which have been considered in these pages. As this analysis has moved from the security problems which challenged Asia's leaders in the region's relatively recent past to the external, internal and transnational security issues facing today's decision-makers, a nagging question has remained which now requires an answer: which of these issues has the greatest impact on Asia's security?

This is not an easy call to make. Part of the answer depends on whether Asia's security can be shaped from anywhere in the region or whether there is a core group of countries whose security experiences are likely to have disproportional effect on the rest. The answer also depends on whose security is at stake. Is the main factor here the security interests of all of the sovereign states in Asia? Or are the security perspectives of particular countries more significant than the others? Is it important instead to be drawn to the security concerns of individual people in the region and of the social groups they are part of? These are more than academic questions. Their answers reveal whether emphasis should be placed on ever-present security problems which may have smaller effects as opposed to the catastrophic security events which have lower chances of occurring. A decision would also need to be made as to whether security challenges that emanate from the deliberate choices of political actors (including terrorism, for example) are bound to be more significant than problems that are closer to being acts of nature (including, for example, the security implications of climate change).

But this analysis is not just about security problems. It also needs to be about responses. The last few chapters have been spent looking at a range of strategies employed in the region to deal with Asia's security problems and to encourage security cooperation of one form or another. The conclusions in this final chapter over what issues should be of greatest concern also need to reflect judgements about which of them is most likely to defy existing strategies. Security in that sense is not the mere absence of threats. It is also about the presence – and in the case of insecurity, the absence – of effective strategies to cope with them.

This leads to the second part of this concluding chapter which will seek to look ahead. If it is found that there are insufficient strategies available in Asia to deal with a very deep security problem it needs to be wondered whether a more effective response might still be developed. It stands to reason that if this serious security problem affects the region as a whole, it is unlikely that a single actor – even if it is the most powerful state in Asia – will have the ability to deal with it effectively on its own. Some sort of cooperative enterprise will be necessary. This volume has already considered various modes of Asian security cooperation. These have included various forms of intervention and the inclusive efforts at security dialogue which have grown up in the last twenty years, largely on the back of ASEAN. They have also included the much more selective alliances, generally led by the United States, whose members are obliged to support one another in the event of an external threat. The previous chapter also drew attention to what appears to be the unlikely possibility of an Asian concert of powers. But something is still missing here: the idea of a new Asian security that rests on a deeper and more enduring basis for cooperation. It might be thought that this is a fool's errand in a region where strategic competition remains so palpable. But it is still important to give that possibility some serious attention.

A pecking order for Asia's security?

In making overall sense of the many regional security themes which have been assessed in this volume, it is useful to return to one of the leading points from the introductory chapter: understanding Asia's security is not a matter of knowing all that one can about each and every one of the security challenges and responses happening in the region and its borderlands. To identify every single one of those factors and appreciate each one of them in all of their complexity would be an almost impossible task that would require several long volumes. But even if this grand task were possible, it would still be an error to conclude that Asia's security is the sum of these multiple parts. One of the main arguments in this book is that a crucial distinction can be made between all of the security issues which are present at least somewhere in Asia on the one hand, and those factors which have a genuinely wider regional effect, on the other hand. This is why the idea of security ripple effects has been used as an informal way of testing which of the issues under consideration stands out because of its wider impact. Regional security in Asia and its borderlands is fundamentally about factors which may initially seem localised but have relevance to the region as a whole because of the ripple effects they create. The title of this book, *Asia's Security*, is itself a reflection of this logic.

It might be thought that this approach has set up an unfair standard which discriminates against some of the security factors which have been considered and the questions which have been asked about them. To select the most obvious example, the main question for Chapter 7 was 'Are Asia's main security problems domestic ones?' Surely, it might be thought, there is only a small chance that the fragmentation of just one of the nation-states of Asia can have a wider regional effect of the sort that is being emphasised. Mustn't these inherently domestic security factors automatically be surpassed by any security challenges which by their nature cross national boundaries? Isn't the prospect of a civil war anywhere in Asia automatically irrelevant on the ripple effects scale in contrast to the possibility of any sort of interstate war in Asia and its borderlands?

But the answer to this question is by no means as clear-cut as this line of thinking would suggest. It is certainly true that some of the internal conflicts and crises which occur Asia and its borderlands remain localised in terms of their main effects. But this depends on their nature, where they are in the region, how big and important a country is under consideration, and how connected to other significant countries it happens to be. Few in the wider Asian region seem particularly bothered by the difficulties which in the last two or three decades have affected the security of Solomon Islands, for example, or of the Papua New Guinean province of Bougainville. But a breakdown of order across the sprawling Indonesian archipelago could certainly have wider regional ramifications. As became perfectly clear in 1999, the violence that flared in East Timor after the pro-independence referendum vote there led to a serious crisis in the relationship between Indonesia, which had controlled East Timor since 1975, and Australia, which led a robustly armed intervention. This is but one example of an internal conflict generating wider effects by raising the chances of an interstate war. Fortunately, in this case, that interstate war was averted.

In a similar vein Afghanistan's internal security conditions were not bound to have a wider impact until the 9/11 terrorist attacks on the United States, which suddenly and substantially concentrated international attention. Over the following years there was great concern that poorly governed spaces in some parts of the region could be the staging ground for wider acts of terror at a very significant distance from Afghanistan itself. This is but one example of the potential for internal and transnational security challenges to be closely connected. This is one reason for the continuing concern about how Pakistan's government manages its own internal security picture. The American drone attacks into Pakistan, which have targeted extremists but also caused understandable tensions in the relationship between the two countries, signal the potential for internal, transnational and interstate security challenges to be intimately linked in some situations. A more worrying example of these connections could come should a major terrorist attack in India, conducted by a group operating from an unevenly governed part of Pakistan, trigger a war between the two South Asian neighbours. Such a war could have a significantly wider regional impact.

There is also no automatic guarantee that any and every dispute between two countries anywhere in the region, while constituting an interstate rather than a purely domestic challenge, will cause significant ripple effects spreading out across Asia and its borderlands. Some minor interstate conflicts do not always carry such a wider significance. The small recurrent problems between Thailand and Cambodia across their border is not the signal of imminent escalation. While serious tensions between the two could contravene the principles behind ASEAN's Treaty of Amity and Cooperation, they are unlikely to trigger wider worries about security in and around mainland Asia. Border tensions which occasionally rise between China and India seem, for the time being at least, to have been contained locally, although they are watched closely for what they suggest about China's intentions as the region's main rising power.

The rarity of interstate wars in today's Asia also works against the idea that this type of problem should dominate one's view of Asia's security. The second chapter of this book was spent endeavouring to understand why Asia seems more peaceful today than it was fifty or sixty years ago. It was found that domestic political changes in a number of Asian countries played a crucial role in the reduced tendency towards war. These changes included the conclusion of sometimes violent struggles for postcolonial independence and national unity, which, as Vietnam's long experience with armed conflict demonstrated, could then be exacerbated by the addition of Cold War competition. Today's Asia is more stable in both of these domestic and international dimensions, which in turn have fewer opportunities to drive one another. And while it could not be concluded from Chapter 4 that the close networks of economic interdependence in Asia are a guarantee against interstate conflict, they certainly add to the costs of war. This is a factor that leaders in the region need to contemplate before they authorise the use of force in anger.

This might encourage a conclusion that interstate armed conflict is important to the wider region largely because of its *potential* effects, rather than because of its *actual* impact today. And if the likelihood of a significant interstate war in Asia is regarded as low, even those potential

Conclusion: Towards a New Asian Security? 273

ripple effects might be discounted, relegating this type of security challenge to the margins of history. Doubts might also be raised about the likelihood of one of Asia's larger powers breaking up in such a manner that truly dramatic wider effects are created across the region. Of course it is unwise to completely rule out the possibility of the serious fragmentation of China, India or Indonesia, to cite the largest examples. But by historical comparisons, these three large countries in Asia, and most of their neighbours and other counterparts, have been enjoying relative internal stability in the last few decades. So once again this may be a matter of *potential* ripple effects from a situation which is also regarded as fairly improbably. Does Asia's international and domestic peace, which exists today in relative if not absolute terms, not mean that regional security analysis needs to concentrate on other sets of problems?

Beyond state-centric confines

It is important to question the proposition that Asia's security (and insecurity) is necessarily about the risk of either significant interstate or domestic armed conflict in the region. By these yardsticks, it might be thought that the subject of this volume is an increasingly obsolete one and that a better focus would be on Asia's prosperity. But, as Chapter 8 showed, there is a range of security issues that, with the exception of terrorism, have little to do with organised violence, but which are nonetheless coming up as significant parts of the regional security debate. These transnational issues by their nature appeared to bypass sovereign states whose dominance of the security picture looks increasingly tenuous to some observers. Moreover in the case of some of these challenges, including climate change and pandemic disease, the emphasis is not on security challenges that come from decisions taken by opposing political actors. These are problems, but they do not emanate from adversaries or enemies.

Transnational challenges appear to have three in-built advantages when it comes to the ripple effects question. First, by their very nature they cross national boundaries. They are therefore not nearly as localised as some of the domestic security issues which are found in parts of the region. Second, a number of these transnational factors seem very difficult to contain by human agency, even when states combine their efforts. The effects of climate change, for example, are felt well beyond the human economic activities that largely give rise to them. Bangladesh and some of the low-lying Pacific Island countries, for example, are already being affected by rising sea levels connected to regional and global carbon emissions occurring thousands of kilometres away.

The displacement of people from trouble spots on one side of the region – from Sri Lanka, Afghanistan and Iran – is felt in the security debate in far away Australia. Thanks to the speed of international air-travel, a pandemic disease originating in southern China can spread rapidly into Southeast Asia and beyond. That these ripple effects are proven and not hypothetical points to the third claim on regional security. Many of these transnational problems are not just possible or even probable events. They are here already, and some are having effects on different parts of the region every day. For example, few parts of Asia and its borderlands are completely immune from the effects of climate change or transnational criminal activities, including drug-trafficking.

But on closer inspection, it is not entirely clear that transnational security challenges complete the quest for the factors which make the best overall sense of Asia's regional security. This is not because they necessarily lack ripple effects in general, but more because it needs to be wondered whether the effects in these cases are necessarily about security. In theory the definition of security from Arnold Wolfers which was used at the outset of this study allows any threat to things which are valued to be a security problem. But the further one moves from violent threats, the less coherent one's understanding of security is likely to become, and the less likely one will find consensus among both scholars and policy-makers that security is actually at stake.

Here transnational issues can be analytically problematic. Transnational terrorism certainly involves organised violence and threats of violence. But despite some of the extraordinary fears that the 9/11 attacks produced, many of the direct effects of terrorism in Asia remain somewhat localised and connected to domestic politics within particular countries in the region. By contrast, climate change will affect every part of the region in some way or another, but whether all of these ripple effects come under the security rubric remains unclear. Is gradual sea level and temperature rise only a security problem when it threatens the survival of states and communities or when, in a rare event, serious violence comes from the competition for scarce water resources? If so, these effects are still specific to some parts of Asia and its borderlands and are less likely to shape the security of the region as a whole.

Readers are likely to come to different conclusions about these problems and whether and how they contribute to regional security. This will depend partly on an assessment of whose security in the region really matters. Groups using terror may be weak actors unable to unseat governments in the region, but their acts can cause significant harm for local communities. Government crackdowns on the same groups may harm some of the citizens for whose security benefit they are sometimes justified, including when these steps are supported by the allies and partners of the country concerned. Even if a resurgent Taliban is of reduced concern to the wider region once the last international troops have left Afghanistan, the consequences for the groups and individuals who will still be in their way will be very serious. Women who risk their lives in seeking education in some parts of South Asia and in Afghanistan cannot be said to enjoy a favourable human security environment. And crop failures caused by environmental change, including depleted soils, failed monsoons, and denuded fisheries, will leave specific communities vulnerable and desperate.

There are perhaps two main ways to try to make sense of what these human security challenges mean for the overall aims of this volume. First, if a particular human security challenge is so ubiquitous that it touches every part of the region (or most of it) then it becomes a common problem which is being experienced in most parts of Asia and its borderlands. Yet human security problems are likely to remain unequally spread. In many situations, it is a case of already vulnerable communities becoming even more vulnerable. Second is the prospect of a changing regional consciousness where, even if they are distant, human security problems in other parts of Asia and its borderlands are regarded as challenges to all communities. There have been some signs of this in the development of interventions for humanitarian purposes, an approach that was considered in Chapter 9. But these actions can challenge the norms of state sovereignty that are held strongly in parts of Asia, and can be more regionally divisive than regionally unifying. Such interventions also have a mixed record in addressing the complex political causes of many of these problems. The appetite for long and costly interventions also has its limits. In short, the change of consciousness required for human security to be put at the centre of Asia's regional security agenda probably involves a fairly significant revolution in political affairs that is far from certain. It may require nothing less than the transformation of the state system itself.

Asia's major power security problem

Such a political transformation is unlikely to begin in Asia, where so many countries in the region place a high value on sovereign statehood. In fact the security dynamic most likely to generate a sense of Asia's security in an authentically regional sense may actually involve a well established theme in the system of states. The variable which most likely to continue generating region-wide ripple effects is the change being brought about as China has been rising, and as the various countries in Asia and its borderlands, including the other major powers, respond to this development. This is the factor which Barry Buzan and Ole Waever (2004) cite as most important in the growing links between the various parts of the region which carry the potential for what they call an Asian regional security super-complex. It is not necessary to adopt quite this same language. But it is important to see that China's rise is helping to bring some of the previously less connected parts of Asia and its borderlands together.

Not all of these aspects of growing regional interdependence are about security per se. The most noticeable impact of China's rise, which has itself stemmed from years of economic growth, lies in the links between the prosperity of almost every country in Asia and its borderlands and a growing China. This provides a common reason for the countries of Asia, including China, to avoid serious security competition, and especially armed conflict, which might upset this mutual enrichment. But as was noted in Chapter 4, this mutual interest in the avoidance of conflict is no guarantee of its continuing absence. Moreover, the changing distribution of power within Asia, in which China's rise is the most obvious feature, may itself also generate a series of security challenges. And at least some of these have wider regional ramifications.

China's re-emergence as an Asian great power challenges America's regional leadership and makes its close neighbour Japan increasingly nervous. Japan's attempts to reduce its self-imposed military limitations have in turn caused China and South Korea to evoke memories of the Second World War period. Any simultaneous rise on the part of India, moreover, will not necessarily make for a more stable regional equilibrium, given that China and India are also strategic competitors. A number of the smaller and medium powers in Asia, including several of the ASEAN countries, are at once being enriched and unsettled by a stronger China. As was clear in Chapter 6, territorial competition in the South China Sea has been heightened as China's political confidence and military ability to support its claims has grown. Japan, China and South Korea are locked in a battle for status over very small territories further north in the East China Sea. And China's neighbour and ally North Korea is a constant worry for South Korea, Japan and the United States.

This mix of strategic nervousness, in which China is a common factor (although by no means the only contributor), has encouraged a number of the surrounding countries to strengthen their own military capabilities. Many of them have also welcomed a stronger American involvement in regional diplomacy, and some now have new security partnerships with Washington. In combination these trends have helped feed growing patterns of arms procurement. As was clear in Chapter 5, this does not mean that Asia is faced with an all-encompassing arms race that will lead the region into war. But there are some increasingly sharp competitions occurring for military advantage in the region. These include the development of advanced maritime capabilities (in terms of platforms, sensors, and missiles) and the applications of cyber-technologies for military purposes. These trends raise risks of misperception which could be particularly destabilising in a crisis.

Asia also remains the main area of intersection for the world's largest and the world's newest nuclear arsenals, held in different quantities and qualities by the United States, Russia, China, India, Pakistan and North Korea. Part of Washington's response to calls from its North Asian allies for more reassurance regarding their long-standing alliance relationships, which was explored in Chapter 11, has been to recommit itself to the extended nuclear deterrence of those countries lying next to nuclear-armed China and North Korea. This makes it more necessary but also more challenging to bring China into US–Russia discussions about nuclear reductions, which have been made more difficult by the breakdown in Washington's relations with Moscow. The growth in US alliances and partnerships, and the closer links between some of these relationships, are due to a range of factors. But chief among them is the response to China's growing political influence and military power.

Interstate war may have been declared obsolete in most parts of the world, although the Russia–Ukraine crisis of 2014 shows that even Europe may not be immune in perpetuity. In Asia it is simply not plausible to make this claim. Tensions between nuclear-armed India and Pakistan remain a point of concern, and old 'flashpoints' such as the Taiwan Strait and the Korean peninsula must not be ignored. But the tense situation in North Asia involving China and Japan demands especially close attention. Given the economic and political importance of these two Asian major powers, a war between them would be likely to have substantial wider effects.

It is simply impossible to know what the ending of a long period of major power peace in Asia would mean for other conflict spots. But even more weight is attached to this possibility when we consider the role of the United States as the leading guarantor of Japan's security and as the leading strategic competitor of China in Asia. A North Asian war which also involved the United States, where there would be pressures to escalate quickly, including the possible use of nuclear weapons, would undoubtedly have important system-wide effects. And even without such a catastrophe, the wider region is already being affected by the increased signs of great power strategic competition. Even the distant countries of the Pacific Islands area, which may seem strategically insignificant, have been noticing an increase in Chinese and American interest.

Asia's security challenges, which are numerous and diverse, would not suddenly all disappear if China and Japan and China and the United States are able to find a way to work through their differences without coming to blows. It would be silly to think that major power relations in Asia and its borderlands are the only issue to focus on. Had this been the case, it would not have been necessary to work through the range of questions which have considered in this book. But by the same token, a serious breakdown in the relations in Asia between these major powers, and the difficult choices this could pose for smaller and medium countries, would test regional security in Asia unlike almost any other challenge that can be envisaged. And despite the mixed feelings that exist about the idea of an Asian concert of powers, perhaps the first order of business for Asia's security is to ensure sufficient cooperation between the big players so that the rest of the region can get on with its existence.

How do we get deeper cooperation?

This still leaves an analytical and practical problem. The preceding two chapters considered the main attempts to build security cooperation in Asia. Chapter 10 investigated the more inclusive form of cooperation built largely around the security diplomacy of a series of ASEAN-centred groupings. The main strength of this approach, its relative inclusiveness – including the desire to engage all the major regional powers in the same setting – has also proven to be a significant limitation. Such practical cooperation as there is tends to be harder to build where it is most needed: in dealing with the more sensitive issues that continue to separate the big players. And the principle of consensus tends to mean that what is agreed avoids the taking of strong positions, let alone the taking of sides.

Chapter 11 then looked at a form of security cooperation that is more exclusive and more aligned: the set of alliances and security partnerships that have grown up in Asia since the early Cold War years, in which the United States continues to play the leading role. Some of the older of these relationships, particularly those involving Japan, South Korea and Australia (Washington's leading Asian allies) provide for extensive military cooperation of a sort that the ASEAN-centred groupings are unlikely ever to offer. But again the advantage of this form of security cooperation is also its weakness: these alliances, and perhaps even some of the newer security partnerships, involve obligations to work together in the event of a common external threat. Unity within the group often depends upon the identification of a common adversary. While it may not have been the original justification for these promises of alliance assistance, China has come to assume the leading role as potential adversary in more recent times, although Japan sometimes appears in turn to be the common adversary that brings together China and South Korea. In either case, what unites the smaller group divides the region.

A common problem afflicts the contemporary evaluation of Asia's more inclusive but insipid plurilateral forums and its more robust but more divisive alliances and security partnerships. In both cases there is too much focus about questions of structure and membership. For Asia's increasing array of forums in which security problems are discussed, but rarely resolved, the issue of who is running the show seems more important than the outcomes of the discussions. The standard answer – that ASEAN is at the centre of these processes – belies the struggle for influence between the major powers which is commonly regarded as the bigger and more important game. But in this structural context a concert between these larger players seems inconceivable, partly because it would be treated as an *organisational* competitor. Meanwhile, the region's alliances offer a competing regional structure where there are two main issues of membership: firstly that the United States remains the one and only essential participant, and secondly, that for each such alliance, outsiders are not welcome to join. The idea that there can be some sort of convergence between Asia's multilateral processes and America's regional alliances, while thoughtfully suggested over a decade ago by William Tow (2002), still seems a very big ask in these circumstances.

Yet while regional countries may always find it difficult to integrate these two forms of security cooperation, the seeds for how they might together think about Asia's security may be found in both of these approaches. ASEAN's greatest effectiveness was in its early years when a group of neighbouring and mutually vulnerable countries sought standards to regulate their behaviour towards one another. Some of these expectations are transmitted in the Treaty of Amity and Cooperation (TAC) which now almost every notable country in Asia and its borderlands has signed up to. Lurking behind these standards, as discussed in Chapter 10, are common goals which might be recognised as important across the region. These include prosperity, freedom from unwanted interference, peace, and stability. Also evident is the goal of security in a comprehensive sense, extending beyond the mere avoidance of armed conflict. And in their more select security relationships, the United States and its allies also routinely commit themselves to common goals. As well as specific security goals including the deterrence of potential adversaries, these goals have also included the promotion of democracy, human rights and the rule of law.

The problems with these goals reflect the shortcomings of the relationships between the groups of countries which have developed them. First, having, in the earlier part of this volume, studied Asia's existing security problems, it cannot be concluded that the goals of the TAC are being universally and consistently respected by the large group of countries who are formally committed to them. What all agree to is not necessarily being followed in individual circumstances. Second, some of the goals that the United States and its allies say they are committed seem designed to exclude some of the region's important countries whose political systems reflect quite different understandings of order and progress. Chief among those excluded is China. What some are following is not acceptable for all.

Notwithstanding these differences, this discussion has moved into some more fertile ground for considering security cooperation in Asia. The focus has shifted instead away from the facile notion that Asia's security depends on how the region is organised to exploring the goals which motivate behaviour (including security cooperation). Elements of this more productive approach have been evident in a number of the issues which have considered in this book. One such example is the idea that a common interest in prosperity can help Asia remain peaceful. Also evident is a widespread assumption that the key to better relations between the major powers in Asia is a common interest in the avoidance of armed conflict. Consideration has also been given in these pages to whether the human security interests of individuals and groups can be given more attention than the national security interests of states and governments. But what if these were more than interests – those objectives that Asia's main political units regarded as useful or advantageous in the quest for their higher ambitions? What if the focus switches to some of the even more important objectives which motivate behaviour? What if the analysis was about varying visions of the good life?

In this case the discussion would be centred on the *values* of states and other political units in Asia and its borderlands. These are goals that are held and felt deeply. They are, as Hedley Bull (1977, p. 67) has argued, goals which are 'valuable in themselves and not merely as a means to an end'. This approach encourages some important and challenging thinking about Asia's security as this volume comes to a close. For example, this notion of values can have significant implications for the widespread argument that a secure Asia is a peaceful region. Is peace here an interest – something that allows the countries of Asia to feel and be secure? Or is it a value which is important in and of itself – something which it is important to work towards in its own right and which may require compromise in other areas? If peace is simply an interest, what happens when a country concludes that its maintenance (and therefore the avoidance of war) is no longer contributing to the goal of a balance of power in Asia? Alternatively, should economic progress be viewed as a means to an end: something that allows states in the region to become more powerful and to achieve the status which they feel they deserve? If this approach is adopted, the possibilities for greater insecurity seem in-built, unless the major powers of Asia are willing to see the power and status of their neighbours grow. Or is economic progress seen as an inherently important value. In this case, Asia's major powers might be able to afford to be less suspicious about the growing power of their regional neighbours.

This returns the conversation to the definition of security with which the present journey began. If security is the absence of threats to acquired values, as Wolfers suggests, then it seems logical to argue that Asia's security depends on how well the countries of Asia understand one another's value systems. Yet this knowledge would not in itself guarantee a positive security environment. I may know your values, but I may well decide that I must reject them. According to the famous historian of ideas, Isaiah Berlin (1997, p. 10) :

> What is clear is that values can clash – that is why civilisations are incompatible. They can be incompatible between cultures, or groups in the same culture, or between you and me. We can discuss each other's point of view, we can try to reach common ground but in the end what you pursue may not be reconcilable with the ends to which I find that I have dedicated my life.

This leaves a conundrum which features in the relationship between the two most important security actors in Asia and its borderlands. These are, of course, China and the United States. China's security depends on the absence of threats to its acquired values, real and perceived. And similarly, America's security depends on the absence of threats to its acquired values, real and perceived. So far, so good, it might be thought. But if their acquired values are incompatible, then it follows that China's security must be at the expense of America's and vice versa. The best outcome in these circumstances would to be to ensure that they had as little as possible to do with one another, something that seems impossible in today's region. Knowledge may actually be a problem here. For each to know the value system of the other would only remind them of the extent to which they were at loggerheads.

But perhaps there are some shared values which can be acquired by both the United States and China, despite their disagreements in other areas. And perhaps these are values that can also be held in common by the other major powers of Asia and its borderlands, and most of the medium and small powers too. Looking back on the twelve chapters we have worked through, it might just be possible to see the beginnings of a set of common regional values and also, possibly, some agreement on the threats to those common values. The question is whether this stock of common values can be rich and deep enough to shape behaviour in Asia when disagreements come.

How important is security for Asia?

This book has been written with the presumption that Asia's security is important. But at the end of this concluding chapter there is one final question to consider: how important is security in comparison to the other values held by decision-makers in Asia and its borderlands? Disregarding security altogether would be hazardous and there may even be some appeal to making security the value through which all other goals in Asia are measured. If one takes security to mean peace through the absence of threats of external war and internal breakdown, for example, it might be thought that Asia simply can't have too much security. The stakes are simply too high. Threats of war alone might harm trade and investment, making the quest for mutual prosperity much more difficult. Ongoing threats of internal conflict can render a government's normal activities next to impossible, challenging the rule of law. Constant exposure to transnational terrorist threats and transnational crime harm prospects for order. Life, liberty and the pursuit of happiness itself, it might be argued, depend on security. China's notion of a harmonious world may similarly depend on a secure environment. Don't all dreams for Asia and its borderlands share this security dependence?

But it may also be dangerous to make too much of security. Putting any single value above or before the others can create a monster out of something that is healthy only in smaller quantities. Any quest for absolute security in Asia, even if it was possible, is bound to involve sacrifices that it would be unwise to make. Asia's security in a wider regional sense may easily come at the expense of claims of justice in parts of the region. Indeed many of the arguments for national self-determination and independence which were made after the Second World War might easily have been rejected on the grounds of Asia's regional security. If security against the threat of severe pandemic disease requires the suspension of civil liberties around the region, is that a price that readers are willing to pay? If the pursuit of security requires the subordination of the goals of smaller powers to the preferences of the larger, is it fair to make that call? And what is the best course of action to take if it becomes clear that an Asian region which accepts the dominance of one major power is better for overall regional security than an ongoing competition between two or more contenders that might mean war? This could well be the crux of Asia's security challenge as the regional distribution of power continues to shift.

As the Asian region changes, and as its global importance continues to grow, the need for people all over the world to understand Asia's security will intensify. Hopefully this book will be a part of that increased awareness. But this is not a case where greater knowledge should lead to greater comfort, simply because some of Asia's security problems are serious and the ripple effects could be very negative, especially when effective regional security cooperation is not easy to arrange. Yet a fascination for Asia's security should not come from predictions of doom and gloom. The subject is forever intellectually fascinating and the debates about its future are of such tremendous significance for decision-makers. Those of us who are students of Asia's security can't compete with the states of Asia economically, diplomatically or militarily. But we may be able to help shape the debates about its future. If this book has readers crafting their own ideas about Asia's security, it will have done its job.

Bibliography

Abrahms, Max (2008) 'What terrorists really want: terrorist motives and counterterrorism strategy', *International Security*, 32, 4, 78–105.
Abuza, Zachary (2003) *Militant Islam in Southeast Asia: Crucible of Terror*, Boulder, CO, Lynne Rienner.
Acharya, Amitav (1992) 'Regional military-security cooperation in the Third World: a conceptual analysis of the relevance and limitations of ASEAN (Association of Southeast Asian Nations)', *Journal of Peace Research*, 29, 1, 7–21.
Acharya, Amitav (1999) 'A concert of Asia?', *Survival*, 41, 3, 84–101.
Acharya, Amitav (2001) *Constructing a Security Community in Southeast Asia: ASEAN and the Problem of Regional Order*, London: Routledge.
Acharya, Amitav (2014) *The End of the American World Order*, Cambridge, Polity Press.
Adamson, Fiona B. (2006) 'Crossing borders: international migration and national security', *International Security*, 31, 1, 165–99.
Afridi, Jamal and Jayshree Bajoria (2010) 'China–Pakistan relations', New York: Council on Foreign Relations. Available at: http://www.cfr.org/china/china-pakistan-relations/p10070?cid=rss-china-china_pakistan_relations-091010.
Ahuja, Partul and Rajat Ganguly (2007) 'The fire within: Naxalite insurgency violence in India', *Small Wars & Insurgencies*, 18, 2, 249–74.
Akashi, Yasuhi (2012) 'An Assessment of the United Nations Transitional Authority in Cambodia (UNTAC)', in Pou Sothirak, Geoff Wade and Mark Hong (eds), *Cambodia: Progress and Challenges Since 1991*, Singapore: Institute for Southeast Asian Studies.
Alagappa, Muthiah (ed.) (2001) *Coercion and Governance: The Declining Political Role of the Military in Asia*, Stanford, CA, Stanford University Press.
Alagappa, Muthiah (2011) 'A changing Asia: prospects for war, peace, cooperation and order', *Political Science*, 63, 2, 155–85.
Alpers, Philip, Robert Muggah and Conor Twyford (2004) 'Trouble in paradise: small arms in the Pacific', *Small Arms Survey 2004: Rights at Risk*, 277–307.
Ambrosio, Thomas (2008) 'Catching the 'Shanghai spirit': how the Shanghai Cooperation Organization promotes authoritarian norms in Central Asia'. *Europe–Asia Studies*, 60, 8, 1321–44.
Anderson, Benedict (1983) *Imagined Communities: Reflections on the Origin and Spread of Nationalism*, London, Verso.
Anderson, Ian, Sue Crengle, Martina Leialoha Kamaka, Tai-Ho Chen, Neal Palafox, and Lisa Jackson-Pulver (2006) 'Indigenous health in Australia, New Zealand, and the Pacific', *The Lancet*, 367, 9524, 1775–85.
Angell, Norman (2010) *The Great Illusion* (first published 1909), New York, Cosimo.
Annan, Kofi (1999) 'Two concepts of sovereignty', *The Economist*, 352, 8137, 49.
Anwar, Dewi Fortuna (1994) *Indonesia in ASEAN: Foreign Policy and Regionalism*, Singapore, Institute of Southeast Asian Studies.
Anwar, Dewi Fortuna (2003) 'Human security: an intractable problem in Asia', in Muthiah Alagappa (ed.), *Asian Security Order: Ideational and Normative Features*, Stanford, CA, Stanford University Press.

Aris, Stephen (2009) 'The Shanghai Cooperation Organisation: "Tackling the three evils": a regional response to non-traditional security challenges or an anti-western bloc?', *Europe–Asia Studies*, 61, 3, 457–82.
Arrow, Kenneth J. (1971) *Essays in the Theory of Risk-Bearing*, Chicago, Markham.
Art, Robert J. (2010) 'The United States and the rise of China: implications for the long haul', *Political Science Quarterly*, 125, 3, 359–91.
Arunatilake, Nisha, Sisira Jayasuriya and Saman Kelegama (2001) 'The economic costs of war in Sri Lanka', *World Development*, 29, 9, 1483–500.
Asian Development Bank (2013) *Asian Economic Integration Monitor*, Manila, Asian Development Bank, October.
Ayoob, Mohammed (2002) 'Humanitarian intervention and state sovereignty', *The International Journal of Human Rights*, 6, 1, 81–102.
Ayson, Robert (2007) 'The "arc of instability" and Australia's strategic policy', *Australian Journal of International Affairs*, 61, 2, 215–31.
Ayson, Robert (2008) 'Strategic studies', in Chris Reus-Smit and Duncan Snidal (eds), *The Oxford Handbook of International Relations*, Oxford, Oxford University Press, 558–75.
Ayson, Robert (2010) 'After a terrorist nuclear attack: envisaging catalytic effects', *Studies in Conflict and Terrorism*, 33, 7, 571–93.
Ayson, Robert (2012) 'Choosing ahead of time? Australia, New Zealand and the US–China contest in Asia', *Contemporary Southeast Asia*, 34, 3, 338–64.
Ayson, Robert and Desmond Ball (2014) 'Can a Sino-Japanese war be controlled?', *Survival*, 56, 6, 135–66.
Baldwin, Richard (2007) 'Managing the noodle bowl: the fragility of East Asian regionalism', *Working Paper Series on Regional Economic Integration*, 7, Asian Development Bank.
Ball, Desmond (1993) 'Arms and affluence', *International Security*, 18, 3, 78–112.
Ball, Desmond, Anthony Milner and Brendan Taylor (2006) 'Track 2 security dialogue in the Asia-Pacific: reflections and future directions', *Asian Security*, 2, 3, 174–88.
Ballard, Chris and Glenn Banks (2003) 'Resource wars: the anthropology of mining', *Annual Review of Anthropology*, 32, 287–313.
Banister, J., D. E. Bloom and L. Rosenberg (2012) 'Population aging and economic growth in China', in Masahiko Aoki and Jinglian Wu (eds), *The Chinese Economy: A New Transition*, Basingstoke, Palgrave Macmillan, 114–50.
Barbieri, Katherine and Jack S. Levy (1999) 'Sleeping with the enemy: the impact of war on trade', *Journal of Peace Research*, 36, 4, 463–79.
Bardhan, Pranab (2006) 'Awakening giants, feet of clay: a comparative assessment of the rise of China and India', *Journal of South Asian Development*, 1, 1, 1–17.
Barno, David W., Nora Bensahel and Travis Sharp (2012) 'Pivot but hedge: a strategy for pivoting to Asia while hedging in the Middle East', *Orbis*, 56, 2, 158–76.
Bateman, Sam (2011) 'Perils of the deep: the dangers of submarine proliferation in the seas of East Asia', *Asian Security*, 7, 1, 61–84.
Bateman, Sam, Joshua Ho, and Jane Chan (2009) *Good Order at Sea in Southeast Asia*, Singapore, S. Rajaratnam School of International Studies.
Beeson, Mark (2009) 'Trading places? China, the United States and the evolution of the international political economy', *Review of International Political Economy*, 16, 4, 729–41.
Beeson, Mark and Li, Fujian (2012) 'Charmed or alarmed? Reading China's regional relations', *Journal of Contemporary China*, 21, 73, 35–51.
Bell, Coral (2007) *The end of the Vasco da Gama era: the next landscape of world politics*, Double Bay, NSW, The Lowy Institute for International Policy.

Bensassi, Sami and Inmaculada Martínez-Zarzoso (2012) 'How costly is modern maritime piracy to the international community?', *Review of International Economics*, 20, 5, 869–83.
Berlin, Isaiah (1997) 'The Pursuit of the Ideal', in Isaiah Berlin, *The Proper Study of Mankind*, ed. Henry Hardy and Roger Hausheer, London, Chatto & Windus, 1–16.
Bisley, Nick (2012) *The Great Powers in the Changing International Order*, Boulder, CO, Lynne Rienner.
Bisley, Nick and Andrew Phillips (2013) 'Rebalance to where? US strategic geography in Asia', *Survival*, 55, 5, 95–114.
Bitzinger, Richard and Barry Desker (2008) 'Why East Asian war is unlikely', *Survival*, 50, 6, 105–28.
Bloom, David E., D. Canning, L. Hu, Y. Liu, A. Mahal and W. Yip (2010) 'The contribution of population health and demographic change to economic growth in China and India', *Journal of Comparative Economics*, 38, 1, 17–33.
Boege, Volker, M. Anne Brown, Kevin P. Clements and Anna Nolan (2008) *States emerging from hybrid political orders: Pacific experiences*, Occasional Paper 11, Brisbane: The Australian Centre for Peace and Conflict Studies, University of Queensland. Available at: http://espace.library.uq.edu.au/eserv/UQ:164904/Occasional_Paper_No_11__Online_final.pdf.
Box, Meredith and Gavin McCormack (2004) 'Terror in Japan', *Critical Asian Studies*, 36, 1, 91–112.
Brands, Henry W. Jr. (1987) 'From ANZUS to SEATO: United States strategic policy towards Australia and New Zealand, 1952–1954', *The International History Review*, 9, 2, 250–70.
Breslin, Shaun (2015) 'Debating human security in China: Towards discursive power?', *Journal of Contemporary Asia*, 45, 2, 243-265.
Brewster, David (2010) The Australia–India security declaration: the quadrilateral redux?, *Security Challenges*, 6, 1, 1–9.
Brewster, David (2011) 'Indian strategic thinking about East Asia', *Journal of Strategic Studies*, 34, 6, 825–52.
Bristow, Damon (2005) 'The Five Power Defence Arrangements: Southeast Asia's unknown regional security organization', *Contemporary Southeast Asia*, 27, 1, 1–20.
Brodie, Bernard (ed.) (1946) *The Absolute Weapon: Atomic Power and the World Order*, New York, Harcourt Brace.
Bull, Hedley (1961) *The Control of the Arms Race: Disarmament and Arms Control in the Missile Age*, London, Weidenfeld & Nicolson for the Institute for Strategic Studies.
Bull, Hedley (1971) 'The New Balance of Power in Asia and the Pacific', *Foreign Affairs*, 49, 4, 669–81.
Bull, Hedley (1977) *The Anarchical Society: A Study of Order in World Politics*, London: Macmillan.
Bull, Hedley (1984) 'Intervention in the Third World', in Hedley Bull (ed.), *Intervention in World Politics*, Oxford, Clarendon Press, 135–56.
Bull, Hedley and Adam Watson (eds) (1985) *The Expansion of International Society*, Oxford, Clarendon Press.
Buszynski, Leszek (2012) 'The South China Sea: oil, maritime claims, and US–China strategic rivalry', *The Washington Quarterly*, 35, 2, 139–56.
Buszynski, Leszek and Iskander Sazlan (2007) 'Maritime claims and energy cooperation in the South China Sea', *Contemporary Southeast Asia*, 29, 1, 143–71.

Buzan, Barry (2000) 'The logic of regional security in the post-Cold War world', in Bjorne Hettne (ed.), *The New Regionalism and the Future of Security and Development*, Vol. 4, Basingstoke: Macmillan.

Buzan, Barry, Ole Waever and Jaap de Wilde (1998) *Security: A New Framework for Analysis*, Boulder, CO, Lynne Rienner.

Buzan, Barry and Ole Waever (2004) *Regions and Powers: The Structure of International Security*, Cambridge, Cambridge University Press.

Byman, Daniel (2013) 'Why drones work: the case for Washington's weapon of choice', *Foreign Affairs*, 32, 4, 32–43.

Byman, Daniel (2014) 'Buddies or burdens? Understanding the Al Qaeda relationship with its affiliate organizations', *Security Studies*, 23, 3, 431–70.

Byman, Daniel and Matthew Waxman (2002) *The Dynamics of Coercion: American Foreign Policy and the Limits of Military Might*, Cambridge, Cambridge University Press.

Caballero-Anthony, Mely (2005) 'SARS in Asia: crisis, vulnerabilities, and regional responses" *Asian Survey*, 45, 3, 475–95.

Cai, Yongshun (2002) 'The resistance of Chinese laid-off workers in the reform period', *The China Quarterly*, 170, 327–44.

Calder, Kent E. (2004) 'Securing security through prosperity: the San Francisco System in comparative perspective', *The Pacific Review*, 17, 1, 135–57.

Campbell, Kurt M. and Ely Ratner (2014) 'Why Washington should focus on Asia', *Foreign Affairs*, 93, 3, 106–16.

Capie, David (2002) *Small Arms Production and Transfers in Southeast Asia*, Canberra, Strategic and Defence Studies Centre, Australian National University.

Capie, David (2008) 'Localization as resistance: the contested diffusion of small arms norms in Southeast Asia', *Security Dialogue*, 39, 6, 637–58.

Capie, David and Paul Evans (2007) *The Asia-Pacific Security Lexicon*, 2nd edn, Singapore, Institute of Southeast Asian Studies.

Cha, Victor D. (2009) 'What do they really want? Obama's North Korea conundrum', *The Washington Quarterly*, 32, 4, 119–38.

Cha, Victor D. (2010) 'Powerplay: origins of the US alliance system in Asia', *International Security*, 34, 3, 158–96.

Cha, Victor D. and David C. Kang (2013) *North Korea: A Debate On Engagement Strategies*, New York, Columbia University Press.

Chambers, Paul (2014) 'Constitutional Change and Security Forces in Southeast Asia: Lessons from Thailand and Myanmar', *Contemporary Southeast Asia*, 36, 1, 101–27.

Chang, Gordon G. (2010) *The Coming Collapse of China*, New York, Random House.

Charney, Jonathan I. (1995) 'Central East Asian maritime boundaries and the law of the sea', *American Journal of International Law*, 89, 4, 724–49.

Chase, Michael S. (2011) 'Chinese suspicion and US intentions', *Survival*, 53, 3, 133–50.

Chellaney, Brahma (2011) *Water: Asia's New Battleground*, Washington, DC, Georgetown University Press.

Chen, Jian (1995) 'China's involvement in the Vietnam War, 1964-69', *The China Quarterly*, 142, 356–87.

Chen, Jie (2008) 'Rapid urbanization in China: a real challenge to soil protection and food security', *Catena*, 69, 1, 1–15.

Chestnut, Sheena (2007) 'Illicit activity and proliferation: North Korean smuggling networks', *International Security*, 32, 1, 80–111.

Chin, Ko-Lin (2009) *The Golden Triangle: Inside Southeast Asia's Drug Trade*, Ithaca, NY, Cornell University Press.
Chomsky, Noam (1969) *American Power and the New Mandarins*, Harmondsworth, Penguin.
Choo, Jaewoo (2005) 'Is institutionalization of the six-party talks possible?', *East Asia*, 22, 4, 39–58.
Chow, Jonathan T. (2005) 'ASEAN counterterrorism cooperation since 9/11', *Asian Survey*, 45, 2, 302–21.
Christensen, Thomas J. (1999) 'China, the US–Japan alliance, and the security dilemma in East Asia', *International Security*, 23, 4, 49–80.
Christensen, Thomas J. (2001) 'Posing problems without catching up: China's rise and problems for U.S. security policy', *International Security*, 25, 4, 5–40.
Christensen, Thomas J. (2006) 'Fostering stability or creating a monster?', *International Security*, 31, 1, 81–126.
Christensen, Thomas J. (2011) 'The advantages of an assertive China: responding to Beijing's abrasive diplomacy', *Foreign Affairs*, 90, 2, 54–67.
Christofferson, Gaye (2010) *China and Maritime Cooperation: Piracy in the Gulf of Aden*, Berlin, Institut für Strategie- Politik- Sicherheits- und Wirtschaftsberatung (ISPSW).
Cirincione, Joseph (2000) 'The Asian nuclear reaction chain', *Foreign Policy*, 118, 120–36.
Clark, Ian (1998) *Globalization and Fragmentation: International Relations in the Twentieth Century*, Oxford, Oxford University Press.
Clarke, Michael (2010) 'China, Xinjiang and the internationalisation of the Uyghur issue', *Global Change, Peace & Security*, 22, 2, 213–29.
Clausewitz, Carl von (1976) *On War*, trans. and ed. Michael Howard and Peter Paret, Princeton, NJ, Princeton University Press.
Clinton, Hillary (2011) 'America's pacific century', *Foreign Policy*, 189, 56–83.
Cohen, Eliot (1996) 'A revolution in warfare', *Foreign Affairs*, 75, 2, 37–54.
Cohen, Stephen P. (2002) *India: emerging power*, Washington, DC, Brookings Institution.
Collins, Kathleen (2002) 'Clans, pacts, and politics in Central Asia', *Journal of Democracy*, 13, 3, 137–52.
Collins, Rosemary and Daud Hassan (2009) 'Applications and Shortcomings of the Law of the Sea in Combating Piracy: A South East Asian Perspective', *Journal of Maritime Law and Commerce*, 40, 1 89–113.
Connell, John (2003) 'New Caledonia: an infinite pause in decolonization?', *The Round Table*, 92, 368, 125–43.
Copeland, Dale C. (1996) 'Economic interdependence and war: a theory of trade expectations', *International Security*, 20, 4, 5–41.
Cotton, James (2001) 'Against the grain: the East Timor Intervention', *Survival*, 43, 1, 127–42.
Crenshaw, Martha (2000) 'The psychology of terrorism: an agenda for the 21st century', *Political Psychology*, 21, 2, 405–20.
Crouch, Harold (2007) *The Army and Politics in Indonesia*, Singapore: Equinox.
Cruz De Castro, Renato (2003) 'The revitalized Philippine–U.S. security relations: a ghost from the Cold War or an alliance for the 21st century?', *Asian Survey*, 43, 6, 971–88.
Cruz De Castro, Renato and Walter Lohman (2012) 'Getting the Philippines Air Force flying again: the role of the U.S.–Philippines alliance', *Backgrounder*, 2733, Washington, DC, The Heritage Foundation.

Cunningham, Fiona and Rory Medcalf (2011) *The Dangers of Denial: Nuclear Weapons in China–India Relations*, Sydney, NSW, Lowy Institute for International Policy.

Dahl, Robert A. (1957) 'The Concept of Power', *Behavioral Science*, 2, 3, 201–15.

Dandeker, Christopher and James Gow (1997) 'The future of peace support operations: strategic peacekeeping and success', *Armed Forces & Society*, 23, 3, 327–47.

Dassú, Marta (1998) 'China and the Asian crisis: pillar of stability or next country at risk?', *The International Spectator*, 33, 3, 29–40.

Davies, Sara E. (2006) 'The Asian rejection? International refugee law in Asia', *Australian Journal of Politics & History*, 52, 4, 562–75.

Deudney, Daniel (1990) 'The case against linking environmental degradation and national security', *Millennium*, 19, 3, 461–76.

Deutsch, Karl W. (1957) *Political Community and the North Atlantic Area: International Organization in the Light of Historical Experience*, Princeton, NJ, Princeton University Press.

DeVotta, Neil (2009) 'The Liberation Tigers of Tamil Eelam and the lost quest for separatism in Sri Lanka', *Asian Survey*, 49, 6, 1021–51.

Dibb, Paul (1986) *The Soviet Union: The Incomplete Superpower*, Urbana, IL and Chicago, University of Illinois Press.

Dibb, Paul (1997) 'The revolution in military affairs and Asian security', *Survival*, 39, 4, 93–116.

Dobbins, James (2012) 'War with China', *Survival*, 54, 4, 7–24.

Drake, Christine (1989) *National Integration in Indonesia: Patterns and Policies*, Hawaii, University of Hawaii Press.

Dupont, Alan (2000) 'ASEAN's Response to the East Timor Crisis', *Australian Journal of International Affairs*, 54, 2, 163–70.

Dupont, Alan (2001) *East Asia Imperilled: Transnational Challenges to Security*, Cambridge, Cambridge University Press.

Dupont, Alan (2008) 'The strategic implications of climate change', *Survival*, 50, 3, 29–54.

Dupont, Alan and Christopher Baker (2014) 'East Asia's maritime disputes: fishing in troubled waters', *The Washington Quarterly*, 37, 1, 79–98.

Earnest, D. C., S. Yetiv and S. M. Carmel (2012) 'Contagion in the transpacific shipping network: international networks and vulnerability interdependence, *International Interactions*, 38, 5, 571–96.

Easley, Leif-Eric (2014) 'Spying on Allies', *Survival*, 56, 4, 141–56.

Emmers, Ralf (2003) 'ASEAN and the securitization of transnational crime in Southeast Asia', *The Pacific Review*, 16, 3, 419–38.

Emmers, Ralf (2010) *Geopolitics and Maritime Territorial Disputes in East Asia*, Abingdon, Routledge.

Enemark, Christian (2006) 'Weapons of mass destruction?', in Robert Ayson and Desmond Ball (eds), *Strategy and Security in the Asia-Pacific*, Crows Nest, NSW, Allen & Unwin, 88–102.

Enemark, Christian (2007) *Disease and Security: Natural Plagues and Biological Weapons in East Asia*, London: Routledge.

Eslake, Saul (2009) 'The global financial crisis of 2007–2009: an Australian perspective', *Economic Papers*, 28, 3, 226–38.

Evans, Michael (2013) American defence policy and the challenge of austerity: some implications for Southeast Asia', *Journal of Southeast Asian Economies*, 30, 2, 164–78.

Evans, Paul (2004) 'Human security and East Asia: in the beginning', *Journal of East Asian Studies*, 4, 263–84.
Facon, Isabelle (2013) 'Moscow's global foreign and security strategy: does the Shanghai Cooperation Organization meet Russian interests?', *Asian Survey*, 53, 3, 461–83.
Fair, C. Christine (2011) 'The Militant Challenge in Pakistan', *Asia Policy*, 11, 105–37.
Fairbank, J. K. and T. T. Ch'en (1968) *The Chinese World Order: Traditional China's Foreign Relations*, Vol. 32, Cambridge, MA, Harvard University Press.
Fearon, James D. and Laitin, David D. (2003) 'Ethnicity, insurgency, and civil war', *American Political Science Review*, 97, 1, 75–90.
Finnemore, Martha (2004) *The Purpose of Intervention: Changing Beliefs about the Use of Force*, Ithaca, NY, Cornell University Press.
Foot, Rosemary (2009) 'China and the United States: between cold and warm peace', *Survival*, 51, 6, 123–46.
Fortna, Virginia Page (2003) 'Scraps of paper? Agreements and the durability of peace', *International Organization*, 57, 2, 337–72.
Fraenkel, Jon, Stewart Firth and Brij V. Lal (eds), (2009) *The 2006 Military Takeover in Fiji: A Coup to End All Coups?*, Canberra, ACT, ANU EPress.
Franckx, Erik (2011) 'American and Chinese views on navigational rights of warships', *Chinese Journal of International Law*, 10, 1, 187–206.
Fravel, M. Taylor (2008) *Strong Borders, Secure Nation: Cooperation and Conflict in China's Territorial Disputes*, Princeton, NJ, Princeton University Press.
Fravel, M. Taylor (2011) 'China's strategy in the South China Sea', *Contemporary Southeast Asia*, 33, 3, 292–319.
Freedman, Lawrence (2007) 'Terrorism as a strategy', *Government and Opposition*, 42, 3, 314–39.
Freedman, Lawrence (2013) *Strategy: A History*, New York, Oxford University Press.
Friedberg, Aaron L. (1993/94) 'Ripe for rivalry: prospects for peace in a multipolar Asia', *International Security*, 18, 3, 5–33.
Friedberg, Aaron L. (2012) *A Contest for Supremacy: China, America, and the Struggle for Mastery in Asia*, New York, W.W. Norton.
Friedberg, Aaron L. (2014) *Beyond Air–Sea Battle: The Debate over US Military Strategy in Asia*, Abingdon, Routledge for The International Institute for Strategic Studies.
Gaddis, John Lewis (1982) *Strategies of Containment: A Critical Appraisal Of Postwar American National Security Policy*, New York, Oxford University Press.
Ganguly, Sumit (2002) *Conflict Unending: India–Pakistan Tensions since 1947*, New Delhi, Oxford University Press.
Ganguly, Sumit (2008) 'Nuclear stability in South Asia', *International Security*, 33, 2, 45–70.
Ganguly, Sumit and Manjeet S. Pardesi (2009) 'Explaining sixty years of India's foreign policy', *India Review*, 8, 1, 4–19.
Gardner, Helen and Christopher Waters (2013) 'Decolonisation in Melanesia', *The Journal of Pacific History*, 48, 2, 113–21.
Gartzke, Erik and Yonatan, Lupu (2012) 'Trading on preconceptions: why World War I was not a failure of economic interdependence', *International Security*, 36, 4, 115–50.
Garver, John W. (2001) *Protracted Contest: Sino-Indian Rivalry in the Twentieth Century*, Seattle, WA University of Washington Press.
Glaser, Bonnie S. (2012) 'Armed clash in the South China Sea, *Contingency Planning Memorandum*, No. 14, New York, Council on Foreign Relations.

Glaser, Bonnie S. and Brittany Billingsley (2012) *Reordering Chinese Priorities on the Korean Peninsula*, Washington, DC, Center for Strategic and International Studies.

Ghosh, Madhuchanda (2008) 'India and Japan's growing synergy: from a political to a strategic focus', *Asian Survey*, 48, 2, 282–302.

Glick, Reuven and Alan M. Taylor (2010) 'Collateral damage: trade disruption and the economic impact of war', *The Review of Economics and Statistics*, 92, 1, 102–27.

Glosny, M. A. (2004) 'Strangulation from the sea? A PRC submarine blockade of Taiwan', *International Security*, 28, 4, 125–160.

Goh, Evelyn (2008) 'Great powers and hierarchical order in Southeast Asia: analyzing regional security strategies', *International Security*, 32, 3, 113–57.

Goh, Evelyn (2011) 'How Japan matters in the evolving East Asian security order', *International Affairs*, 87, 4, 887–902.

Goh, Evelyn (2013) *The Struggle for Order: Hegemony, Hierarchy, and Transition in Post-Cold War East Asia*, Oxford, Oxford University Press.

Goldstein, Avery (2013) 'First things first: the pressing danger of crisis instability in U.S.–China relations', *International Security*, 37, 4, 49–89.

Goldstein, Lyle and William Murray (2004) 'Undersea dragons: China's maturing submarine force', *International Security*, 28, 4, 161–196.

Goldstein, Lyle (2011) 'Chinese naval strategy in the South China Sea: an abundance of noise and smoke, but little fire', *Contemporary Southeast Asia*, 33, 3, 320–47.

Gompert, David C. and Martin Libicki (2014) 'Cyber warfare and Sino-American crisis instability', *Survival*, 56, 4, 7–22.

Goodhand, Jonathan (2008) 'Corrupting or consolidating the peace? The drugs economy and post-conflict peacebuilding in Afghanistan', *International Peacekeeping*, 15, 3, 405–23.

Gries, Peter Hays (2003) *China's New Nationalism: Pride, Politics, and Diplomacy*, Berkeley, CA, University of California Press.

Grønning, Bjørn Elias Mikalsen (2014) 'Japan's shifting military priorities: counterbalancing China's rise', *Asian Security*, 10, 1, 1–21.

Guan, Ang Cheng (2009) 'Singapore and the Vietnam war', *Journal of Southeast Asian Studies*, 40, 2, 353–84.

Hack, Karl (2013) *Defence and Decolonisation in Southeast Asia: Britain, Malaysia and Singapore*, Abingdon, Routledge.

Haacke, Jürgen (2008) 'ASEAN and political change in Myanmar: towards a regional initiative?', *Contemporary Southeast Asia*, 30, 3, 351–78.

Haacke, Jürgen (2009) 'The ASEAN Regional Forum: from dialogue to practical security cooperation?', *Cambridge Review of International Affairs*, 22, 3, 427–49.

Haas, Ernst B. (1953) 'The balance of power: prescription, concept, or propaganda?', *World Politics*, 5, 4, 442–77.

Hartfiel, Robert and Brian L. Job (2007) 'Raising the risks of war: defence spending trends and competitive arms processes in East Asia', *The Pacific Review*, 20, 1, 1–22.

He, Yinan (2007) 'History, Chinese nationalism and the emerging Sino–Japanese conflict', *Journal of Contemporary China*, 16, 50, 1–24.

Hegarty, David and Anna Powles (2006) 'South Pacific security', in Robert Ayson and Desmond Ball (eds), *Strategy and Security in the Asia-Pacific*, Crows Nest, NSW, Allen & Unwin, 257–69.

Hemmer, Christopher and Peter J. Katzenstein, (2002) 'Why is there no NATO in Asia? Collective identity, regionalism, and the origins of multilateralism', *International Organization*, 56, 3, 575–607.

Herman, Paul F. Jr and Gregory F. Treverton (2009), 'The political consequences of climate change', *Survival*, 51, 2, 137–48.

Hess, Steve (2013) 'From the Arab Spring to the Chinese Winter: the institutional sources of authoritarian vulnerability and resilience in Egypt, Tunisia, and China', *International Political Science Review*, 34, 3, 254–72.

Hill, Hal (1998) 'The Indonesian economy: the strange and sudden death of a tiger', in Geoff Forrester and R. J. May (eds), *The Fall of Soeharto*, Bathurst: Crawford House Publishing, 93–104.

Hill, Matthew (2010) 'A velvet glove: coercion, and the Australasian response to the 2006 Fijian coup', *Security Challenges*, 6, 2, 105–22.

Hobbes, Thomas (1651 [1968]) *Leviathan*. ed. C. B. McPherson, Aylesbury, Pelican.

Holsag, Jonathan (2009) 'The persistent military security dilemma between China and India', *Journal of Strategic Studies*, 32, 6, 811–40.

Huang, Chin-Hao (2011) 'Principles and praxis of China's peacekeeping', *International Peacekeeping*, 18, 3, 257–70.

Hughes, Christopher W. (2009) *Japan's Remilitarisation* Abingdon, Routledge for the International Institute for Strategic Studies.

Huntington, Samuel (1996) *The Clash of Civilizations and the Remaking of World Order* New York, Simon & Schuster.

Hutchison, Emma (2010) 'Trauma and the politics of emotions: constituting identity, security and community after the Bali bombing', *International Relations*, 24, 1, 65–86.

Ikenberry, G. John (2008) 'The rise of China and the future of the West: can the liberal system survive?', *Foreign Affairs*, 87, 1, 23–37.

Ikenberry, G. John (2011) 'The future of the liberal international order: internationalism after America', *Foreign Affairs*, 90, 3, 56–68.

Immerzeel, Walter W., Ludovicus P. H van Beek, and Marc F. P. Bierkins (2010) 'Climate change will affect the Asian water towers', *Science* 328, 5984, 1382–5.

International Commission on Intervention and State Sovereignty (2001) *The Responsibility to Protect* Ottawa, ON, Canada, ICISS.

International Crisis Group (2012) *Indonesia: Dynamics of Violence in Papua*, Asia Report 232, International Crisis Group.

Jervis, Robert (1976) *Perception and Misperception in International Politics*, Princeton, NJ, Princeton University Press.

Jervis, Robert (1978) 'Cooperation under the security dilemma', *World Politics*, 30, 2, 167–214.

Jervis, Robert (1997) *System Effects: Complexity in Political and Social Life*, Princeton, NJ, Princeton University Press.

Johnson, Chalmers (2004) *Blowback: The Costs and Consequences of American Empire*, New York, Henry Holt.

Johnson, Thomas H. (2006) 'Afghanistan's post-Taliban transition: the state of state-building after war', *Central Asian Survey*, 25, 1–2, 1–26.

Johnson, Thomas H. and M. Chris Mason (2007) 'Understanding the Taliban and insurgency in Afghanistan', *Orbis*, 51, 1, 71–89.

Johnson, Thomas H. and M. Chris Mason, (2008) 'No sign until the burst of fire: understanding the Pakistan–Afghanistan frontier', *International Security*, 32, 4, 41–77.

Johnston, Alistair Iain (2003) 'Is China a status quo power?', *International Security*, 27, 4, 5–56.

Jones, David Martin and M. L. R. Smith (2007) 'Making Process, not progress: ASEAN and the evolving East Asian regional order', *International Security*, 32, 1, 148–84.

Jones, Lee (2010) 'Still in the "drivers' seat", but for how long? ASEAN's capacity for leadership in East-Asian international relations', *Journal of Current Southeast Asian Affairs*, 29, 3, 95–113.
Jones, Seth G. (2006) 'Averting failure in Afghanistan', *Survival*, 48, 1, 111–28.
Jones, Seth G. (2007) 'Pakistan's dangerous game', *Survival*, 49, 1, 15–32.
Jones, Sidney (2005) 'The changing nature of Jemaah Islamiyah', *Contemporary Southeast Asia*, 59, 2, 169–78.
Jones, Simon (2008) 'India, Pakistan, and counterinsurgency operations in Jammu and Kashmir', *Small Wars & Insurgencies*, 19, 1, 1–22.
Kabutaulaka, Tarcisius Tara (2005) 'Australian foreign policy and the RAMSI intervention in Solomon islands', *The Contemporary Pacific*, 17, 2, 283–308.
Kahler, Miles (2004) 'Economic security in an era of globalization, definition and provision', *The Pacific Review*, 17, 4, 485–502.
Kahler, Miles and Scott L. Kastner (2006) 'Strategic uses of economic interdependence: engagement policies on the Korean Peninsula and across the Taiwan Strait', *Journal of Peace Research*, 43, 5, 523–41.
Kalhan, Anil, Gerald P. Conroy, Mamta Kaushal, Sam Scott Miller and Jed S. Rakoff (2006) 'Colonial continuities: human rights, terrorism, and security laws in India', *Columbia Journal of Asian Law*, 20, 93–234.
Kandiyoti, Deniz (2007) 'Old dilemmas or new challenges? The politics of gender and reconstruction in Afghanistan', *Development and Change*, 38, 2, 169–99.
Kang, David C. (2003) 'Getting Asia wrong: the need for new analytical frameworks', *International Security*, 27, 4, 57–85.
Kang, David C. (2007) *China Rising: Peace, Power and Order in East Asia*, New York, Columbia University Press.
Kaplan, R. D. (2010) *Monsoon: The Indian Ocean and the Future of American Power*, New York: Random House.
Kapur, S. Paul (2008) 'Nuclear terrorism: prospects for Asia,' in M. Alagappa (ed.), *The Long Shadow: Nuclear Weapons and Security in 21st Century Asia*, Stanford, CA, Stanford University Press, pp. 323–46.
Kapur, S. Paul and Sumit Ganguly (2007) 'The transformation of US–India relations: an explanation for the rapprochement and prospects for the future', *Asian Survey*, 67, 4, 642–56.
Katsumata, Hiro (2003) 'Reconstruction of diplomatic norms in Southeast Asia: the case for strict adherence to the "ASEAN way"', *Contemporary Southeast Asia*, 25, 1, 104–21.
Kennedy, Paul M. (1987) *The Rise and Fall of the Great Powers*, New York, Random House.
Khong, Yuen Foong (2004) 'Coping with strategic uncertainty: the role of institutions and soft-balancing in Southeast Asia's post-Cold War strategy', in J. J. Suh, Peter J. Katzenstein and Allen Carlson (eds), *Rethinking Security in East Asia: Identity, Power and Efficiency*, Stanford, CA, Stanford University Press, 172–208.
Kiernan, Ben (2003) 'The demography of genocide in Southeast Asia: the death tolls in Cambodia, 1975–79, and East Timor, 1975–80', *Critical Asian Studies*, 35, 4, 585–97.
Kilcullen, David (2006) 'Counter-insurgency redux', *Survival*, 48, 4, 111–30.
Kivimäki, Timo (2011) 'East Asian relative peace and the ASEAN way', *International Relations of the Asia-Pacific*, 11, 57–85.

Kochhar, Kaplana, and Ejaz Ghani (2013) 'What can India and Pakistan do to maximize the benefits from trade?', in Michael Kugelman and Robert M. Hathaway (eds), *Pakistan–India Trade: What Needs to Be Done?*, Washington, DC, Woodrow Wilson Center, 97–115.
Kohli, Harinder, Ashok Sharma and Anil Sood (eds) (2011) *Asia 2050: Realizing the Asian Century*, Washington, DC, Emerging Markets Forum.
Koo, Min Gyo (2009) 'The Senkaku/Diaoyu dispute and Sino-Japanese political-economic relations: cold politics and hot economics?', *The Pacific Review*, 22, 2, 205–32.
Kreps, Sarah and Micah Zenko (2014) 'The next drone wars; preparing for proliferation', *Foreign Affairs*, 93, 2, 68–79.
Kuik, Cheng-Chwee (2005) 'Multilateralism in China's ASEAN policy: its evolution, characteristics and aspiration', *Contemporary Southeast Asia*, 27, 1, 102–22.
Ladwig III, Walter C. (2009) 'Delhi's Pacific ambition: naval power, "Look East", and India's emerging influence in the Asia-Pacific', *Asian Security*, 5, 2, 87–113.
Laksmana, Evan A. (2011) 'Variations on a theme: dimensions of ambivalence in Indonesia–China relations', *Harvard Asia Quarterly*, 13, 1, 24–31.
Lampton, David M. (2013) 'A new type of major-power relationship: seeking a durable foundation for U.S.–China ties', *Asia Policy*, 16, 51–68.
Lang, Hazel J. (2002) *Fear and Sanctuary: Burmese Refugees in Thailand*, Ithaca, NY, Cornell University Press.
Lanteigne, Marc (2008) 'China's maritime security and the "Malacca dilemma"', *Asian Security*, 4, 2, 143–61.
Layne, Christopher (1997) 'From preponderance to offshore balancing: America's future grand strategy', *International Security*, 22, 1, 86–124.
Lee, Jong-Wha and Innwon Park (2005) 'Free trade areas in East Asia: discriminatory or non-discriminatory?', *The World Economy*, 28, 1, 21–48.
Lee, Sheryn (2013) 'Burying the hatchet? the sources and limits of Japan–South Korea security cooperation', *Asian Security*, 9, 2, 93–110.
Leifer, Michael (1999) 'The ASEAN peace process: a category mistake', *The Pacific Review*, 12, 1, 25–38.
Leung, Guy (2011) 'China's energy security: Perception and Reality', *Energy Policy*, 39, 3, 1330-1337.
Lewis, John Wilson and Xue Litai (1988) *China Builds the Bomb*, Stanford, CA, Stanford University Press.
Levite, Ariel (2002) 'Never say never again: nuclear reversal revisited', *International Security*, 27, 3, 59–88.
Li Mingjiang (2008) 'China debates soft power', *Chinese Journal of International Politics*, 2, 2, 287–308.
Lim, Robyn (1998) 'The ASEAN Regional Forum: building on sand', *Contemporary Southeast Asia*, 20, 2, 115–36.
Lind, Jennifer (2009) 'The perils of apology: what Japan shouldn't learn from Germany', *Foreign Affairs*, 88, 3, 132–46.
Liow, Joseph Chinyong (2006) *Muslim resistance in Southern Thailand and Southern Philippines: religion, ideology, and politics*, Policy Studies 24, Washington, DC, East–West Center.
Lobell, David B. Marshall B. Burke, Claudia Tebaldi, Michael D. Mastrandrea, Walter P. Falcon and Rosamond N. Naylor (2008) 'Prioritizing climate change adaptation needs for food security in 2030', *Science*, 319, 5863, 607–10.
Lüthi, Lorenz M. (2008) *The Sino-Soviet Split: Cold War in the communist world*, Princeton, NJ, Princeton University Press.

MacFarlane, S. Neil (2002) *Intervention in Contemporary World Politics*, Adelphi Paper 350, London, International Institute for Strategic Studies.
MacFarlane, S. Neil, Carolyn J. Thielking and Thomas G. Weiss (2004) 'The responsibility to protect: is anyone interested in humanitarian intervention?', *Third World Quarterly*, 25, 5, 977–92.
MacFarlane, S. Neil and Stina Torjeson (2005) '"Awash with weapons"? The case of small arms in Kyrgyzstan', *Central Asian Survey*, 24, 1, 5–19.
Mackinder, Halford John (1904) 'The geographical pivot of history', *The Geographical Journal*, 23, 4, 421–37.
MacMillan, Margaret (2007) *Nixon and Mao: The Week that Changed the World*, New York, Random House.
MacMillan, Margaret (2013) 'The rhyme of history: lessons of the Great War', Washington, DC, Brookings Institution. Available at: http://www.brookings.edu/research/essays/2013/rhyme-of-history.
Mahbubani, Kishore (2009) *The New Asian Hemisphere: The Irresistible Shift of Global Power to the East*, New York, PublicAffairs.
Maley, William (2013) 'Statebuilding in Afghanistan: challenges and pathologies', *Central Asian Survey*, 32, 3, 255–70.
Mandiant (2013) *APT1: Exposing One of China's Cyber Espionage Units*, Alexandria, VA: Mandiant. Available at: http://intelreport.mandiant.com/Mandiant_APT1_Report.pdf.
Manicom, James (2014) 'China and American seapower in East Asia: is accommodation possible?', *Journal of Strategic Studies*, 37, 3, 345–71.
Mao, Zedong (2005) *On Guerilla Warfare*, trans. Samuel B. Griffith, Mineola, NY, Dover Publications.
Marten, Kimberley (2012) *Warlords: Strong Arm Brokers in Weak States*, Ithaca, NY, Cornell University Press.
Mastanduno, Michael (2014) 'Realism and Asia', in Saadia M. Pekkanen, John Ravenhill and Rosemary Foot (eds), *The Oxford Handbook of the International Relations of Asia*, Oxford, Oxford University Press, 25–44.
Matheson, Michael J. (2001) 'United Nations governance of postconflict societies', *American Journal of International Law*, 95, 1, 76–85.
May, Ronald (2010) *The Melanesian Spearhead Group: Testing Pacific Island Solidarity*, Canberra, ACT, Australian Strategic Policy Institute.
May, Ronald (2012) 'Papua New Guinea: issues of external and internal security', *Security Challenges*, 8, 4, 47–60.
McCrisken, Trevor (2013) 'Obama's drone war', *Survival*, 55, 2, 97–122.
Mearsheimer, John J. (2001) *The Tragedy of Great Power Politics*, New York, W.W. Norton.
Mearsheimer, John J. (2014) 'Can China rise peacefully?', *The National Interest*. Available at: http://nationalinterest.org/commentary/can-china-rise-peacefully-10204.
Mehta, Pratap Bhanu (2009) 'Still under Nehru's shadow? The absence of foreign policy frameworks in India', *India Review*, 8, 3, 209–33.
Menkhaus, Ken (2009) 'Dangerous waters', *Survival*, 51, 1, 21–5.
Meyer, Paul (2012) 'Diplomatic alternatives to cyber-warfare', *The RUSI Journal*, 157, 1, 14–19.
Mietzner, Marcus (2007) 'Local elections and autonomy in Papua and Aceh: mitigating or fueling secessionism?', *Indonesia*, 84, 1–39.
Mitchell, Derek J. (2002) 'A blueprint for U.S policy toward a unified Korea', *The Washington Quarterly*, 26, 1, 123–37.

Mochizuki, Mike M. (2007) 'Japan's shifting strategy toward the rise of China', *Journal of Strategic Studies*, 30, 4–5, 739–76.
Mohan, C. Raja (2006) 'India and the balance of power', *Foreign Affairs*, 85, 4, 17–32.
Mohan, C. Raja (2013) *Samudra Manthan: Sino-Indian Rivalry in the Indo-Pacific*, New Delhi, Oxford University Press.
Moore, Gregory J. (2008) 'How North Korea threatens China's interests: understanding Chinese "duplicity" on the North Korean nuclear issue', *International Relations of the Asia-Pacific*, 8, 1, 1–29.
Morgan, Matthew J. (2004) 'The origins of the new terrorism', *Parameters*, 34, 1, 29–43.
Morrison, Wayne M. (2014) *China–US Trade Issues*, Washington, DC, Congressional Research Service.
Morrow, Daniel and Michael Carriere (1999) 'The economic impacts of the 1998 sanctions on India and Pakistan', *The Nonproliferation Review*, 6, 4, 1–16.
Mortreux, Colette and Jon Barnett (2009) 'Climate change, migration and adaptation in Funafuti, Tuvalu', *Global Environmental Change*, 19, 1 105–12.
Mountz, Alison (2011) 'The enforcement archipelago: detention, haunting, and asylum on islands', *Political Geography*, 30, 3, 118–28.
Mowatt-Larssen, Rolf (2009) 'Nuclear security in Pakistan: reducing the risks of nuclear terrorism', *Arms Control Today*, 39, 6, 6–11.
Mulvenon, James (2013) 'The new central military commission', *China Leadership Monitor* 40, Hoover Institution. Available at: http://www.hoover.org/sites/default/files/uploads/documents/CLM40JM.pdf.
Narang, Vipin (2010) 'Posturing for peace? Pakistan's nuclear postures and South Asian stability', *International Security*, 34, 3, 38–78.
Narine, Shaun (1998) 'ASEAN and the management of regional security', *Pacific Affairs*, 71, 2, 195–214.
Narushige, Michishita and Peter van der Hoest (2013) 'Another Cold War in Asia?', *The ASAN Forum*. Available at: http://www.theasanforum.org/another-cold-war-in-asia/.
Nicholls, Robert J. and Anny Cazenave (2010) 'Sea-level rise and its impact on coastal zones', *Science*, 328, 5985, 1517–20.
Nixon, Richard M. (1967) 'Asia after Viet Nam', *Foreign Affairs*, 46, 1, 111–125.
Nye, Joseph (1990) 'The Changing Nature of World Power', *Political Science Quarterly*, 105, 2, 177–92.
Nye, Joseph (2004) *Soft Power: The Means to Success in World Politics*, New York, PublicAffairs.
O'Neil, Andrew (2007) *Nuclear Proliferation in Northeast Asia: The Quest for Security*, Basingstoke, Palgrave Macmillan.
O'Neil, Andrew (2011) 'Extended nuclear deterrence in East Asia: redundant or resurgent?', *International Affairs*, 87, 6, 1439–457.
Pan, Chengxin (2014) 'The 'Indo-Pacific' and geopolitical anxieties about China's rise in the Asian regional order', *Australian Journal of International Affairs*, 68, 4, 453–69.
Pandey, Gyanendra (2001) *Remembering Partition: Violence, Nationalism and History in India*, Cambridge, Cambridge University Press.
Paoli, Letizia, Irina Rabkov, Victoria Greenfield and Peter Reuter (2007) 'Tajikistan: the rise of a narco-state', *Journal of Drug Issues*, 37, 4, 951–79.
Pape, Robert (2005) *Dying to Win: The Strategic Logic of Suicide Terrorism*, New York, Random House.

Park, John S. (2005) 'Inside multilateralism: the six-party talks,' *Washington Quarterly*, 28, 4, 73–91.
Parnini, Syeda Naushin (2013) 'The crisis of the Rohingya as a Muslim minority in Myanmar and bilateral relations with Bangladesh', *Journal of Muslim Minority Affairs*, 33, 2, 281–97.
Pei, Minxin (2006) *China's Trapped Transition: The Limits of Developmental Autocracy*, Cambridge, MA: Harvard University Press.
Perlo-Freeman, Sam and Carlina Solmirano (2014) *Trends in World Military Expenditure, 2014*, SIPRI Fact Sheet, Stockholm: Stockholm International Peace Research Institute.
Prins, Gwyn (1993) *Threats Without Enemies: Facing Environmental Insecurity*, London: Earthscan.
Rabasa, Angel, Robert D. Blackwill, Peter Chalk, Kim Cragin, C. Christine Fair, Brian A. Jackson, Brian Michael Jenkins, Seth G. Jones, Nathaniel Shestak, and Ashley J. Tellis (2009) *The Lessons of Mumbai*, Santa Monica, CA, RAND Corporation.
Raghavan, Srinath (2010) *War and Peace in Modern India*, Basingstoke, Palgrave Macmillan.
Rajagopalan, Rajesh and Varun Sahni (2008) 'India and the Great Powers: strategic imperatives, normative necessities', *South Asian Survey*, 15, 1, 5–32.
Ramakrishna, Kumar (2001) '"Transmogrifying" Malaya: the impact of Sir Gerald Templer (1952–54)', *Journal of Southeast Asian Studies*, 32, 1, 79–92.
Ramanathan, Veerabhadran and Gregory Carmichael (2008) 'Global and regional climate changes due to black carbon', *Nature Geoscience*, 1, 4, 221–7.
Ravenhill, John (2001) *APEC and the Construction of Pacific Rim Regionalism*, Cambridge, Cambridge University Press.
Reuveny, Rafael (2007) 'Climate change-induced migration and violent conflict', *Political Geography*, 26, 6, 656–73.
Rid, Thomas (2012) 'Cyberwar will not take place', *Journal of Strategic Studies*, 35, 1, 5–32.
Riedel, Bruce (2008) 'Pakistan and terror: the eye of the storm', *Annals of the American Academy of Political and Social Science*, 618, 31–45.
Roberts, Adam (2009) 'Doctrine and reality in Afghanistan', *Survival*, 51, 1, 29–60.
Robock, Alan and Owen Brian Toon (2009) 'Local nuclear war', *Scientific American*, 302, 1, 74–81.
Rolfe, Jim (2001) 'Peacekeeping the pacific way in Bougainville', *International Peacekeeping*, 8, 4, 38–55.
Rosenberg, David (2009) 'The political economy of piracy in the South China Sea', *Naval War College Review*, 62, 3, 43–58.
Ross, Robert R. (2009) 'Here be dragons: is China a military threat? Myth', Debate with Aaron Friedberg, *National Interest*, 103, 19, 25–31, 33–4.
Rotberg, (2003) 'Failed states, collapsed states and weak states: causes and indicators', in Robert I. Rotberg (ed.), *State Failure and State Weakness in a Time of Terror*, Washington DC: Brookings Institution Press.
Roy, Denny (1994) 'North Korea and the "Madman Theory" ', *Security Dialogue*, 25, 3, 307–16.
Roy, Denny (2005) 'Southeast Asia and China: balancing or bandwagoning?', *Contemporary Southeast Asia*, 27, 2, 305–22.
Roy, Olivier (2007) *The New Central Asia: Geopolitics and the Birth of Nations*, London, I.B. Tauris.
Rudd, Kevin (2013) 'Beyond the pivot: a new road map for US–Chinese relations', *Foreign Affairs*, 92, 2, 9–15.

Rutland, Peter (2008) 'Russia as an Energy Superpower', *New Political Economy*, 13, 2, 203–10.
Sáez, Lawrence (2011) *The South Asian Association for Regional Cooperation (SAARC): An Emerging Collaboration Architecture*, London, Routledge.
Sagan, Scott D. (1988) 'The Origins of the Pacific War', *The Journal of Interdisciplinary History*, 18, 4, 893–922.
Saikal, Amin (2006) 'Securing Afghanistan's Border', *Survival*, 48, 1, 129–42.
Samad, Paridah Abd and Darusalam Abu Bakar (1992) 'Malaysia–Philippines relations: the issue of Sabah', *Asian Survey*, 32, 6, 554–67.
Samore, Gary (2003) 'The Korean nuclear crisis', *Survival*, 45, 1, 7–24.
Santoro, David (2012) 'The pessimistic nuclear weapons states: France, Russia and China', in Tanya Ogilvie-White and David Santoro (eds), *Slaying the Nuclear Dragon: Disarmament Dynamics in the Twenty-First Century*, Athens, GA, University of Georgia Press, 118–50.
Schelling, Thomas C. (1960) *The Strategy of Conflict*, Cambridge, MA, Harvard University Press.
Schelling, Thomas C. (1966) *Arms and Influence*, New Haven, CT, Yale University Press.
Schreer, Benjamin (2013) *Planning the Unthinkable War: 'Air Sea Battle' and Its Implications for Australia*, Canberra, ACT, Australian Strategic Policy Institute.
Schreer, Benjamin and Brendan Taylor (2011) 'The Korean Crises and Sino-American Rivalry', *Survival*, 53, 1, 13–19.
Schroeder, Paul W. (1986) 'The 19th-century international system: changes in the structure', *World Politics*, 39, 1, 1–26.
Scott, David (2008) *China and the International System 1840–1949: Power, Presence, and Perceptions in a Century of Humiliation*, New York, New York State University Press.
Seekins, Donald M. (2005) 'Burma and US sanctions: punishing an authoritarian regime', *Asian Survey*, 45, 3, 437–52.
Segal, Adam M. (2014) 'Cyberspace: the new strategic realm in US–China relations', *Strategic Analysis*, 38, 4, 577–81.
Selth, Andrew (2008) 'Even paranoids have enemies: Cyclone Nargis and Myanmar's fears of invasion', *Contemporary Southeast Asia*, 30, 3, 379–402.
Selth, Andrew (2013) 'Burma's security forces: performing, reforming or transforming?', *Regional Outlook Paper* 45, Brisbane, QLD, Griffith University Asia Institute.
Shah, Sikander Ahmed (2010) 'War on terrorism: self defense, operation enduring freedom, and the legality of U.S. drone attacks in Pakistan', *Washington University Global Studies Law Review*, 9, 1, 77–129.
Shambaugh, David (2004/5) 'China engages Asia: reshaping the regional order', *International Security*, 29, 3, 64–99.
Shambaugh, David (2008) *China's Communist Party: Atrophy and Adaptation*, Berkeley, CA, University of California Press.
Shawcross, William (1979) *Sideshow: Kissinger, Nixon, and the Destruction of Cambodia*, New York, Simon & Schuster.
Simpson, Brad (2005) 'Illegally and beautifully': the United States, the Indonesian invasion of East Timor and the international community, 1974–76, *Cold War History*, 5, 3, 281–315.
Simpson, Emile (2013) *War from the Ground Up: Twenty-First-Century Combat as Politics*, Oxford, Oxford University Press.
Simpson, John (1994) 'Nuclear non-proliferation in the Post-Cold War era', *International Affairs*, 70, 1, 17–39.

Singer, J. David. (1958) 'Threat-perception and the armament-tension dilemma', *Journal of Conflict Resolution*, 2, 1, 90–105.
Singer, Peter W. and Allan Friedman (2014) *Cybersecurity and Cyberwar: What Everyone Needs to Know*, Oxford, Oxford University Press.
Singh, L. P. (1962) 'The Thai–Cambodian temple dispute', *Asian Survey*, 2, 8, 23–6.
Sisson, Richard and Leo E. Rose (1990) *War and Secession: Pakistan, India, and the Creation of Bangladesh*, Berkeley, CA, University of California Press.
Smith, Paul (2000) 'Transnational security threats and state survival: a role for the military', *Parameters* 30, 3, 78–91.
Smith, Sheila (2013) 'A Sino-Japanese Clash in the East China Sea', *Contingency Planning Memorandum* No. 18, New York, Council on Foreign Relations.
Snitwongse, Kusuma (1998) 'Thirty years of ASEAN: achievements through political cooperation', *The Pacific Review*, 11, 2, 183–94.
Snyder, Timothy (2010) *Bloodlands: Europe between Hitler and Stalin*, New York, Basic Books.
Spykman, Nicholas J. (1944) *The Geography of the Peace*, New York, Harcourt, Brace.
Steinhoff, Patricia G. (1989) 'Hijackers, bombers, and bank robbers: managerial style in the Japanese Red Army', *The Journal of Asian Studies*, 48, 4, 724–40.
Stern, Paul C. and Druckman, Daniel (2000) 'Evaluating interventions in history: the case of international conflict resolution', *International Studies Review*, 2, 1, 33–63.
Stevenson, David (1996) *Armaments and the Coming of War: Europe 1904–1914*, Oxford, Clarendon Press.
Stockwell, A. J. (1992) 'Southeast Asia in war and peace: the end of European colonial empires', in Nicholas Tarling (ed.), *The Cambridge History of Southeast Asia, Vol II: The Nineteenth and Twentieth Centuries*, Cambridge, Cambridge University Press.
Storey, Ian (1999) 'Creeping assertiveness: China, the Philippines and the South China Sea dispute', *Contemporary Southeast Asia*, 21, 1, 95–118.
Storey, Ian (2007) *The United States and ASEAN–China Relations: All Quiet on the Southeast Asia Front*, Carlisle, PA, US Army War College Strategic Studies Institute.
Stuart, Douglas T. (2001) 'Reconciling non-intervention and human rights', *UN Chronicle*, 38, 2, 32–3.
Stubbs, Richard (2002) 'ASEAN plus three: emerging East Asian regionalism?', *Asian Survey*, 42, 3, 440–55.
Stueck, William (1995) *The Korean War: An International History*, Princeton, NJ, Princeton University Press.
Sturgeon, Timothy J. and Olga Memdovic (2011) 'Mapping global value chains: intermediate goods trade and structural change in the world economy', *United Nations Industrial Development Organisation Working Paper*, 05/2010, Vienna, UNIDO.
Subritzky, John (2000) 'Britain, Konfrontasi, and the end of empire in Southeast Asia, 1961–65', *The Journal of Imperial and Commonwealth History*, 28, 3, 209–27.
Sun Tzu (1994) *The Art of War*, trans. Ralph D. Sawyer, Boulder, CO, Westview Press.
Taagepera, Rein (1997) 'Expansion and contraction patterns of large polities: context for Russia', *International Studies Quarterly*, 41, 3, 475–504.
Tan, See Seng (2012) '"Talking their walk"? The evolution of defense regionalism in Southeast Asia', *Asian Security*, 8, 3, 232–50.

Tankel, Stephen (2009) 'Lashkar-e-taiba: from 9/11 to Mumbai', *Developments in Radicalisation and Political Violence*, London, International Centre for the Study of Radicalisation and Political Violence.
Tannewald, Nina (2008) *The Nuclear Taboo: The United States and the Non-Use Of Nuclear Weapons since 1945*, Cambridge, Cambridge University Press.
Taylor, Brendan (2010) *Sanctions as Grand Strategy*, Abingdon, Routledge for International Institute for Strategic Studies.
Tellis, Ashley J. (2013) 'Balancing without containment: a U.S. strategy for confronting China's rise', *The Washington Quarterly*, 36, 4, 109–24.
Tkacik, Michael (2014) 'Chinese nuclear weapons enhancements – implications for Chinese employment policy', *Defence Studies*, 14, 2, 161–91.
Tow, William T. (1978) 'The Janzus option: a key to Asian/Pacific security', *Asian Survey*, 18, 12, 1221–34.
Tow, William T. (2002) *Asia-Pacific Strategic Relations: Seeking Convergent Security*, Cambridge, Cambridge University Press.
Trampuz, Andrej, R. M. Prabhu, T. F. Smith and L. M. Baddour (2004) 'Avian influenza: a new pandemic threat?', *Mayo Clinic Proceedings*, 79, 4, 523–30.
Twining, Daniel (2007) 'America's grand design in Asia', *Washington Quarterly*, 30, 3, 79–94.
United Nations (1945) *Charter of the United Nations and Statute of the International Court of Justice*, San Francisco, United Nations. Available at: https://treaties.un.org/doc/publication/ctc/uncharter.pdf.
United Nations Development Program (1994) *Human Development Report 1994: New Dimensions of Human Security*, New York, Oxford University Press.
Van Creveld, Martin (1991) *The Transformation of War*, New York, The Free Press.
Vatikiotis, Michael (2009) 'Managing armed conflict in Southeast Asia: the role of mediation', *Southeast Asian Affairs 2009*, 28–35.
Wainwright, Elsina (2003) 'Responding to state failure – the case of Australia and Solomon Islands', *Australian Journal of International Affairs*, 57, 3, 485–98.
Walker, William (1992) 'Nuclear weapons and the former Soviet republics', *International Affairs*, 68, 2, 255–77.
Walt, Stephen M. (1997) 'Why alliances endure or collapse', *Survival*, 39, 1, 156–79.
Wang, Jisi (2005) 'China's search for stability with America', *Foreign Affairs*, 84, 5, 39–48.
Weber, Max (1948) *From Max Weber: Essays in Sociology*, trans. and ed. by H. H. Gerth and C. Wright Mills, London, Routledge & Kegan Paul.
Weiss, Jessica Chen (2013) 'Authoritarian signaling, mass audiences, and nationalist protest in China', *International Organization*, 67, 1, 1–35.
Wezeman, Siemon T. and Wezeman, Pieter D. (2014) *Trends in International Arms Transfers, 2013*, Stockholm, SIPRI. Available at: http://books.sipri.org/files/FS/SIPRIFS1403.pdf.
Wheeler, Nicholas J. (2010) '"I had gone to Lahore with a message of goodwill but in return we got Kargil": the promise and perils of "leaps of trust" in India–Pakistan relations', *India Review*, 9, 3, 319–44.
Wheeler, Nicholas J. and Tim Dunne (2001) 'East Timor and the new humanitarian interventionism', *International Affairs*, 77, 4, 805–27.
White, Hugh (2011) 'Power shift: rethinking Australia's place in the Asian century', *Australian Journal of International Affairs*, 65, 1, 81–93.
White, Hugh (2012) *The China Choice: Why America Should Share Power*, Collingwood, Victoria, Black.

Wickramasinghe (2013) 'Sri Lanka in 2013: Postwar Oppressive Stability', *Asian Survey*, 54, 1, 199–205.
Williams, Brian Glyn (2010) 'The CIA's covert Predator drone war in Pakistan, 2004–2010: the history of an assassination campaign'. *Studies in Conflict & Terrorism* 33, 10, 871–892.
Wohlforth, William C. (2009) 'Unipolarity, status competition, and great power war', *World Politics*, 61, 1, 28–57.
Wolfers, Arnold (1952) '"National security" as an ambiguous symbol', *Political Science Quarterly*, 67, 4, 481–502.
Xiang, Lanxin (2012) 'China and the "pivot"', *Survival*, 54, 5, 113–28.
Yan, Xuetong (2011) *Ancient Chinese Thought, Modern Chinese Power*, ed. Daniel A. Bell and Sun Zhe, Princeton, NJ, Princeton University Press.
Yan, Xuetong (2014) 'From keeping a low profile to striving for excellence', *The Chinese Journal of International Politics*, 7, 2, 153–84.
Yang, Kuisong (2000) 'The Sino-Soviet Border Clash of 1969: from Zhenbao Island to Sino-American rapprochement', *Cold War History*, 1, 1, 21–52.
Yahuda, Michael (2011) *The International Politics of the Asia-Pacific*, 3rd edn, Abingdon, Routledge.
Yang, Jian (2011) *The Pacific Islands in China's Grand Strategy: Small States, Big Games*, New York, Palgrave Macmillan.
Yasutomo, Dennis T. (1989) 'Why aid? Japan as an "aid great power"', *Pacific Affairs*, 62, 4, 490–503.
You, Ji (2001) 'China and North Korea: a fragile relationship of strategic convenience', *Journal of Contemporary China*, 10, 28, 387–98.
Young, Adam J. and Mark J. Valencia (2003) 'Conflation of piracy and terrorism in Southeast Asia: rectitude and utility', *Contemporary Southeast Asia*, 25, 2, 269–83.
Younossi, Obaid, Peter Dahl Thruelsen, Jonathan Vaccaro, Jerry M. Sollinger, and Brian Grady (2009) *The Long March: Building an Afghan National Army*, Santa Monica, RAND.
Zartman, I. William (2001) 'The timing of peace initiatives: hurting stalemates and ripe moments', *Global Review of Ethnopolitics*, 1, 1, 8–18.
Zhang, Shu Guang (1992) *Deterrence and Strategic Culture: Chinese–American Confrontations, 1949–1958*, Ithaca, NY, Cornell University Press.
Zhang, Xiaoming (2005) 'China's 1979 War with Vietnam: A Reassessment', *The China Quarterly*, 184, 851–74.
Zhao, Suisheng (2005) 'China's pragmatic nationalism: is it manageable?', *The Washington Quarterly*, 29, 1, 131–44.
Zhao, Suisheng (2008) 'China's global search for energy security: cooperation and competition in Asia-Pacific', *Journal of Contemporary China*, 17, 55, 207–27.
Zheng, Bijian (2005) 'China's "peaceful rise" to great power status', *Foreign Affairs*, 84, 5, 18–24.
Zhu, Feng (2006) 'Shifting tides: China and North Korea', in Ronald Husiken (ed.), *The Architecture of Security in the Asia-Pacific*, Canberra, ACT, ANU E Press, 45–57.
Zoellick, Robert B. (1997) 'Economics and security in the changing Asia-Pacific', *Survival*, 39,4, 29–51.
Zou, Keyuan (2012) 'China's U-shaped line in the South China Sea revisited', *Ocean Development & International Law*, 43, 1, 18–34.

Index

Abe, Shinzo, 54, 70, 258
Abrahms, Max, 177
Abu Bakar, Darusalam, 137
Abu Sayyaf Group, 179, 222
Abuza, Zachary, 179
Aceh, 149, 170–1, 220
Acharya, Amitav, 56, 231, 242, 264
Accidental war, 109
Accommodation, 73
Adamson, Fiona, 165
Afghanistan, 3, 6, 11, 19, 26, 36, 65, 122–3, 124, 130, 137–8, 145, 150, 156–9, 164–5, 166, 171, 172, 173, 177, 179, 180, 182, 186–7, 188, 190, 191, 202, 208, 209–14
Africa, 6, 36, 42, 45, 100, 137, 138, 180, 183, 185, 203, 216, 217, 222, 238, 243, 255, 259, 271, 274, 275
Afridi, Jamal, 254
Ahtisaari, Martti, 220
Ahuja, Partul, 99, 158–9
Aid, overseas, 68, 210, 228, 240, 261
Air Defence Information Zone, 139, 257
Air–sea battle, 117, 119, 126
Akashi, Yasui, 206
Alagappa, Muthiah, 45, 154, 168
Alaska, 68
Alpers, Philip, 122
Al Qaeda, 111, 124, 156, 177, 178, 179–80, 182, 210–11
Alliances, 5, 12, 20, 25, 33, 36, 37, 39, 42, 46, 48, 55, 56, 62, 63, 66, 69, 73, 74, 75, 76, 77, 88, 98, 107, 112, 120, 139–40, 141, 154, 156, 158, 172, 202, 217, 228, 230–1, 232, 239, 245, 246–268, 270, 276, 277, 278–9
Ambrosio, Thomas, 216
American Samoa, 170

Anderson, Benedict, 142
Anderson, Ian, 106
Angell, Norman, 88–9
Anglo-Malayan/Malaysian Defence Agreement, (AMDA) 42, 250
Angola, 44
Annan, Kofi, 214
Anti-access/area denial, (A2AD) 62, 117, 119, 126
Anti-communism, 46
Anti-Terrorist Regional Structure (RATS), 239
Anwar, Dewi Fortuna, 41, 161
Apocalyptic worldviews, 178, 181
Arab Spring, 65, 167
Aris, Stephen, 239
Armed conflict, see War
Armed forces, 17, 35, 44, 52, 53, 54, 55, 73, 140, 142, 152–5, 158, 159, 161, 162, 166, 183, 185, 187, 192, 193, 197, 201, 202, 204, 205, 206, 207, 208–9, 210–11, 213, 217, 222, 226, 231, 233, 235–6, 239, 243–4, 245, 250, 251, 259, 275
Arms build-up, 15, 23–4, 35, 36, 48, 60–1, 70, 71, 97, 103–26, 127, 130, 134, 221, 246, 248, 253, 254, 256, 257, 263, 266, 276–7
Arms races, 103, 106, 113–16, 276
Arms smuggling, 186, 188–9, 209
Arrow, Kenneth, 81
Art, Robert, 73
Arunachal Pradesh, 131
Arunatilake, Nisha, 85
ASEAN+3, 234
ASEAN Defence Ministers Meeting Plus (ADMM+), 235–6, 238, 245
ASEAN Regional Forum (ARF), 233–4, 236, 238, 239
Asia Major, 6

Index

Asia Minor, 6
Asia-Pacific Community, 238, 240
Asia-Pacific Economic Cooperation (APEC), 57, 233
Asia, region of 1, 6–13
Asian Development Bank, 80, 82
Asian Financial Crisis, 82, 84
Association of Southeast Asian Nations (ASEAN), 5, 8, 10, 20, 25, 40–1, 46, 63, 72, 74, 75, 77, 93, 132, 135, 146–7, 148, 154, 172, 190, 192, 202, 207, 218, 222, 224–5, 230–8, 240, 241, 242–3, 244, 245, 246, 252, 261, 266, 270, 272, 276, 278, 279
Asylum seekers, *see* Migration
Asymmetry, 91, 110, 246
Atomic weapons, *see* Nuclear weapons
Aum Shinrikyo, 181
Aung Sun Suu Kyi, 155, 218
Australia 12, 17, 21, 29, 36, 39, 42, 43, 56, 70, 75, 84, 98, 100, 105, 113, 122, 125, 132, 160, 165, 170, 171, 173, 181, 183, 187, 188, 191–2, 197, 205, 207–9, 219, 221, 222, 224, 226, 228, 232, 233, 235, 238, 241, 244, 248, 249, 250, 251, 255, 256, 257, 258, 259–60, 261, 266, 271, 274, 278
Australia, New Zealand, United States Treaty (ANZUS), 36, 39, 75, 248, 250, 255, 266
Austria, 19, 89, 264
Authoritarianism, 44, 121, 161, 202, 216, 218, 264
Autonomous weapons, 125
Autonomy, 170, 171, 173, 220, 221, 222
Ayoob, Mohammed, 40
Ayson, Robert, 75, 111, 146, 160, 200, 243

Bainimarama, Frank, 219
Bajoria, Jayshee, 254
Baker, Christopher, 140
Balance of power, 37, 72, 77–8, 88, 117, 131, 235, 242–3, 253, 259, 260, 261, 280
Balance of terror, 83, 91

Ball, Desmond, 115, 146, 237, 243
Ballard, Chris, 100
Bali, 176, 181, 234
Bandwagoning, 73
Bangladesh, 4, 6, 11, 29, 40, 47, 100, 149, 151, 154, 159, 169, 188, 189, 190, 195, 196, 197, 201, 205, 243, 273
Banks, Glenn, 100
Bannister, J., 84
Barbieri, Katherine, 90
Bardhan, Pranab, 99
Bargaining, 92, 229
Barnett, Jon, 195
Barno, David, 66
Bateman, Sam, 113, 185
Bay of Bengal, 190
Beek, Ludovicus van , 195
Beeson, Mark, 59, 83
Bell, Coral, 35
Bensassi, Sami, 184
Berlin, Isaiah, 281
Bhutan, 11, 130–1, 243
Bhutto, Benazir, 158
Bierkins, Marc, 195
Biketawa Declaration, 224–5
Bilateral cooperation, 229, 235, 256–7
Billingsley, Brittany, 94
bin Laden, Osama, 179–80, 210
Biological weapons, 105
Bird flu, 193–4
Bisley, Nick, 52, 97
Bitzinger, Richard, 117
Bloom, David, 71
Bombings, 159, 176, 179, 181–2, 234
Boege, Volker, 161
Borneo, 41
Bougainville, 44, 99, 151, 169, 221, 228, 271
Box, Meredith, 181
Brahmaputra Basin, 195
Brands, Henry, 249
Breslin, Shaun, 163
Brewster, David, 71
Bristow, Damon, 42, 250
Britain, 6, 15, 37, 41, 42, 43, 45, 55–6, 70, 87, 88, 98, 105, 113, 114, 138, 154, 157, 169, 177, 212, 222, 232, 233, 241, 248, 249, 250, 263–4

British India, 6, 29, 131, 137, 169
Brodie, Bernard, 30
Brunei, 7, 10, 46, 93, 132, 205, 233
Buddhism, 145, 164
Bull, Hedley, 7, 52, 113, 203, 242, 265, 280
Burma, *see* Myanmar
Burnham, 221
Bush, George H.W., 205, 214
Bush, George W., 179, 215
Buszynski, Leszek, 96, 128
Buzan, Barry, 12, 22, 193, 276
Byman, Daniel, 94, 124, 180

Caballero-Anthony, Mely, 192
Calder, Kent, 248
Caliphate, 178, 179, 180
Cam Ranh Bay, 256
Cambodia, 10, 32, 34, 40, 45, 63, 131–2, 148, 150, 161, 190, 196, 205–6, 209, 211, 222, 232, 233, 241, 272
Campbell, Kurt, 65
Canada, 12, 83, 173, 192, 238, 241
Capie, David, 122, 188, 230, 231
Capitalism, 143
Carbon emissions, 194, 273
Caribbean countries, 138
Carmichael, Gregory, 100
Carriere, Michael, 95
Casualties of war and violence, 30, 32, 38–9, 41, 43, 79, 82, 86, 103, 104, 105, 108–9, 123, 124, 129, 158, 161, 176–7, 178, 181, 185, 188, 205, 207, 213, 214, 217
Cazenave, Anny, 194
Central America, 66
Central Asia, 6, 11, 12, 43, 44, 65, 67, 82, 98, 122–3, 169, 190, 216, 238–9, 255
Central Military Commission (China), 153
Cha, Victor, 39, 77, 172, 246
Chambers, Paul, 154
Chan, Jane, 185
Chang, Gordon, 20
Charney, Jonathan, 133
Chase, Michael, 261
Ch'en, T.T., 7
Chechenya, 239

Chellaney, Brahma, 196
Chemical weapons, 105, 181, 213
Chen, Jian, 32
Chen, Jie, 167
Chestnut, Sheena, 187
Chiang Kai-Shek, 138, 249
Chile, 12, 93
China, 2, 3, 4, 6, 7, 8, 9, 10, 11, 12, 14, 17, 19, 25, 29, 30–1, 32, 33, 36, 38, 39, 40, 42, 43, 46, 47, 48, 49, 50–79, 81, 82–4, 85, 86, 93, 94, 96–9, 100, 105, 106, 112, 113, 114, 116–17, 118, 119–21, 125, 130–1, 132, 134–5, 137, 138–40, 143–4, 145, 146–7, 148–9, 151, 153, 158, 163, 164, 165, 166–8, 169, 171, 183, 189, 192, 193, 195, 196, 201, 202, 205–6, 216, 218, 219, 226, 228, 233, 234, 235, 236, 238–40, 243, 246–7, 249, 251, 252–67, 272, 273, 274, 275–6, 277, 278, 280, 281, 282
China–India relations, 11, 29–30, 31, 38, 71, 72–3, 83, 106, 110, 131, 149, 226, 227, 253, 258, 265, 272, 276
China–Japan relations, 2, 20, 21, 22, 39, 69, 70, 73, 83, 88, 89, 90, 91–2, 104, 126, 128, 125, 137, 139, 143–4, 145, 146, 227, 228, 237, 240, 243, 247, 253–4, 258, 263, 276, 277
China–North Korea relations, 63, 73, 141–2
China–Russia relations, 63, 67, 68, 120–1, 254
China–Southeast Asia relations, 63–4, 74, 77
China–Soviet relations, 33, 36, 38
China–Taiwan relations, 31, 48, 61, 62, 86, 91, 138–9, 249
China–US relations, 2, 9, 10, 12, 14, 15, 23, 31, 33–4, 36, 37–8, 48, 50, 51–67, 69, 73–4, 83–4, 92, 93, 96, 97, 104, 119–20, 126, 134, 139–41, 148–9, 227–8, 235, 236, 237, 238, 240, 243, 246–7, 250, 252–67, 276, 277, 280, 281
China–Vietnam War, 26, 226

Index   305

China–Vietnam relations, 36, 74, 118, 128, 146–7, 260
Chomsky, Noam, 47
Choo, Jaewoo, 240
Chow, Jonathan, 234
Christensen, Thomas 2, 36, 59, 255
Christofferson, Gaye, 183
Cirincione, Joseph, 106
Civil–military relations, 233
Civil war, *see* Internal conflict
Clark Air Force Base, 249
Clark, Ian, 189
Clarke, Michael, 159
Clausewitz, Carl von, 28, 85, 127, 179
Climate change, 3–4, 14, 17, 100–1, 109, 168, 176, 194–7, 198, 269, 273–4
Clinton, Bill, 207
Clinton, Hillary, 256
Cobden, Richard, 88
Code of conduct, 146–7
Coercion, 15, 17, 56, 79, 91, 94–5, 105, 106, 140, 146, 201, 202, 207, 214, 218, 219–20, 222, 249
Cohen, Eliot, 112
Cohen, Stephen, 70
Cold War, 8, 11, 15, 27, 29, 31, 32–3, 42, 46, 47, 48, 49–50, 57, 58, 67, 68, 71, 77, 79, 83, 92, 104, 105, 107, 108, 110, 157, 170, 202, 204, 209, 214, 229, 230, 245, 250, 252, 254, 260–1, 266, 272, 278
Collective security, 205
Collins, Kathleen, 44
Collins, Rosemary, 186
Colonialism, 6, 7–8, 28, 29, 42–3, 44, 45, 48, 55, 70, 77, 87, 137–8, 142, 143, 148, 150, 153, 157, 159, 161, 169–70, 172, 183, 204, 207, 218, 221, 225, 230, 237, 246, 250, 252, 266, 272
Command and control systems, 110
Common interests, *see* Interests
Commonwealth, British, 42
Communication, 110
Communist Party of China, 3, 9, 65, 99, 138, 143–4, 153, 158, 163, 167–8, 253, 254, 255
Comprehensive security, 232, 279

Concert of powers, 262–7, 270, 278
Confidence-building measures, 238
Connell, John, 170
Consensus, 232–3, 236, 238, 244, 278
Constructivism, 215–16, 241
Containment, 79, 90, 92, 93, 258, 261
Cook Islands, 132
Cooperation, 37, 94, 97, 121, 128, 135, 147, 183, 185, 189–90, 192, 197, 200–1, 203, 207, 209, 224–245, 246, 252, 260, 263–7, 270, 278–81
Copeland, Dale, 90
Copenhagen School, 193
Cotton, James, 201
Council for Security Cooperation in the Asia-Pacific (CSCAP), 236
Counter-insurgency, 210, 212–14, 216
Credibility, 140, 251
Crenshaw, Martha, 178
Crime, 14, 122, 176, 186–9, 191, 193, 198, 209, 234, 239, 274, 282
Crimea, 67, 264
Crimes against humanity, 206
Crisis management, 24, 75, 110, 137, 139, 146, 207, 226–7, 233, 236, 259–60, 277
Crouch, Harold, 153
Cruz De Castro, Renato, 115, 249
Cuban Missile Crisis, 105
Cultural revolution (China), 34
Culture, 20, 221, 230, 264–5, 281
Cunningham, Fiona, 110
Cyberspace issues, 23, 55, 104, 118–121, 125, 144, 277
Cyclones, 197, 220, 236

Dahl, Robert, 53
Daioyu, 135, 137, 139, 257
Dandeker, Christopher, 203
Darwin, 256
Dassú, Marta, 83
Davies, Sara, 190
Decision-making, 127, 144, 152, 168, 187, 197, 215, 229, 263, 272, 273, 282

Index

Democracy, 5, 44, 46, 63, 72, 82, 121, 154, 164, 167, 173, 218, 233, 258, 264, 279
Demographic factors, 52, 56, 67, 69, 70, 71, 75, 84, 98, 108, 132, 154, 158, 166–7, 195, 197, 209
Deng Xiaoping, 9, 34, 37, 38, 42
Desker, Barry, 117
Détente, 33, 36, 37, 47, 157
Deterrence, 30, 31, 76, 92, 105, 107, 108, 110, 122, 227–8, 248, 251, 256, 261, 277, 279
Deudney, Daniel, 198
Deutsch, Karl, 241–2
DeVotta, Neil, 129
Dialogue, 233, 236–7, 240, 257, 268, 270
Dibb, Paul, 68, 116
Diplomacy, 40–1, 56–7, 63–4, 68, 71–2, 77, 147, 220, 230, 236–7, 255, 276, 278
Disarmament, 43, 103, 107–8, 122, 205–6
Disaster relief, *see* Natural disasters
Disease, 162, 164, 176, 192–4, 198, 234, 245, 273, 274, 282
Dobbins, James, 62
Dokdo, 135, 137
Domestic politics and security, 39–40, 41, 44, 46, 48, 64–5, 66
Domino effect, 8, 172
Drake, Christian, 142–3
Drones, 104, 123–5, 217, 272
Drought, 164, 195
Druckman, Daniel, 201
Drug trafficking, 3, 157, 186–8, 209, 239, 274
Dunne, Tim, 207
Dupont, Alan, 140, 196, 198, 207
Durand line, 137–8
Dutch East Indies, 204, 220

Earnest, D., 86
Easley, Leif-Eric, 120
East Asia, 6, 7, 9, 10, 11, 12, 19, 20, 21, 24, 42, 46, 55, 57, 58, 61, 67, 71, 82, 93, 96, 98, 116, 117, 118, 128, 141, 148, 193, 205, 207, 235, 248, 251, 255, 258, 265
East Asia Summit (EAS), 57, 234–5, 238, 256, 276
East China Sea, 2, 14, 22, 24, 62, 79, 92, 128, 125, 139, 145, 227, 256, 257
East Timor, 10, 37, 44, 132, 143, 149, 150, 161, 169, 170, 206–8, 209, 214, 215, 224, 226, 228, 271
Economic growth, 4–5, 9, 10, 12, 23, 32, 34–5, 37–8, 41, 47–8, 52, 53, 54, 55–6, 58–9, 60, 64, 65, 66, 69, 72, 73, 74, 77, 79, 80–102, 115, 118–19, 160, 162–4, 167, 184, 185, 192, 195, 197, 218, 232, 233, 255, 259, 273, 276, 279, 280–1
Economic interdependence, *see* Interdependence
Economic sanctions, *see* Sanctions
Economics–security relationship, 34–5, 37–8, 41, 48, 80–102, 259, 261, 272
Egypt, 167
Embargo, 95–6
Emmers, Ralf, 148, 191, 193
Endeavour Accord, 221
Enemark, Christian, 105, 193
Energy resources, 4, 67, 75, 84, 85, 95–8, 128, 132, 146, 148, 184, 240
English-speaking countries, 241
Environmental degradation, 14, 84, 167, 196, 198, 275
Equidistance, 132
Equilibrium, 64, 75, 148–9, 259, 276
Escalation, 20, 62, 85, 92, 96, 104, 110, 139, 143, 146, 148, 182, 272
Eslake, Saul, 84
Ethnicity, 164–5, 169, 173, 211
Eurasia, 6
Europe, 6, 28, 33, 36, 42, 43, 52, 55, 58, 67, 79, 84, 87, 88, 95, 120, 137, 148, 169–70, 184, 186, 190, 220, 225–6, 241, 254, 256, 263–6, 277
European Coal and Steel Commission, 28

European Union, 28, 95, 218, 219, 230, 238, 241
Evans, Michael, 66
Evans, Paul, 163, 230, 231
Exclusive Economic Zones (EEZs), 97, 115, 130, 132, 134

Facon, Isabelle, 239
Failed states, 160, 208–9, 210
Fair, Christine, 158
Fairbank, John, 7
Fear, 17, 64, 77, 90, 177, 181–3, 197, 256
Fearon, James, 173
Federalism, 230
Fiji, 12, 43, 154, 155, 219, 244
Financial crisis (1997), 184, 235
Finland, 220
Finnemore, Martha, 215, 241
Firth, Stewart, 154
First Track dialogue, 236–7
First World War, 2, 19, 88–9, 104–5, 113, 193, 264
Fisheries, 128, 140, 145–6, 184, 244, 275
Five Power Defence Arrangements, 42, 232, 250
Floods, 195, 197
Food production, 100, 195, 196, 275
Foot, Rosemary, 260
Force, *see* Violence
Fortna, Virginia Page, 204
Forum Fisheries Agency, 244
Fraenkel, Jon, 154
France, 28, 29, 42, 43, 45, 88, 98, 105, 131, 169, 170, 241, 244, 249, 263–4
Franckx, Erik, 134
Fravel, M. Taylor, 131, 147
Free Aceh Movement, 220
Freedman, Lawrence, 183, 263
Freedom of navigation, 134, 146, 149
Free Trade Agreements, 93
French Polynesia, 170
Friedberg, Aaron, 10, 62, 117, 226, 229, 262
Friedman, Allan, 119
Fukushima, 70

G2, 265–6
Gaddafi, Muammar, 221
Gaddis, John Lewis, 79
Gama, Vasco da, 35
Ganges River, 196
Ganguly, Rajat, 99, 158–9
Ganguly, Sumit, 29, 71, 95, 109
Gardner, Helen, 43
Gartzke, Eric, 88
Garver, John, 29
Gender, 164, 275
Genocide, 161, 206
Geographic factors, 20, 59, 63, 67–8, 70, 107, 120, 122–3, 129–130, 141, 143, 149, 157, 158, 176, 180, 188, 226, 228, 243
Germany, 28, 88–9, 105, 113, 241, 264
Ghani, Ejaz, 87
Ghosh, Madhuchanda, 97
Glaser, Bonnie, 94, 140
Glick, Reuven, 86
Global Financial Crisis, 66, 84, 255
Globalisation, 90–1, 101, 141, 176, 189
Glosny, Michael A., 86
Goh, Evelyn, 69, 78, 91, 253
Golden triangle, 187
Goldstein, Avery, 110
Goldstein, Lyle, 116, 134
Gompert, David, 119
Goodhand, Jonatham 186
Governance, 180, 184, 186, 196, 205, 208
Gow, James, 203
Great powers, 32, 52–4, 60, 63, 64–7, 68–9, 71, 73, 76, 77, 78, 79, 80, 97, 106, 134, 137, 149, 185, 194, 226, 230, 232, 233–4, 237, 238, 240, 258, 262, 263–8, 273, 277–8, 279, 280, 282
Greece, 89
Gries, Peter Hayes, 143
Grønning, Bjørn, 114
Guadacanal, 208
Guam, 11, 35, 66, 257
Guam Doctrine, 35
Guan Ang Chen, 37
Guerilla strategy, 121, 158, 179

Haacke, Jürgen, 218, 234
Haas, Ernst B., 78
Hack, Karl, 55
Hainan, 134
Hamas, 159
Hartfiel, Robert, 116
Hassan, Daud, 186
Hawaii, 8, 12, 35, 66, 96, 201
Hazaras, 164–5, 173
He Yinan, 145
Health problems, 106, 109
Hearts and minds, 212, 216
Hegarty, David, 43
Hegemony, 142, 251, 253, 262, 263, 264, 282
Hemmer, Christopher, 252
Herman, Paul, 196
Herodotus, 6
Heroin, 186–7
Hess, Steve, 167
Hezbollah, 159
Hierarchy, 253
Hijacking, 178, 183, 184
Hill, Hal, 82
Hill, Matthew, 219
Himalayas, 196
Hinduism, 144, 164
Hiroshima, 30, 105
Hitler, Adolf, 38
HIV/AIDS, 194
Ho Chi Minh, 32, 46, 153
Ho, Joshua, 185
Hobbes, Thomas, 16
Hokkaido, 135
Holsag, Jonathan, 131
Hong Kong, 6, 9, 43, 118, 138, 169, 192
Hu Jintao, 153
Huang, Chin-Hao, 206
Hub and spokes, 250, 256–7
Hughes, Christopher, 70
Human rights, 161, 171, 190, 202, 214–15, 218, 261, 279, 282
Human security, 16–17, 18, 24, 123, 151–2, 161–4, 175, 192, 197, 206, 215–16, 274–5, 280
Humanitarian intervention, 207, 214–16, 275

Huntington, Samuel, 145, 173
Hussein, Saddam, 117
Hutchison, Emma, 181–2

Identity, 70, 137, 138, 142–3, 170, 185, 231, 242, 243
Ideology, 142
Ikenberry, G. John, 58–9, 88
Immerzeel, Walter, 195
Impartiality, 206, 222
Independence, 29, 41, 43, 44, 45, 46, 47, 67, 70, 91, 114, 127, 129–130, 132, 142–3, 150, 153, 154, 155, 156, 160, 167, 169–71, 172, 177, 201, 206, 207, 220, 225, 226, 233, 239, 244, 253, 254, 260, 266, 272, 282
India, 2, 3, 4, 5, 6, 7, 8, 9, 10, 11, 14, 19, 23, 31, 36, 40, 43, 47, 49, 55, 60, 62, 63, 68, 70–2, 75, 78, 84, 95, 97, 98, 99, 100, 106, 107, 108–9, 111, 112, 113, 116, 118, 130–1, 137, 144–5, 149, 151, 158–9, 164, 166, 168, 169, 173, 180, 181, 182–3, 185, 195, 196, 197, 201, 204, 205, 226, 235, 238, 243, 252–3, 258, 260, 263, 265, 266, 272, 273, 276, 277
India–Pakistan relations, 2, 11, 14, 21, 22, 26, 29, 31, 40, 43, 45, 46, 47, 71, 87, 88, 106, 107, 108–9, 111, 118, 126, 128, 131, 137, 144–5, 146, 148, 155, 164, 166, 180, 181, 182–3, 201, 204, 226, 227, 242, 243–4, 266, 272, 277
India–US relations, 71–3, 95, 252–3, 261, 265
Indian Ocean, 9, 10, 71, 97, 118, 183, 184, 220
Indochina, 45
Indonesia, 3, 5, 6, 7, 8, 10, 12, 17, 21, 27, 37, 41–2, 43, 44, 74–5, 82, 113, 114–15, 129, 142–3, 145, 149, 151, 153–4, 159, 160, 161, 169, 170–1, 173, 176, 179, 181, 184–5, 188, 191, 204, 205, 206–8, 218, 220, 221, 222, 226, 231, 232, 233,

Index

238, 252, 253, 256, 257, 259, 266, 271, 273
Indus Basin, 195
Information technology, 104, 112, 124
Insecurity, 21, 49, 76, 81, 97, 129, 173, 198, 280
Institutions, 17, 25, 56, 157, 161, 166, 168, 186, 205, 208–9, 211, 236–7, 241, 242, 246
Insurgency, 17, 26, 34, 41, 46, 47, 115, 137, 151, 156–9, 171, 177, 179, 186, 187, 188, 189, 202–3, 211–14, 217, 222, 232
Integration, 22, 59, 81–2, 91, 142–3, 183, 234, 256, 259
Intelligence, 75, 119–120, 137, 142, 158, 251
Interdependence, 5, 9, 12, 20, 21, 23, 37, 59, 71, 75, 79, 80, 86–90, 97, 101–2, 122, 127, 130, 143, 165, 167, 228, 241, 242–3, 272, 276
Interests, 33, 37, 40, 80, 87, 90, 97, 111, 127, 130, 141, 145, 152, 159, 161, 185, 197, 200, 206, 209, 215, 224, 225, 227, 229–30, 235, 242, 249, 251, 252, 253, 266, 269, 276, 280
Interference, *see* Intervention
Internal conflict, 14, 15, 18, 19, 24, 26, 40, 42, 43, 44, 46, 48, 49, 65, 85, 99–100, 103, 104, 114, 122, 123, 128–9, 151, 154, 155, 188, 190, 191, 209, 210, 213, 221, 225, 282
Internal security, 3, 149, 150–174, 175, 177, 179, 196, 198–9, 202, 204, 207, 218, 232, 255, 259, 261, 268, 271–3, 274
International Contact Group, 222
International Court of Justice, 146
International Crisis Group, 171
International Force for East Timor (INTERFET), 207
International Institute for Strategic Studies, 235
International law, 17, 97, 186, 189, 190–2, 217
International security, 18, 203
Internet, *see* Cyberspace issues

Interstate conflict, 4, 26, 27–8, 48, 75, 85, 118, 168, 196, 198, 202, 205, 207, 214, 224, 225–6, 231, 239, 243, 263, 264–7, 271–3, 277–8, 282
Intrastate conflict, *see* Internal conflict
International Commission on Intervention and State Sovereignty, 215
International Monetary Fund, 56, 95
Intervention, 24, 40, 42, 130, 162, 172, 173–4, 188, 197, 199, 201–23, 224–5, 232, 244, 257, 270, 271
Intimidation, *see* Coercion
Investment, 56, 82, 88, 92, 98, 99
Iran, 66, 98, 107, 120, 186–7, 191, 274
Iraq, 6, 28, 65, 66, 117, 159, 180, 205, 212
Iraq War, 28, 211, 213, 215–16, 255
Irian Jaya, *see* West Papua
Irish Liberation Army, 177
Islam, 5, 7, 17, 137, 144–5, 164, 167, 169, 179, 180, 190, 222
Islamic State, 66, 180, 213
Israel, 66, 120, 177, 178, 220

Jaffna Peninsula, 155
Jammu and Kashmir, *see* Kashmir
Japan, 2, 4, 7–8, 9, 10, 11, 12, 14, 23, 28, 30, 31, 35, 36, 38–9, 48, 49, 52, 54, 55, 59, 60, 61, 62, 68–70, 72, 75, 76, 77, 81, 84, 91, 95, 96, 97, 98, 105, 107, 108, 110, 113, 114, 116, 117, 118, 120, 125, 135, 137, 141, 143–4, 145, 148, 158, 168, 169, 177–8, 181, 183, 185, 201, 202, 205, 206, 222, 227, 228, 234, 238, 239, 243, 247, 250, 251, 253–4, 255, 256, 257, 258, 259, 263, 264, 265, 266, 276, 278
Japan Self-Defence Force, 70
Japanese Red Army, 177–8
Jemaah Islamiyah, 179, 181
Jervis, Robert, 15, 21, 125
Job, Brian, 116
Johnson, Chalmers, 189
Johnson, Thomas, 157, 180, 211
Johnston, Alastair Iain, 253

Jones, David Martin, 232
Jones, Lee, 266
Jones, Seth, 159, 210
Jones, Sidney, 179
Jones, Simon, 137
Juche, 142
Justice, 215, 264, 282

Kabul, 157, 210
Kabutaulaka, Tarcisius, 209
Kachin Independence Army, 155
Kahler, Miles, 91, 101
Kalashnikov, 121, 158, 188
Kalhan, Anil, 164
Kamchatka peninsula, 135
Kandiyoti, Deniz, 164
Kang, David, 10, 39, 253
Kaplan, Robert 9, 97
Kapur, S. Paul, 95, 111
Kargil crisis, 26, 109
Karzai, Hamid, 156, 210–13
Kashmir, 2, 14, 26, 29, 47, 128, 131, 137, 141, 148, 161, 180, 182, 204
Kastner, Scott, 91
Katsumata, Hiro, 172
Katzenstein, Peter, 252
Kazakhstan, 11, 43, 108, 130, 169, 238
Kennedy, Paul, 56
Khmer Rouge, 32, 34, 40, 161, 205–6, 211, 215, 241
Khong, Yuen Foong, 233
Kilcullen, David, 156
Kiernan, Ben, 161
Kim-il Sung, 8
Kim Jong-Un, 39
Kiribati, 224
Kissinger, Henry, 34
Kivimäki, Timo, 26
Kochhar, Kaplana, 87
Kohli, Harinder, 80
Konfrontasi, 41–2
Koo, Min Gyo, 92
Korea–Japan relations, 39, 70, 89, 128, 135, 137, 139, 143, 257
Korean peninsula, 6, 7, 14, 33, 39–40, 48, 68, 73, 94, 141, 248–9, 253, 260, 277

Korean War, 8, 26, 28–9, 30–1, 32–3, 40, 105, 141, 203, 205, 226, 240, 248
Kosovo, 161
Kreps, Sarah, 125
Kuik, Cheng-Chwee, 63
Kuomintang, 153
Kuril islands, 135
Kuwait, 28, 205
Kyrgyzstan, 11, 43, 44, 123, 130, 190, 238

Ladakh, 131
Ladwig, Walter, 118
Laitin, David, 173
Laksmana, Evan, 74
Lal, Brij, 154
Lampton, David, 57, 265
Lang, Hazel, 166
Lanteigne, Marc, 55
Laos, 10, 45, 63, 130, 190, 196, 233
Lashkar-e-Taiba (LeT), 180, 181
Latin America, 36, 66, 120, 203
Laws of conflict, 124
Layne, Christopher, 57–8
Lebanon, 178
Lee Jhong-Wa, 93
Lee Kuan Yew, 37
Lee, Sheryn, 257
Legitimacy, 161, 167–8, 183, 196, 223, 263
Leifer, Michael, 224
Lennon, John, 129
Leung, Guy, 96
Levite, Ariel, 108
Levy, Jack S., 90
Lewis, John Wilson, 106
Li Fujian, 59
Li Mingjiang, 64
Liberalism, 87–8, 92–3, 102, 189, 218, 233
Liberty, 99, 162, 164, 192, 206, 264, 282
Libicki, Martin, 119
Libya, 65, 213, 221
Lim, Robyn, 234
Limited war, 109, 226–7
Lind, Jennifer, 39
Lohman, Walter, 115

Lombok Treaty, 208
Lupu, Yonatan, 88
Lüthi, Lorenz, 33

Macau, 6
Mackinder, Halford, 6
MacFarlane, S. Neil, 123, 202, 215–16
MacMillan, Margaret, 264
Mahbubani, Kishore, 259
Malacca Strait, 6, 9, 128, 184, 185
Malaita, 208
Malaya, *see* Malaysia
Malaysia, 3, 6, 7, 8, 10, 21, 41, 42, 46, 56, 113, 132, 135, 137, 143, 145, 151, 179, 183, 184, 185, 205, 207, 212, 221, 222, 231, 232, 250, 253, 2562
Maldives, 243
Maley, William, 157
Manchuria, 7
Mandiant, 120
Manicom, James, 149
Mao Zedong, 8, 29–31, 33, 34, 38, 42, 138, 153, 158, 163
Major powers, *see* Great powers
Marco Polo, 6
Maritime power, 35, 36, 37, 61, 71, 104, 113, 116, 118, 140, 149, 183, 236, 260, 277
Marshall Islands, 170
Marten, Kimberley, 157
Martinez-Zarzoso, Immaculada, 184
Mason, M. Chris, 157, 180
Mastanduno, Michael, 89
Matheson, Michael, 205
May, Ronald, 194, 244
McChristen, Trevor, 217
McCormack, Gavin, 181
Mearsheimer, John, 60, 92
Medcalf, Rory, 110
Media, 166, 181
Mediation, 219–22
Mehta, Pratap Bhanu, 72
Mekong River, 196
Melanesia, 160–1
Melanesian Spearhead Group, 244
Memdovic, Olga, 81–2
Menkhaus, Ken, 183
Methamphetamines, 187

Mexico, 12, 83
Meyer, Paul, 121
Micronesia, 66, 106, 194
Middle East, 6, 9, 42, 45, 55, 65, 66, 97, 98, 159, 167, 180, 255, 256
Mietzner, Marcus, 171
Migration, 14, 17, 100, 164–6, 167, 175, 189–92, 195, 196, 274
Militarism, 7, 38–9
Military bases, 35, 75, 247–9, 252, 253, 256, 257
Military cooperation, 42, 70, 261, 278
Military medicine, 236
Military power, 3, 5, 23, 45, 48, 49, 58, 68, 71, 73, 77, 94, 96–7, 103, 105
Military rule, 5, 46
Military spending, 54–5, 61, 66, 69, 71
Military support, 248–9, 250, 255, 259–60, 261–2, 278
Military technology, 24, 104, 260
Milner, Anthony, 237
Milosevic, Slobodan, 161
Mindinao, 156, 222
Mischief Reef, 147
Missile defence, 114
Missiles, 48, 61, 91, 104–5, 116, 117, 240, 277
Mitchell, Derek, 141
Mochizuki, Michael, 73
Modi, Narendra, 258
Mohan, C. Raja, 72, 265
Mongolia, 10, 130, 192, 194
Monroe Doctrine, 66
Monsoon, 195, 275
Moore, Gregory, 63
Morgan, Matthew, 178
Moro Islamic Liberation Front (MILF), 129, 169, 179, 188, 221–2
Moro National Liberation Front (MNLF), 188, 221
Morrison, Wayne, 83
Morrow, Daniel, 95
Mortreux, Colette, 195
Mountz, Alison, 192
Mowatt-Larssen, Rolf, 166
Mozambique, 44
Muggah, Robert, 122
Mujahadeen, 122, 188

Multilateral cooperation, 229, 230, 237, 240, 261
Mulvenon, James, 153
Mumbai, 180, 181
Murray, Williamson, 116
Myanmar, 3, 7, 10, 26, 63, 98, 122, 123, 130, 150, 155, 159, 164, 165–6, 173, 187, 188, 190, 196, 197, 218, 219, 233

Nagasaki, 20, 105
Napoleon, 263–4
Narang, Vipin, 109
Narine, Shaun, 41, 232
Narushige, Michishita, 79
Nation-building, 206, 210–11
Nation states, 17–18, 22, 88, 142
National security, 18, 44, 46, 63, 152, 191, 215, 216, 280
National Security Agency (US), 120
Nationalism, 24, 29, 89, 128, 142–5, 148, 159, 167, 168, 169, 177, 179, 255, 258
Natural disasters, 4, 162, 195, 197, 234, 236, 237, 245, 249
Nauru, 191
Naxalites, 3, 99, 158–9
Nazi Germany, 36, 38
Nehru, Jawaharlal, 226
Nepal, 11, 130, 243
Netherlands, 43, 171, 204
Netizens, 144
Neutrality, 77, 86, 232, 252
New Caledonia, 170, 244
New wars, 178–9
New world order, 214
New Zealand, 12, 19, 29, 36, 39, 42, 54, 56, 75, 93, 132, 170, 173, 183, 196, 205, 207–9, 219, 221, 228, 232, 233, 235, 241, 244, 248, 249, 250, 255
Nicholls, Robert, 194
Niue, 132, 170
Nixon, Richard, 32, 34, 35, 47, 56
Non-aligned movement, 8
Non-alignment, 37, 71, 74, 252–3
Non-intervention, principle of, 40, 46, 147, 204, 207, 215–16, 218, 222, 231–2, 279

Non-state actors, 17, 24, 76, 111, 121, 160, 166, 176, 177, 179, 181, 182, 185, 187, 207, 208, 209, 211, 234
Norms, 67, 202, 203, 215–16, 223, 225, 231, 233, 242, 279–80
North Asia, 10, 23, 36, 48, 128, 131, 135–6, 143, 154, 184, 202, 226, 233, 235, 240, 243, 244, 251, 260, 277
North Atlantic Treaty Organization (NATO), 33, 36, 65, 67, 156–7, 209, 211–13, 217, 231, 246, 249, 254, 260
North Korea, 2, 3, 5, 8, 10, 11, 17, 21, 36, 39, 49, 57, 61, 63, 73–4, 76, 77, 88, 93, 94, 107, 109, 114, 130, 141–2, 154, 178, 187, 203, 219, 226, 228, 239–40, 249, 251, 254, 257, 259, 276, 277
Northern Tribal Alliance, 210
Northern Ireland, 177
Northern Marianas, 170
Nuclear Non-Proliferation Treaty (NPT), 106, 111, 252–3
Nuclear proliferation, 31, 66, 72, 94, 98, 106–7, 110–11
Nuclear tests, 105–6, 228, 240, 250
Nuclear war, 108–111
Nuclear weapons, 3, 27, 30, 31, 43, 46, 48, 53, 55, 61, 63, 68, 69–70, 71, 73–4, 76, 79, 95, 103–11, 114, 119, 142, 146, 148, 166, 178, 226–7, 239–40, 249, 250, 251, 254, 256, 257, 266, 277
Nuclear winter, 108
Nye, Joseph, 54, 79

Obama, Barack, 65, 235, 255–6, 265
Oil, *see* Energy
Okinawa, 35
O'Neil, Andrew, 107, 251
Opium, 186
Order, 69, 73, 132, 150, 154, 155–6, 160, 184, 191, 192, 193–4, 195, 199, 209, 282
Organisasi Papua Merdeka (OPM), 171
Organisation of Islamic Cooperation, 221

Index

Pacific island countries, 4, 12, 43, 75, 99, 100, 105, 122, 138, 160, 165, 170, 188, 191, 194, 197, 205, 208–9, 219, 221, 222, 224–5, 228, 244, 248, 250, 261, 273
Pacific Islands Forum, 209, 224–5, 244
Pacific Ocean, 9, 10, 12, 68, 97, 184
Pakistan, 2, 3, 6, 11, 14, 26, 31, 43, 47, 49, 63, 71, 76, 95, 98, 106, 107, 108–9, 111, 112, 113, 118, 122, 123, 124, 130, 137–8, 144–5, 149, 150, 154, 155, 157–8, 159, 166, 169, 171, 173, 176, 179–81, 182–3, 186, 189, 190, 201, 204, 205, 214, 216–17, 226, 243, 249, 254, 266, 272, 277
Palau, 170
Palestine, 177–8, 220
Palestinian Liberation Organisation (PLO), 177
Pan, Chengxin, 66
Pandey, Gyanendra, 169
Panetta, Leon, 256
Paoli, Letizia, 187
Pape, Robert, 177
Papua New Guinea, 3, 7, 12, 19, 43, 44, 99–100, 122, 132, 151, 160, 169, 170, 171, 173, 188, 191, 194, 220–1, 244, 271
Pardesi, Manjeet, 71
Park, Innwon, 93
Park, John, 239–40
Parnini, Syeda, 190
Partnerships, 247, 252, 255, 256, 258, 260, 263, 276
Peace, 23, 24, 26–50, 77, 89, 101, 103, 113, 126, 127–8, 143, 144, 146, 168, 204, 209, 221, 224, 226, 228, 231, 242, 243, 245, 247, 266, 267, 272–3, 279, 280
Peacekeeping, 24, 201, 203–6, 222, 236
Pearl Harbor, *see* Hawaii
Pei, Minxin, 99
People's Liberation Army (PLA), 31, 60, 62, 120, 131, 168
Perceptions of security, 15, 17, 125, 185, 191, 247, 260–2, 267, 277

Philippines, 3, 7, 8, 10, 12, 14, 17, 26, 29, 35, 36, 41, 42, 46, 62, 115, 128, 129, 132, 134–5, 137, 139–40, 147, 149, 150, 156, 159, 169, 179, 188, 192, 197, 205, 207, 221–2, 231, 232, 233, 236, 249, 252, 256, 259, 266
Phillips, Andrew, 97
Piracy, 3, 14, 176, 183–5, 186
Pivot to Asia, 256–8
Plurilateral cooperation, 229, 235, 237, 279
Polar ice, 194
Police, 192, 197, 208, 213
Popular Front for the Liberation of Palestine, 178
Portugal, 6, 44, 161
Poverty, *see* Economic growth
Power, 23, 37–8, 52–3, 63, 69, 71, 76, 96, 148–9, 216, 233, 242, 280–1
Power, distribution of, 25, 27, 37, 42, 45, 52, 57, 60, 68, 73–4, 78, 80, 118, 122, 140, 147, 148, 156, 235, 243–4, 247, 250, 251, 252, 257, 261, 263, 265, 275–7, 280–1, 282
Power, projection, 61, 62, 78, 104, 115, 116–17, 142, 244, 248, 251, 255, 260, 261
Powles, Anna, 43
Preventive diplomacy, 233, 236
Prins, Gwyn, 197
Production chains, 4, 81
Proliferation, *see* Arms build-up
Provincial Reconstruction Teams, 210
Proximity, 21–2
Proxy wars, 47
Prussia, 263
Psychological factors, 177, 182, 191, 212, 260
Putin, Vladimir, 67, 95, 239

Quadrilateral of democracies, 258

Rabasa, Angel, 181
Raghavan, Srinath, 226
Rainfall patterns, 195
Rakhine, 190
Ramakrishna, Kumar, 212
Ramanathan, Veerabhadran, 100

RAND, 213
Rationality, 90
Ratner, Ely, 65
Ravenhill, John, 57
Reagan, Ronald, 250
Realism, 89
Rebalancing, *see* Pivot to Asia
Referent objects, 163
Refugees, *see* Migration
Regional Assistance Mission to Solomon Islands (RAMSI), 208, 211, 225
Regional Comprehensive Economic Partnership (RCEP), 93
Regional cooperation, 25, 28, 41–2, 57, 63, 209, 223, 224–245, 246, 258, 266–7, 278–9, 283
Regional security, 20–2, 130, 150, 151, 167, 173, 175–6, 186, 270–1, 274, 276, 282
Regional security architecture, 237–8, 279
Religion, 7, 144–5, 178, 212
Resilience, 231
Resources, 9, 14
Responsibility to Protect (R2P), 214–16
Restraint, 76, 87, 129–30, 147, 204, 226–7, 246
Reuveny, Rafael, 195
Revolution in Military Affairs (RMA), 112
Revolutions, 152, 178
Rid, Thomas, 119
Riedel, Bruce, 180–1
Ripeness, 221
Ripple effects, 20, 21, 45, 50, 53, 59, 62, 65, 76, 79, 85, 103, 120, 151, 158, 159, 165, 166, 167–8, 171–2, 175, 176, 180, 181, 182, 185, 186, 191, 193, 200, 246, 270–8, 283
Risk, 84
River systems, 195–6
Roberts, Adam, 210
Robock, Alan, 108
Rohyingas, 164, 173, 190
Rolfe, Jim, 221
Rose, Leo 47
Rosenberg, David, 184
Ross, Robert, 61

Rotberg, Robert, 160
Roy, Denny, 39, 252
Roy, Olivier, 169
Rudd, Kevin, 238, 240
Rule of law, 279, 282
Rules, 32, 64, 148–9, 167, 184, 193, 204, 266
Russia, 3, 6, 7, 11, 43, 55, 61, 63, 65, 67–8, 69, 70, 89, 95, 104, 107, 113, 120–1, 130, 135, 157, 168, 170, 216, 227, 233, 235, 238–9, 251, 254, 264, 277
Russo-Japanese War, 7
Rutland, Peter, 67
Rwanda, 206, 214

Sabah, 41, 136–7
Saddam Hussein, 28, 205, 211
Sáez, Lawrence, 243–4
Sagan, Scott, 95–6
Saikal, Amin, 138
Samad, Paridah Abd, 137
Samore, Gary, 94
San Francisco system, 247–50, 253–4, 262, 266
Sanctions, 56, 94–6, 98, 202, 207, 217–20
Santoro, David, 107
Sarawak, 41
Sarin, 181
Saudi Arabia, 205, 213, 222
Sazlan, Iskander, 128
Scarborough Shoal, 135, 140, 256
Schelling, Thomas, 105, 115, 144, 200, 227, 229
Schreer, Benjamin, 63, 75
Schroeder, Paul, 264
Scott, David, 138
Sea-lanes, 37, 55, 86, 96, 184–5
Sea levels, 4, 194–5, 273, 274
Sea lines of communication, *see* Sea-lanes
Secession, 149, 169, 202, 222
Second Track dialogue, 236–7
Second World War, 7–8, 27, 28, 30, 38, 49, 68, 88, 95–7, 104, 105, 108, 114, 122, 135, 137, 141, 150, 151, 168, 188, 201, 205, 225, 227, 241, 247, 248, 257, 276, 282

Index

Securitisation, 193
Security assurances, 76–7
Security community, 241–5
Security, conceptions of, 14–20, 23, 35, 49, 51, 123, 128, 150–2, 160, 161, 162–3, 165, 175–6, 193, 194, 197–8, 200, 215, 232, 254, 269–70, 273, 274, 281–3
Security dilemma, 125–6, 247, 258–262
Seekins, Donald, 218
Segal, Adam, 120
Selth, Andrew, 155, 197
Senkaku, 135, 137, 139, 257
September 2001 attacks (9/11), 124, 156, 160, 177, 178, 180–2, 209–11, 215, 217, 238, 257, 271, 274
Serbia, 89
Severe Acute Respiratory Syndrome (SARS), 192
Shah, Sakinder Ahmed, 217
Shambaugh, David, 65, 255
Shanghai Cooperation Organisation (SCO), 11, 67, 216, 238–9, 254
Shangri-La Dialogue, 235
Shawcross, William, 32
Signalling, 92, 227, 235, 239
Simpson, Brad, 37
Simpson, Emile, 212
Simpson, John, 76
Singapore, 6, 8, 9, 10, 21, 37, 41, 42, 56, 62, 74, 93, 113, 118, 125, 143, 179, 183, 184, 185, 192, 207, 212, 222, 231, 232, 233, 235, 250, 252, 256, 259
Singer, J. David, 189–90
Singer, Peter W., 119
Singh, L.P., 131
Sisson, Richard, 47
Six Party Talks, 239–40
Small arms, 104, 121–3, 158, 188–9
Smith, M.L.R., 232
Smith, Paul, 175
Smith, Sheila, 139
Snitwongse, Kusuma, 77
Snowden, Edward, 120
Snyder, Timothy, 38
Social contract, 99, 253
Social groups, 163, 177, 215, 242, 275, 280

Socialisation, 242
Soft power, 54, 65
Solomon Islands, 3, 12, 99–100, 122, 150, 160, 165, 188, 208–9, 211, 224–5, 244, 271
Somalia, 183–4, 208
South Asia, 6, 7, 10, 12, 23, 24, 29, 31, 43, 71–2, 76, 82, 84, 87, 100, 111, 131, 144–5, 148, 154, 181, 183, 190, 195, 204, 233, 243–4, 263, 266, 272, 275
South Asian Association for Regional Cooperation (SAARC), 243–4
South China Sea, 14, 24, 46, 47, 62, 70, 74, 77, 79, 96, 118, 128, 132–5, 138, 140, 146–7, 227, 234, 255, 258, 259, 260, 261, 266, 276
South Korea, 5, 8, 9, 10, 12, 17, 21, 29, 35, 36, 39, 46, 49, 55, 61, 73–4, 76, 94, 107, 113, 114, 116, 118, 120, 135, 137, 139, 142, 154, 183, 203, 222, 234, 239, 243, 246, 248–9, 251, 253, 255, 256, 257, 259, 261, 266, 276, 278
South Pacific, *see* Pacific Island countries
South Sudan, 170
Southeast Asia, 3, 7, 8, 10, 20, 22, 26–7, 29, 32, 34, 36, 37, 40–1, 43, 47, 56, 63, 66–7, 68, 70, 71, 74, 76, 77, 113, 115, 122, 128, 132, 145, 147, 154, 161, 171, 179, 181, 184–5, 188, 190, 191, 195, 196, 202, 205, 213, 218, 222, 224, 230–6, 238, 240, 241, 242–3, 245, 250, 252, 258, 261, 266, 274
Southeast Asia Treaty Organisation (SEATO), 249
Sovereignty, 17–18, 44, 45–6, 47, 67, 124, 128–9, 132–3, 142, 146, 149, 152, 157, 160, 170, 172, 176, 185, 202, 203, 207, 209, 214, 217, 218, 230–1, 239, 248, 264, 275
Soviet Union, 8, 11, 15, 29, 30, 32, 33, 34, 36, 43, 46, 47, 48, 49, 58–9, 67, 78, 79, 83, 92, 105, 110, 114, 122, 123, 135, 138, 157, 169–70, 179, 187, 189, 190, 201, 203, 245, 249, 250, 251, 252, 254, 261, 266

Spain, 137
Spanish influenza, 193
Spheres of influence, 66
Spratly Islands, 147
Spykman, Nicholas, 6
Srebrenica, 206
Sri Lanka, 3, 6, 11, 85, 129, 149, 150, 155–6, 158, 159, 169, 176, 188, 190, 191, 243, 274
Stabilisation, 211–12, 216
Stability, 37, 48, 58, 88, 99, 110, 232, 233, 253, 272, 273, 276, 279
Stalemate, 221
Stalin, Joseph, 8, 38
State building, 29, 44–5
States, 17–18, 150, 152, 154, 157–8, 160–1, 163, 170, 172, 175–6, 181, 188–9, 196–7, 204, 208–9, 211, 214, 220, 225, 226, 229–30, 241, 245, 269, 270, 273, 274–5, 283
States system, 8, 9, 17, 45, 128–130, 150, 170, 176, 202, 215, 225, 245, 275, 280
Status, 89–90, 106, 257, 276, 280–1
Steinhoff, Patricia, 177–8
Stern, Paul, 201
Stevenson, David, 113
Stockholm International Peace Research Institute (SIPRI), 54
Stockwell, A.J., 29
Storey, Ian, 74, 147
Strategy, 173, 178–9, 183, 198, 200–1, 210, 211, 234, 251, 258, 269–70
Stuart, Douglas, 214
Stubbs, Richard, 235
Stueck, William, 30
Sturgeon, Timothy, 81–2
Subic Bay, 35, 249
Submarines, 61, 112–13, 114, 116, 117, 134
Subritzky, John, 41
Suez Canal, 55
Suicide terrorism, *see* Terrorism
Suharto, 37, 41–2, 44, 82, 153, 161, 204, 206, 220
Sukarno, 41, 142, 153, 204
Sumatra, 184
Summers, Larry, 83
Sun Tzu, 118

Survival, 16–17, 18, 47
Syria, 65, 66, 180, 213
Systems, 21, 44, 45, 51, 53, 58, 78, 130, 197, 225, 281

Taagepera, Rein, 166
Tacit agreements, 227
Taiwan, 2, 6, 7, 9, 10, 36, 46, 48, 62, 79, 86, 91, 113, 116, 132, 135, 138–9, 146, 153, 154, 167, 192, 201, 249
Taiwan Strait, 31, 62, 91, 128, 139, 201, 277
Tajikistan, 11, 43, 44, 123, 130, 151, 186–7, 238
Takeshima, 135, 137
Taliban, 122, 137, 156–7, 164–5, 186, 210–13, 216, 217, 275
Tamil Tigers, 85, 129, 155, 159, 169, 188
Tan, See Seng, 236
Tankel, Stephen, 180
Tannewald, Nina, 108
Taylor, Alan M., 86
Taylor, Brendan, 63, 95, 237
Tellis, Ashley, 261
Templer, Gerard, 212
Territorial disputes, 2, 14, 17, 20, 22, 24, 29, 41, 45–6, 62, 66, 72–3, 74, 96, 112, 118, 128–141, 145–9, 196, 222, 225, 227, 234, 249, 255, 260, 261, 266, 276
Terrorism, 3, 14, 15, 76, 86, 111, 121, 124, 157, 158, 160, 162, 164, 176–83, 185, 198, 209–11, 215, 222, 234, 239, 245, 254, 257, 269, 271–2, 273, 274, 282
Thailand, 3, 8, 10, 12, 14, 29, 36, 41, 42, 82, 113, 131–2, 148, 151, 154, 155, 156, 159, 165–6, 169, 187, 188, 192, 196, 205, 207, 222, 231, 232, 241, 249, 251, 252, 259, 272
Thielking, Carolyn, 215–16
Third parties, 220–2
Third world, 203
Threats, 14–15, 16, 31, 35, 37, 49, 57, 59, 74, 86, 94, 96, 101, 105, 107, 129, 142, 161, 163, 164, 190, 197–8, 201, 203, 207, 212, 219,

225, 231–2, 233, 239, 244, 245, 246, 247, 249, 251, 257, 260–2, 270, 274, 278, 281, 282
Thucydides, 89
Tiananmen Square, 167, 216, 219
Tibet, 30, 131, 138, 149, 167, 196
Timor Leste, *see* East Timor
Tkacik, Michael, 61
Tokyo, 181
Toon, Owen Brian, 108
Tokelau, 132, 170
Torjeson, Stina, 123
Tow, William, 257, 279
Trade, 7, 17, 20, 64, 75, 82, 83, 85–8, 89–90, 92–3, 184–5, 228, 239, 255, 256, 261
Trampuz, Andrej, 193
Trans-Pacific Partnership, 93, 256, 261
Transnational crime, *see* Crime
Transnational security, 3, 14, 24, 137–8, 165–6, 174, 175–99, 239, 268, 271–2
Treaty of Amity and Cooperation (TAC), 135, 231, 235, 242, 245, 272, 273–4, 279–80
Treverton, Gregory, 196
Tripolarity, 33
Truce Monitoring Group, 221
Turkey, 6, 222
Turkmenistan, 11, 43, 44
Tuvalu, 100, 187, 194–5
Twining, Daniel, 253
Twyford, Conor, 122

Ukraine, 11, 65, 67, 95, 277
United Kingdom, *see* Britain
United Nations, 8, 29, 57, 75, 98, 129–30, 138, 203–8, 214, 216, 229, 231
United Nations Convention on the Law of the Sea (UNCLOS), 115, 132–4, 186
United Nations Development Programme, 162
United Nations Economic, Scientific and Cultural Organisation (UNESCO), 189
United Nations Military Observer Group in India and Pakistan (UNMOGIP), 204

United Nations Transitional Authority in Cambodia (UNTAC), 205
United States of America 2, 3, 5, 8, 10, 11–12, 14, 15, 20, 25, 28, 30–1, 32–3, 35–8, 39, 42, 45, 46, 47, 48, 49, 51–79, 83–4, 86, 88, 92, 93, 95, 96, 98, 105–6, 110, 113, 114, 116–17, 119–21, 122, 124–5, 134, 139–42, 148–9, 154, 169, 170, 172, 177, 178, 180, 182, 183, 185, 188, 189, 192, 201, 202, 209–11, 212, 213, 216–17, 218, 219, 222, 226–7, 232, 233, 235, 236, 238, 239, 241, 243, 245, 246–67, 270, 271–2, 276, 277, 278–9, 280, 281
United States–Southeast Asia relations, 64, 66–7, 74
United States–Japan relations, 35, 68–9, 73, 95–6, 107, 110, 139, 141, 228, 246, 247–8, 253–4, 256, 261, 262
United States Navy, 12, 35, 55, 74, 116, 252, 256, 257, 261
Use of force, *see* Violence
Uyghurs, 159, 167
Uzbekistan, 11, 43, 44, 169, 190, 194, 238

Valencia, Mark, 185
Values, 14–15, 23, 35, 49, 54, 70, 79, 235, 274, 280–1
Van Creveld, Martin, 179
van der Hoest, Peter, 79
Vanuatu, 43, 160, 244
Vatikiotis, Michael, 220
Vietnam, 5, 8, 10, 22, 32, 34, 40, 42, 45, 46, 62, 74, 96, 113, 118, 130, 132, 134, 153, 168, 169, 188, 190, 192, 196, 197, 205, 218, 226, 233, 241, 249, 253, 254, 256, 260, 266, 272
Vietnam–Cambodia war, 34, 40, 41, 205, 233, 241
Vietnam War, 8, 29, 32, 36, 37, 47, 105, 202, 212–13, 249
Violence, 15, 101–2, 151–3, 155, 157, 159–60, 161, 163, 164, 168, 169, 170, 171–3, 176–7, 178, 180–1, 184, 187, 190, 198, 201, 204, 206,

207, 209, 211, 213, 215, 224, 226, 228, 241, 243, 271, 272, 273, 274
Vulnerability, 18, 31, 76, 101, 106, 198, 214, 216, 219, 220, 221, 222, 275, 279

Wa State Army, 187
Waever, Ole, 12, 193, 276
Wainwright, Elsina, 208–9
Walker, William, 43
Walt, Stephen, 248
Wang Jisi, 228
War, 16, 19, 23, 26–50, 60, 73, 79, 80, 81, 85, 86–7, 88–9, 90, 103, 104, 105, 109–110, 113–14, 118, 119, 123, 124, 125, 126, 127, 130, 131–2, 137, 143, 144, 146, 148, 151, 156, 158, 168–9, 173, 177, 178–9, 182, 195–6, 201–2, 203, 204, 212, 225, 228, 241, 244, 247, 248, 264–8, 276, 279, 280, 282
War crimes, 206
War on terror, 179
Warlords, 157
Warsaw Pact, 33, 254, 260
Water supplies, 100–1, 195–6, 274
Waters, Christopher, 43
Watson, Adam, 7
Waxman, Matthew, 94
Weapons of mass destruction, 105, 227
Weber, Max, 152–3
Weiss, Jessica Chen, 92
Weiss, Thomas, 215–16
West Papua, 43, 129, 149, 159, 171, 173, 204, 205, 207, 257
West Timor, 207
Western countries, 15, 33, 35, 42, 147, 164, 173, 207, 210, 218, 222, 237, 239, 248, 249, 259
Western Pacific, 257
Westphalia (system), 7
Wezemon, Pieter, 112
Wezeman, Siemon, 112
Wheeler, Nicholas, 144, 207

White, Hugh, 45, 60, 73, 98, 110, 260, 263, 265
Whole of government, 211–12
Wickramasinghe, Nira, 155–6
Wilde, Jaap de, 193
Williams, Brian, 124
Wilson, Woodrow, 88
Wohlforth, William, 90
Wolfers, Arnold, 15–16, 123, 160, 198, 215, 274, 281
World Trade Organization, 93

Xi Jinping, 57, 153, 265
Xiang, Lanxin, 93
Xinjiang, 3, 159, 167, 239
Xue Litai, 106

Yahuda, Michael, 33, 67
Yan Xuetong, 53, 262
Yang, Jian, 75
Yang Kuisong, 33
Yasukuni Shrine, 257
Yasutomo, Dennis, 68
Yellow River, 196
Yemen, 213
Yokosuka, 35, 116
Yongshun, Cai, 98
You Ji, 141
Young, Adam, 185
Younossi, Obaid, 213
Yugoslavia, 161, 206, 214
Yunnan, 196

Zartmann, I. William, 221
Zenko, Micah, 124
Zero-sum games, 78, 145, 227–8, 243, 259
Zhang Shu Guang, 31
Zhang Xiaoming, 34
Zhao, Suisheng, 97, 168
Zheng, Bijian, 60
Zhu Feng, 228
Zoellick, Robert, 88–9
Zone of Peace, Freedom and Neutrality (ZOPFAN), 77, 232, 252
Zou Keyuan, 135

www.ingramcontent.com/pod-product-compliance
Ingram Content Group UK Ltd.
Pitfield, Milton Keynes, MK11 3LW, UK
UKHW041113160825
461921UK00007B/151